Jim Schlaefer
1029 Fiedler Lane
Apt #1
Madison, WI
 255-2437
1978

Medical
Virology

MEDICAL VIROLOGY

Second Edition

FRANK FENNER
Centre for Resource & Environmental Studies
The Australian National University
Canberra, A.C.T., Australia

DAVID O. WHITE
School of Microbiology
University of Melbourne
Melbourne, Victoria, Australia

ACADEMIC PRESS   NEW YORK   SAN FRANCISCO   LONDON

A Subsidiary of Harcourt Brace Jovanovich, Publishers

ACADEMIC PRESS, INC.
111 Fifth Avenue, New York, New York 10003

United Kingdom Edition published by
ACADEMIC PRESS, INC. (LONDON) LTD.
24/28 Oval Road, London NW1

Library of Congress Cataloging in Publication Data

Fenner, Frank John,   (date)
    Medical virology.

    Includes bibliographies and index.
    1.   Virus diseases.   I.   White, David O., joint
author.   II.   Title.   [DNLM: 1.   Virus diseases.
2.   Viruses.   QZ65 F336m]
RC114.5.F45   1975   616.01'94   75-33981
ISBN 0−12−253060−8

Contents

Part I. PRINCIPLES OF ANIMAL VIROLOGY

1. Structure and Classification of Viruses

2. Cultivation and Assay of Viruses

v

3. Viral Multiplication

4. Viral Genetics

5. Effects of Viruses on Cells

6. Pathogenesis of Viral Infections

7. Host Responses to Viral Infections

8. Persistent Infections

9. Oncogenic Viruses

10. Epidemiology of Viral Infections

11. Evolutionary Aspects of Viral Diseases

12. Immunization against Viral Diseases

13. Chemotherapy of Viral Diseases

14. Laboratory Diagnosis of Viral Disease

Part II. VIRUSES OF MAN

15. Adenoviridae

16. Herpetoviridae

17. Poxviridae

18. Picornaviridae

19. Arboviruses: Togaviridae, Bunyaviridae, Reoviridae

20. Orthomyxoviridae

21. Paramyxoviridae

22. Other Families of Viruses

23.   Hepatitis, Rubella, "Slow" Viruses

24.   Common Viral Syndromes

Preface

Certain aspects of virology have expanded greatly since the first edition of this book was published in 1970, notably studies on the molecular biology of viruses in general and oncogenic viruses in particular. In medical virology there have been steady rather than dramatic advances. The major explosion in knowledge of medically important viruses that followed the introduction of cultured cells came to an end in the mid-1960's, and most of the viruses that resisted cultivation then (e.g., the hepatitis viruses) remain a problem today. However, the advances that have been made in viral classification, in immunology and immunopathology, in laboratory methods for rapid diagnosis, and in the etiology of several viral diseases, notably hepatitis, enteritis, and slow infections of the brain, justify the production of a second edition.

We have not changed the plan of the book but have essentially updated it to 1975, and have expanded certain sections that are important for medical students and practitioners, in particular the chapters on laboratory diagnosis and common viral syndromes. Photographs of patients have been introduced; despite their obvious limitations, black-and-white reproductions have been used for economic reasons. We have also expanded the lists of "Further Reading" at the end of each chapter with the aim of simplifying the entry of interested readers to the scientific literature.

As before, "Medical Virology" is intended primarily as a textbook for medical students, but Part I has been written as a self-sufficient synopsis of the principles of animal virology in the hope that it will serve as a useful introduction to the subject for students of science, dentistry, or veterinary science, and indeed for graduate students embarking on virological research.

We are indebted once more to those scientists acknowledged in the Preface to the First Edition, especially Mr. Ian Jack, and to Dr. Ian Gust and the coauthors of the second edition of "The Biology of Animal Viruses," which we have used as a basis for the revision of Part I. Our secretaries, Mrs. Margaret Mahoney, Mrs. Patricia Morrison, and Miss Clare Moore, have once again provided sterling help in the typing and preparation of the book, and the staff of Academic Press have cooperated to ensure its rapid production and low price.

We thank the authors and publishers of books or journals listed in the legends to the plates and figures for permission to use the illustrations thus acknowledged.

Frank Fenner
David O. White

Preface

To First Edition

The course of infectious diseases that affect man has changed remarkably in the last century and even during the last twenty years. In the more affluent countries and, recently, also in developing countries, improved hygiene and chemotherapy have greatly reduced mortality and, to a lesser extent, morbidity due to bacterial and protozoal diseases, but have had little effect on morbidity due to viral infections. Nevertheless, the pattern of viral diseases has also changed. Immunization and other preventive measures have led to the virtual disappearance of yellow fever and smallpox; in technologically advanced countries poliomyelitis and measles are also disappearing, and rubella might soon follow. "Civilization" led first to the appearance of epidemic poliomyelitis and then to its control by vaccination. We are now witnessing a comparable increase in the incidence of infectious hepatitis, and await the laboratory isolation of the causative viruses as an essential prelude to the development of a vaccine. Urbanization and rapid intercontinental travel have lead to widespread dissemination of the respiratory viruses. The morbidity due to these viruses and the variety of viruses involved will probably increase still more in the future.

Those who will become doctors in the next few decades need to understand the nature of these changes and to know enough about the molecular biology of viral replication to take advantage of antiviral chemotherapy when practical procedures are developed and to appreciate the possible role of viruses as causative agents in cancer. The aim of this book is to provide the student of medicine with a background that will enable him to appreciate viral diseases as they afflict human beings in urbanized western society, both as problems in management at the level of the individual patient and as problems in public health. Part I summarizes the principles of animal virology in relation to human infection and disease. It is derived, in part, from a condensation and extensive reorientation of a larger monograph, "The Biology of Animal Viruses," written by the senior author. Part II deals with the viruses of man and the diseases they cause. No attempt has been made to supplant the descriptions of disease and differential diagnosis supplied in medical textbooks. The aim has been to apply to each group of viruses the principles of human virology outlined in Part I of the book.

We considered, but finally rejected, the proposition to include chapters on chlamydiae, rickettsiae, and mycoplasmas. These agents are now known to be bacterial rather than viral in nature, and we believe that the time has come to break with tradition and exclude them from textbooks of virology.

The chapters in Part I conclude with a brief summary, but those of Part II do not lend themselves so readily to this approach since they contain much detailed information arranged in a standard format. We have selected illustrations and diagrams that will assist medical students in obtaining a clearer understanding of human viruses and viral diseases. No photographs of human patients have been included because of the difficulty of depicting signs other than skin rashes by black-and-white photography. We suggest that teachers supplement our illustrations with the excellent color photographs supplied in Swain and Dodds' "Clinical Virology."

We are grateful to the following colleagues in Australia for their comments on individual chapters: Drs. R. L. Doherty, A. A. Ferris, J. Forbes, J. R. L. Forsyth, I. H. Holmes, I. Jack, I. D. Marshall, and D. H. Watson. We are also indebted to all those, too numerous to mention here, who responded generously to our appeals for illustrative material; acknowledgments accompany the legends to the figures and plates. We owe a special debt to Ian Jack, of the Royal Children's Hospital, Melbourne, for providing so many of the photographs used.

The staff of Academic Press has given us much assistance with the production of the book and with the preparation of the figures. We are grateful to our secretaries, Mrs. Margaret Mahoney and Miss Elizabeth Duff, for their devotion and skill in preparing the manuscript.

Frank Fenner
David O. White

Supplementary Reading

The text contains no references, but a short "Further Reading" list is provided at the end of each chapter. For the convenience of teacher and student we list below additional books on general virology and virological techniques as well as periodical publications in which review articles on virology appear. The latter are a continuing source of authoritative papers. Review articles on viral diseases in these and other periodicals can be readily identified in the Monthly Bibliography of Medical Reviews issued both separately and as a section of *Index Medicus* by the National Library of Medicine in the United States.

Books

Andrewes, C. H., and Pereira, H. G. (1972). "Viruses of Vertebrates," Third Edition. Balliere, London. A comprehensive catalogue of animal viruses.

Fenner, F. J., McAuslan, B. R., Mims, C. A., Sambrook, J. F., and White, D. O. (1974). "The Biology of Animal Viruses," Second Edition. Academic Press, New York. Provides detailed information and references for Part I (Chapters 1–14).

Fraenkel-Conrat, H., and Wagner, R. R., Eds. (1974 onwards). "Comprehensive Virology." Plenum, New York. A multivolume treatise that deals with all viruses. Early volumes have accounts of the reproduction of several families of animal viruses.

Grist, N. R., Ross, C. A. C., and Bell, E. J. (1975). "Diagnostic Methods in Clinical Virology," Second Edition. Blackwell, Oxford.

Habel, K., and Salzman, N. P. (1969). "Fundamental Techniques in Virology." Academic Press, New York. An authoritative account of laboratory methods used in animal virology.

Horsfall, F. L., Jr., and Tamm, I., Eds. (1965). "Viral and Rickettsial Infections of Man," Fourth Edition. Lippincott, Philadelphia, Pennsylvania. Still the best source for information and references for Part II (Chapters 15–24).

Lennette, E. H., and Schmidt, N. J. Eds. (1969). "Diagnostic Procedures for Viral and Rickettsial Infections," Fourth Edition. American Public Health Association, New York. Standard reference book for diagnostic virology.

Luria, S. E., and Darnell, J. E. (1967). "General Virology," Second Edition. Wiley, New York. Best account of general principles of virology, but now rather out of date.

Maramorosch, K., and Koprowski, H. Eds. (1967–1971). "Methods in Virology," Volumes, I–V. Academic Press, New York. Comprehensive summary of virological techniques.

Swain, R. H. A., and Dodds, T. C. (1967). "Clinical Virology." Livingstone, Edinburgh and London. Good source of color illustrations.

Review Publications

Advances in Virus Research (K. M. Smith, M. A. Lauffer, and F. B. Bang, eds.). Academic Press, New York. Published annually since 1953.

Annual Review of Microbiology (M. P. Starr, ed.). Annual Reviews, Stanford, California. Published annually since 1947.

Bacteriological Reviews. Williams & Wilkins, Baltimore, Maryland. Published quarterly.

Cold Spring Harbor Symposia on Quantitative Biology. Cold Spring Harbor Laboratory of Quantitative Biology, New York. Annual publications on various subjects; several volumes deal with advanced research in virology or immunology.

Current Topics in Microbiology and Immunology (Ergebnisse der Mikrobiologie und Immunitatsforschung). Springer, Vienna. Published several times a year since 1967.

Modern Trends in Medical Virology (R. B. Heath and A. P. Waterson, eds.). Butterworth, London and Washington, D.C. Published irregularly since 1967.

Monographs in Virology (J. L. Melnick, ed.). Karger, Basel. Published irregularly since 1968.

Perspectives in Virology (M. Pollard, ed.). Proceedings of biennial symposia, mostly on animal virology. from 1959 onwards. Now published by Academic Press, New York.

Progress in Medical Virology (J. L. Melnick, ed.). Karger, Basel. Published annually since 1958.

Recent Advances in Medical Microbiology (A. P. Waterson, ed.). Churchill, London. Published irregularly since 1967.

Symposia of the Society of General Microbiology. Cambridge University Press, London and New York. Annual publications; several volumes contain review articles relevant to animal virology.

Virology Monographs (S. Gard, C. Hallauer, and K. F. Meyer, eds.). Springer, Vienna. Published irregularly since 1968.

WHO Technical Report Series. World Health Organization, Geneva, Switzerland. Irregular reports include authoritative articles on practical problems in medical virology.

PART I

Principles of
Animal Virology

CHAPTER 1

Structure and Classification of Viruses

INTRODUCTION

The unicellular microorganisms can be arranged in order of decreasing size and complexity: protozoa, yeasts and certain fungi, bacteria, mycoplasmas, rickettsiae, and chlamydiae. Then there is a major discontinuity, for the viruses cannot really be regarded as microorganisms at all. True microorganisms, however small and simple, are cells. They always contain DNA as the repository of their genetic information, and they also have their own machinery for producing energy and macromolecules. Microorganisms grow by synthesizing their own macromolecular constituents (nucleic acid, protein, carbohydrate, and lipid), and they multiply by binary fission.

Viruses, on the other hand, contain only one type of nucleic acid which may be either DNA or RNA, double-stranded or single-stranded. Furthermore, since viruses have no ribosomes, mitochondria, or other organelles, they are completely dependent upon their cellular hosts for the machinery of protein synthesis, energy production, and so on. Unlike any of the microorganisms, many viruses can, in suitable cells, reproduce themselves from a single nucleic acid molecule. The key differences between viruses and microorganisms are listed in Table 1-1.

Years of debate have centered around the question of whether viruses should be considered as living things. The controversy was first brought to a head when Stanley crystallized tobacco mosaic virus in 1935, and then again in 1956 when Gierer and Schramm demonstrated that purified viral nucleic acid can be infectious. Work emanating from the laboratories of Spiegelman and Kornberg in the late 1960's demonstrated that purified viral RNA or DNA will "multiply" in a test tube (in the presence of the requisite precursors and enzymes). It has also been known since the early 1960's that viral RNA

TABLE 1-1

Properties of Microorganisms and Viruses

	GROWTH ON NONLIVING MEDIA	BINARY FISSION	DNA AND RNA	RIBO-SOMES	SENSITIVITY TO ANTIBIOTICS	SENSITIVITY TO INTERFERON
Bacteria	+	+	+	+	+	−
Mycoplasmas	+	+	+	+	+	−
Rickettsiae	−	+	+	·I	+	−
Chlamydiae	−	+	+	+	+	+
Viruses	−	−	−	−	−	+

can be translated into viral coat protein by bacterial ribosomes *in vitro,* and that typical viral particles can be assembled *in vitro* from their constituent nucleic acid and protein. For scientists, the question of whether viruses are "living" or "nonliving" is now little more than one of semantics, but doubtless it is a question that will engage the attention of the lay press when a virus really is synthesized in a test tube for the first time.

Several important practical consequences flow from the differences between microorganisms and viruses; for example, some viruses (but no microorganisms) may persist indefinitely by the integration of their DNA (or a DNA copy of their RNA) with that of the host cell, and viruses, being metabolically inert, are not susceptible to antibiotics that act against specific steps in the metabolic pathways of microorganisms.

MORPHOLOGY OF VIRUSES

Physical Methods for Studying Viruses

For many years it has been known that viruses are smaller than microorganisms. In 1898 Loeffler and Frosch demonstrated that foot-and-mouth disease of cattle could be transferred by material which could pass through a filter of average pore diameter too small to admit the passage of bacteria. The new group of "organisms" became known as the "filterable viruses." For a time they were also called "ultramicroscopic," since all but the poxviruses were beyond the limit of resolution of light [200 nanometers (nm) = 200 millimicrons $(m\mu) = 2000$ angstroms (Å)]. Only with the advent of the electron microscope was it possible to study the morphology of viruses properly. It became apparent that they ranged in size from the poxviruses, which at 300 × 250 × 100 nm are about the size of the smallest micro-organisms (mycoplasmas), down to the parvoviruses, which at 20 nm in diameter are about the size of the largest protein molecules, e.g., hemocyanins. Then in 1959, our knowledge of viral ultrastructure was transformed when Brenner and Horne applied negative staining to the electron microscopy of viruses. Potassium phosphotungstate, which is electron dense, fills the interstices of the viral surface, giving the resulting electron micrograph a degree of fine detail not previously possible. The technique of X-ray diffraction had already provided evidence that the protein molecules of the viral coat are packed around the nucleic acid molecule in a symmetrical arrangement; negative staining confirmed and extended this finding in a most revealing way.

Viral Structure

In the simpler viruses the mature virus particle, the *virion*, consists of a single molecule of nucleic acid surrounded by a protein coat, the *capsid;* the capsid and its enclosed nucleic acid together constitute the *nucleocapsid* (Fig. 1-1). In some of the more complex viruses the capsid surrounds a protein *core*. The viral nucleic acid is not just stuffed inside the virion; it probably has a specific relationship with the polypeptides of either capsid or core. The capsid is composed of a large number of *capsomers*, which are held together by noncovalent bonds and are visible by electron microscopy. Each capsomer is in turn made up of one or more polypeptide chains. X-Ray diffraction reveals the symmetry of arrangement of these molecules ("chemical units"); electron microscopy reveals the arrangement of the capsomers ("morphological units").

For reasons of genetic economy, capsids are composed of repeating units of one or a small number of different kinds of polypeptides. In order to form a complete shell to protect the viral nucleic acid from nucleases, molecules of these polypeptides must pack together symmetrically, and only two kinds of symmetry have been recognized in capsids: icosahedral and helical (Fig. 1-1).

Icosahedral Symmetry. The icosahedron is one of the classical "Platonic solids" of geometry; it has 12 vertices and 20 faces, each an equilateral triangle. It has axes of two-, three-, and fivefold rotational symmetry, passing through the edges, faces, and vertices, respectively (see Fig. 1-2). The mathematical problems involved in distributing subunits symmetrically over the surface of such a solid were solved long ago, and in recent times they have been applied not only in studies of viral structure (Plate 1-1) but also in the architect Buckminster Fuller's

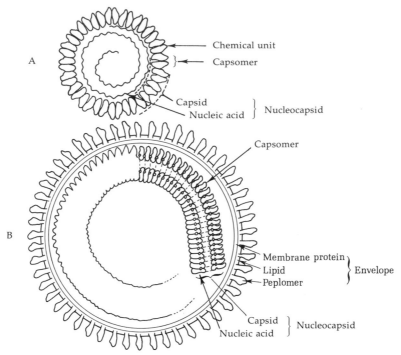

FIG. 1-1. *Schematic diagrams of the structure of a simple nonenveloped virion with an icosahedral capsid (A) and an enveloped virion with a tubular nucleocapsid with helical symmetry (B). The capsids are composed of capsomers, which are in turn made up of one or more chemical subunits (polypeptide chains). Many icosahedral viruses have a "core" (not illustrated), which consists of protein(s) directly associated with the nucleic acid, inside the icosahedral capsid. In viruses of type B the envelope is a complex structure consisting of an inner virus-specified protein shell (membrane protein) and a lipid layer derived from cellular lipids, in which are embedded virus-specified glycoprotein subunits (peplomers), each of which, like a capsomer, consists of one or more polypeptide chains. [Modified from D. L. D. Caspar et al., Cold Spring Harbor Symp. Quant. Biol. **27**, 49 (1962).]*

geodesic domes. In each case the object is to enclose a maximal volume within a strong structure.

In theory, the triangular facets of the icosahedron can be further subdivided into equilateral triangles. Only certain subdivisions are possible: the number of new triangles per facet is called the triangulation number T, and $T = H^2 + HK + K^2$ where H and K are any pair of integers. Three chemical units can be placed on each small triangle, but they can cluster together in different ways, as shown in Fig. 1-2. If, as is common, the chemical units cluster into pentamer and

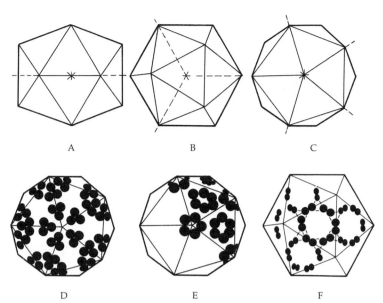

FIG. 1-2. *Features of icosahedral structure. Regular icosahedron viewed along twofold (A), threefold (B), and fivefold (C) axes. Various clusterings of structural subunits give characteristic appearances of capsomers in electron micrographs. With T = 3 the structural subunits may be arranged as 20T trimers (D), capsomers being then difficult to define, as in poliovirus; or they may be grouped as 12 pentamers and 20 hexamers (E) which form bulky capsomers as in parvoviruses, or as dimers on the faces and edges of the triangular facets (F), producing an appearance of a bulky capsomer on each face, as in caliciviruses.*

hexamer capsomers (Fig. 1-2E), then the number of capsomers per capsid is given by the simple formula $10T + 2$. There are always 12 pentamers, but the total number of capsomers may be 12, 32, 72, 162, 252, and so on, depending on the virus group. The total number of polypeptide molecules per capsid is always $60T$. In some viruses the pentamers and hexamers are composed of different species of polypeptide.

Recently various kinds of icosahedral virions have been found to contain an additional inner protein shell; in the case of the reoviruses, this "inner capsid" also has icosahedral symmetry. In other cases, e.g., adenoviruses, poxviruses, and perhaps herpetoviruses, the inner DNA-containing protein "core" has not been shown to have a definite symmetrical structure.

Helical (Screw) Symmetry. The nucleocapsids of certain RNA viruses show a different type of symmetry: the capsomers and nucleic acid

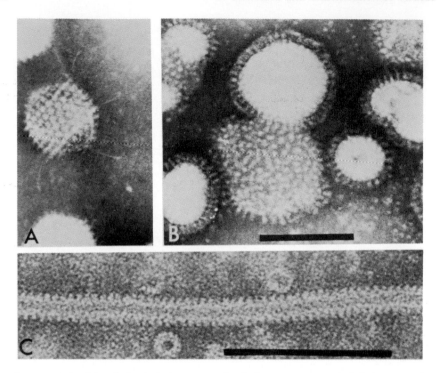

PLATE 1-1. *Morphological features of viral structure revealed by negative staining (bars = 100 nm). (A) Icosahedral structure of adenovirus capsid. At each of the 12 vertices there is a penton base capsomer from which projects a fiber with a small terminal knob; each of the 20 triangular facets contains 12 identical hexon capsomers. The capsid encloses a protein core with which the DNA is associated. (Courtesy Dr. N. G. Wrigley.) (B) Envelope of influenza virus (family: Orthomyxoviridae). The projections (peplomers) are of two morphological types: the hemagglutinin subunits which are rectangular rods of protein and the neuraminidase subunits which have a more complex structure (see Fig. 20-1). Both are embedded in lipid, and this lipoprotein envelope encloses a coiled ribonucleoprotein tube. (Courtesy Dr. N. G. Wrigley.) (C) Nucleocapsid of parainfluenza virus 1 (family: Paramyxoviridae). The RNA is wound within and protected by a helix composed of identical protein subunits. The complete nucleocapsid is 1 μm long, but in the intact particle is coiled up within a roughly spherical envelope 0.15 μm in diameter. (Courtesy Dr. A. J. Gibbs.)*

molecule are wound together in a helix or spiral (Plate 1-1). With all helical nucleocapsids analyzed so far, each capsomer consists of a single polypeptide molecule. The rod-shaped plant viruses consist solely of such a naked nucleocapsid. However, in viruses of vertebrates this tubular nucleocapsid is wound into a coil, and surrounded by a loosely fitting *envelope* of lipoprotein. The envelope usually consists of an inner "membrane protein" that is virus-specified, and an outer lipoprotein complex (Fig. 1-1), which is derived directly from the cellular cytoplasmic membrane during the release of the nucleocapsid from the infected cell by a process of "budding." However, the outer lipoprotein complex is not an unmodified cellular membrane; the cell protein in the membrane has been replaced by virus-specified glycoprotein subunits called *peplomers* (*peplos* = envelope). These peplomers can be clearly seen in electron micrographs as "spikes" projecting from the envelope (Plate 1-1).

Envelopes are not restricted to viruses of helical symmetry; some icosahedral viruses have them also, as do the rhabdoviruses, in which the envelope is closely applied to a bullet-shaped shell that encloses a tubular nucleocapsid with helical symmetry.

CHEMICAL COMPOSITION OF VIRUSES

Methods of Purification of Viruses

An essential preliminary to the chemical analysis of viruses has been the development of adequate methods of purifying virus particles. Special problems are created by the close association of viruses with the cells they parasitize; it is not an easy matter to rid virions of associated cell debris, or even from viral proteins synthesized in excess in the infected cell. Furthermore, the infectivity of virions is very sensitive to inactivation by heat, acid, alkali, and sometimes lipid solvents or osmotic shock. Accordingly, throughout all purification protocols the virus is maintained at near neutral pH and 4°.

Liberation of Virus from Cells. The first step in the purification process consists of liberating virus from the cells in which it has grown. The supernatant fluid bathing an infected cell culture provides the cleanest starting source of virions, but some viruses must be released from the cells by such methods as sonic vibration, homogenization, or repeated freeze-thawing.

Chemical Methods of Purification. The surface of the virion consists of protein, sometimes together with associated lipid. Accordingly, chemical methods of purifying viruses must avoid conditions leading to the denaturation of proteins. Viruses can be adsorbed to columns of DEAE-cellulose, calcium phosphate, aluminum hydroxide gel, ion-exchange resins, Sephadex, or erythrocytes, and then eluted with buffers of different pH and ionic strength, or at a different temperature. Alternatively, nonenveloped viruses can be separated from lipids and soluble proteins by partitioning them into the aqueous phase of a fluorocarbon-water emulsion.

Physical Methods of Purification. After preliminary purification and concentration by chemical methods, virus particles can be separated from soluble contaminants by physical procedures such as centrifugation. Differential centrifugation consists of alternate cycles of low- and high-speed centrifugation to deposit, first large contaminating particles, then virions. Rate zonal (velocity gradient) centrifugation through a preformed gradient of a dense solute such as sucrose permits the virions to move through the gradient at a rate determined by their sedimentation coefficient (a function of their density, size, and shape). Equilibrium (isopycnic) gradient centrifugation, on the other hand, separates virions according to their buoyant density, in dense solutes like cesium chloride or potassium tartrate (or sucrose in the case of enveloped viruses of low density); after prolonged ultracentrifugation at very high gravitational forces the virions will come to rest in a sharp band in that part of the tube where the solution is at the same density as the virus, i.e., usually within the range 1.15–1.4.

Concentration of Viruses. Following (or preceding) purification, viruses can be concentrated into a small volume by ultracentrifugation, freeze-drying, pervaporation, or dialysis against a hydrophilic agent such as polyethylene glycol.

Nucleic Acid

Viruses contain only a single species of nucleic acid. This may be DNA or RNA; indeed, the RNA viruses comprise the only instance in nature in which RNA is the sole carrier of genetic information. Such RNA can be double-stranded or single-stranded, as can viral DNA. The DNA of the papovaviruses is circular, and has a "supercoiled" configuration. As yet no animal viral nucleic acid has been found to be methylated, or to contain novel bases of the type encountered in bacterial viruses or mammalian transfer RNA's. However, the base composition of DNA from mammalian viruses covers a far wider range than that of the mammals themselves (e.g., the guanine plus cytosine content can vary from 35 to 75%). In the evolutionary hierarchy, therefore, viruses comprise a considerably more diverse group than mammals. This is supported by nucleic acid hybridization tests, which show that viruses from separate families differ more from one another than, say, a mouse does from an elephant.

The molecular weight of viral nucleic acid varies over a broad range from 1.6–160 million daltons, corresponding to a range of about 3–160 cistrons. On the whole, the larger viruses contain more nucleic acid. The nucleic acid can be extracted from the virion with detergents (such as sodium dodecyl sulfate) or phenol. The released molecules are fragile, but the relatively short nucleic acid molecules derived from some small viruses are demonstrably infectious when inoculated into cell culture in the presence of histones, bentonite, DEAE-dextran or dimethyl sulfoxide to protect them against nucleases, and high osmotic pressure to facilitate their entry into cells. In other cases, the isolated nucleic acid is not infectious even though it contains all the necessary genetic information, for its transcription depends on a virion-associated transcriptase without which multiplication cannot proceed.

All DNA viruses have genomes that consist of a single molecule of nucleic acid, but the genomes of many RNA viruses consist of several different molecules (sometimes corresponding with individual genes) which are probably loosely linked together in the virion. In viruses whose genome consists of single stranded nucleic acid, the viral nucleic acid is either the "positive" strand (in RNA viruses, equivalent to messenger RNA) or the "negative" (complementary) strand. Preparations of some viruses with genomes of single-stranded DNA consist of particles that contain either the positive or the complementary strand.

Viral preparations often contain some particles with an atypical content of nucleic acid. Host cell DNA is found in some papovaviruses, and cellular ribosomes are incorporated in some arenaviruses. Several copies of the complete viral genome may be enclosed within a single particle (as in paramyxoviruses), or viral particles may be formed that contain no nucleic acid ("empty" particles) or that have an incomplete genome, lacking part of the nucleic acid that is needed for infectivity.

Viral nucleic acid molecules can be clearly seen and measured under the electron microscope (Plate 1-2).

Protein

The major constituent of all viruses is protein. Its primary role seems to be to provide the viral nucleic acid with a protective coat. The proteins on the surface of the virion have a special affinity for complementary receptors present on the surface of susceptible cells. They also contain the antigenic determinants against which the body directs its immunological response to infection. Most species of virion contain several different polypeptides separable by polyacrylamide gel electrophoresis; the number ranges from one or two in the case of some simple viruses to over thirty in the case of the highly complex poxviruses. Some histone-like polypeptides comprise an inner core in-

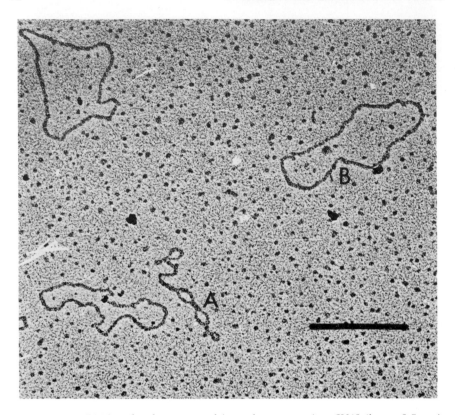

PLATE 1-2. *DNA molecules extracted from the papovavirus SV40 (bar = 0.5 μm).*
Molecules of SV40 DNA exist in two major forms. When it is isolated from the virus
particles most of the DNA occurs in the configuration shown in A, as double-
stranded closed-circular molecules containing superhelical twists. If one of the DNA
strands is broken, the superhelical twists are relieved and the molecule assumes a
relaxed circular configuration (B). (Courtesy Dr. P. Sharp.)

timately associated with the nucleic acid, but most make up one or
more concentric shells of capsomers comprising the capsid of the
virion. Each capsomer is composed of one to six molecules of a single
species of polypeptide; however, capsomers from different regions of
the virion may be composed of different polypeptides. Yet other polypeptides, in the form of glycoproteins, make up the peplomers projecting from the envelope of enveloped viruses, in which cellular proteins are no longer present. Some virions carry a limited number of
enzymes for specialized purposes, the nucleocapsid-associated transcriptase being the most important example. An interesting feature of
the capsid proteins is that they are usually assembled in the virion in

such a way that the virions are resistant to digestion by proteolytic enzymes.

Lipid and Carbohydrate

Certain viruses derive an *envelope* in the process of "budding" through the cytoplasmic membrane (or more rarely, nuclear membrane) of the infected cell, the lipids of the envelope being characteristic of the species of cell. Enveloped viruses are susceptible to destruction by lipid solvents, such as ether, chloroform, and bile salts. Poxviruses contain lipids in the virion itself (i.e., not in an envelope derived from the cell).

Enveloped viruses also contain a small amount of carbohydrate in addition to the sugars of the nucleic acid molecule. Some of this carbohydrate occurs as cellular glycolipid, but most is in the form of glycoprotein in the peplomers of the viral envelope.

INACTIVATION OF VIRUSES

Knowledge of the physical and chemical agents that inactivate the infectivity of the virion is relevant not only to antisepsis and disinfection, but also to vaccine production. Furthermore, the preservation of viruses for laboratory studies, including the isolation of virus from clinical specimens, demands the avoidance of such inactivating conditions.

We are not concerned here with *virostatic* agents, which inhibit the growth of viruses, but only with *virocidal* agents, which inactivate the infectivity of the virion by direct action upon its nucleic acid, protein, or lipid *in vitro*.

Physical Agents

Heat. Viruses are notoriously heat-labile. Within a few minutes at temperatures of 55°–60° the capsid protein is denatured, with the result that the virion is no longer infectious, presumably because it is no longer capable of normal cellular attachment and/or uncoating. Even at body temperature some loss of infectivity occurs, although this loss is greatly outweighed by the rate at which viable particles continue to multiply *in vivo*. At ambient temperatures the rate of decay is slower but nevertheless significant mainly due to changes in the nucleic acid. Accordingly, viruses must be stored at low temperatures; 4° is usually satisfactory for a day or so, but longer-term preservation requires temperatures well below zero, e.g., −70°, the temperature of solid carbon dioxide, or $-196°$, the temperature of liquid nitrogen. As a rule of thumb it could be said that the half-life of most viruses can be measured in seconds at $60°$, minutes at $37°$, hours at $20°$, days at $4°$, months at $-70°$, and years at $-196°$. The enveloped viruses are more heat-labile than naked icosahedra. Some enveloped respiratory viruses, notably the respiratory syncytial virus, tend to be inactivated by the actual process of freezing and subsequent thawing, probably as a result of disruption of the virion by ice crystals. This places the physician in something of a dilemma; perhaps the most useful advice is that specimens taken from patients should be dispatched to the laboratory as rapidly as practicable in a thermos flask of ice, but not frozen. Laboratory research workers, on the other hand, often need to preserve stocks of viable virus for years on end. This they achieve in one of two ways: (a) "Rapid-freezing" of small aliquots of virus suspended in media containing protective protein, or better still, dimethyl sulfoxide, followed by storage at $-70°$ or $-196°$; (b) Freeze-drying (lyophilization), i.e., dehydration of a frozen virus suspension under vacuum; the resulting powder can thereafter be stored indefinitely at $4°$.

Ionic Environment and pH. On the whole viruses prefer an isotonic environment at a physiological pH, but their limits of tolerance are fairly wide. The infectivity of certain viruses may actually be protected against heat inactivation by suspending them in molar Mg^{2+}, but for other viruses the situation is reversed. The reason for the differences is unknown; they are of interest mainly to laboratory research workers and vaccine manufacturers.

Radiation. Viruses are inactivated by ionizing (e.g., X-rays, γ-rays) and nonionizing (e.g., ultraviolet) radiations by different mechanisms. The efficiency of inactivation by ionizing radiation of viruses with single-stranded nucleic acids is very high, for almost every ionization causes a lethal break in the polynucleotide chain. With double-stranded nucleic acids only a proportion of the one-strand breaks is lethal.

Ultraviolet irradiation produces its damaging effects on DNA by causing adjacent pyrimidine bases on the same polynucleotide chain to form dimers; uracil dimers are responsible for most of the UV damage to RNA viruses.

Chemical Agents

Lipid Solvents. Enveloped viruses are readily destroyed by lipid solvents, such as ether, chloroform or sodium deoxycholate. Indeed, sensitivity to lipid solvents is employed as a preliminary screening test in the identification of new viral isolates. A corollary is that enveloped viruses usually do not survive exposure to the bile in the alimentary tract.

Antiseptics and Disinfectants. The nonenveloped viruses are resistant to many of the chemicals conventionally employed as antibacterial disinfectants. Anionic detergents such as sodium dodecyl sulfate are effective against all enveloped viruses and some, but not all, nonenveloped icosahedral viruses. Formaldehyde, dilute hydrochloric acid, or sodium hypochlorite will satisfactorily disinfect contaminated articles, but are too irritant to be brought into contact with man. There is a definite need for a virocidal chemical sufficiently innocuous to be used for routine skin antisepsis.

Chlorination of drinking-water supplies does not always succeed in ridding them of the more resistant viruses, such as the hepatitis viruses and the enteroviruses. The effectiveness of chlorination is greatly influenced by the amount of organic matter present in the water, since proteins, in particular, tend to protect viruses against inactivation.

Heterotricyclic "vital" dyes, such as neutral red or acridine orange, may become intercalated between the bases of replicating viral nucleic acid molecules. This is not lethal in itself, but becomes so upon subsequent exposure to light—a process known as photodynamic inactivation.

CLASSIFICATION OF VIRUSES

Virtually every kind of organism can be parasitized by viruses: vertebrate and invertebrate animals, plants, fungi, protozoa, mycoplasmas, and bacteria. There are even some "satellite" viruses which are in a sense "parasites" on other viruses, although in such cases both agents are parasitic on their host cells. All viruses, whatever their cellular hosts, share the features summarized in Table 1-1, hence, taxonomists concerned with viral classification have now accepted a scheme of classification that embraces all viruses. From the operational point of view, however, we can divide viruses into those that affect vertebrate animals, insects, plants, and bacteria. In this book we shall be concerned solely with the viruses of vertebrate animals, primarily those that affect man. Some of these viruses, the arthropod-borne or arboviruses, will also multiply in insects or other arthropods.

There are several hundred recognized viruses of man alone, and there must be many more that have not been discovered yet. In order to simplify their study we need to sort them into groups that share certain common properties.

Classification Based on Physicochemical Criteria

The classification of viruses is still in its infancy, but some of the newer techniques upon which their classification can now be based are introducing a degree of refinement to the science of taxonomy that had not previously existed. These depend on the fact that all phenotypic characters reflect the nucleotide sequence of the genetic material, and in viruses this is usually a single small molecule which can be purified and analyzed in a variety of ways. The ultimate test of resemblance, short of determining the complete nucleotide sequence as was first accomplished for the RNA bacteriophage MS2 in 1975, is the nucleic acid hybridization test, which depends upon precise homology between short nucleotide sequences. This method is very useful for studying viral messenger RNA's (mRNA) (see Chapter 3) and can be used for estimating the degree of relationship of closely related viruses. It is too sensitive to be useful with distantly related viruses, hence, cruder methods of comparison, like polynucleotide maps, molecular weight, and nearest neighbor nucleotide doublet frequencies of the viral nucleic acid give more meaningful results. The primary importance of the nucleic acid is reflected in the major binary division of all viruses into those containing DNA and those containing RNA.

The capsid proteins are a direct expression of viral genes and the amino acid sequences of the constituent polypeptides are a direct reflection of the nucleotide sequences of these genes. Cruder measurements like peptide maps and even amino acid composition give valuable evidence of relationships between viruses. Antigenicity depends upon short sequences of amino acids in critical parts of the folded polypeptide chain. As serological methods are so simple, antigenic relationships form the cornerstone of viral classification within genera, and sometimes within families.

Looking into the future, an important feature that may prove of great value at the "family" level of classification is what may be called the "strategy of the viral genome." This denotes the way in which the genetic information is translated into proteins, and thus encompasses such features as the nature and structure of the viral genome, the "polarity" of single-stranded RNA (messenger or complementary), and whether translation of the genome into individual polypeptides results from fragmentation of the genome or from posttranscriptional or post-

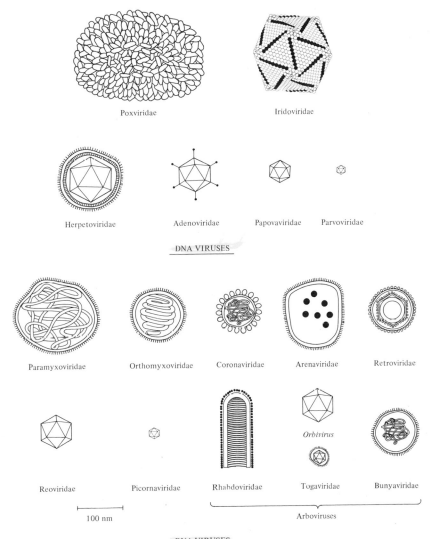

Poxviridae   Iridoviridae

Herpetoviridae   Adenoviridae   Papovaviridae   Parvoviridae

DNA VIRUSES

Paramyxoviridae   Orthomyxoviridae   Coronaviridae   Arenaviridae   Retroviridae

Reoviridae   Picornaviridae   Rhabdoviridae   Orbivirus / Togaviridae   Bunyaviridae

100 nm   Arboviruses

RNA VIRUSES

FIG. 1-3. *Diagram illustrating the shapes and relative sizes of animal viruses of the major families (bar = 100 nm).*

translational cleavage (see Chapter 3 for a fuller discussion of these terms).

Less directly related to the nucleotide sequence of the viral nucleic acid, but still ultimately dependent on it, is the morphology of the virion. Since this is more readily determined than any other viral

TABLE 1-2

Morphology of the Families of Animal Viruses

FAMILY	SHAPE	DIAMETER (nm)	ENVELOPE	SYMMETRY	NUMBER OF CAPSOMERS (IF ICOSAHEDRAL)
DNA viruses					
Parvoviridae	Spherical	20	−	Icosahedral	32
Papovaviridae	Spherical	45–55	−	Icosahedral	72
Adenoviridae	Spherical	70–80	−	Icosahedral	252
Herpetoviridae	Spherical	150	+	Icosahedral	162
Poxviridae	Brick-shaped	100 × 240 × 300	−	−	−
RNA viruses					
Picornaviridae	Spherical	20–30	−	Icosahedral	? 60
Togaviridae	Spherical	40–60	+	Icosahedral	?
Bunyaviridae	Spherical	90–100	+	Helical	−
Arenaviridae	Spherical	85–120	+	? Helical	−
Coronaviridae	Spherical	80–120	+	Helical	−
Retroviridae	Spherical	100–120	+	Helical	−
Orthomyxoviridae	Spherical or filamentous	80–120	+	Helical	−
Paramyxoviridae	Spherical or filamentous	100–200	+	Helical	−
Rhabdoviridae	Bullet-shaped	70 × 180	+	Helical	−
Reoviridae	Spherical	50–80	−	Icosahedral	?

TABLE 1-3

Chemical Composition of the Families of Animal Viruses

| FAMILY | NUCLEIC ACID | | PROTEINS[c] | TRANSCRIPTASE |
	CONFIGURATION[a]	MW[b]		
DNA viruses				
Parvoviridae	SS	2	3	−
Papovaviridae	DS	3–5	6	−
Adenoviridae	DS	20–25	9	
Herpetoviridae	DS	100	12–24	−
Poxviridae	DS	160	>30	+
RNA viruses				
Picornaviridae	SS	2–3	4	−
Togaviridae	SS	4	3	−
Bunyaviridae	SS	6	3	+
Arenaviridae	SS	6	?	?
Coronaviridae	SS	9	16	+
Retroviridae	SS	10–12	7–8	+
Orthomyxoviridae	SS	5	7	+
Paramyxoviridae	SS	7	6	+
Rhabdoviridae	SS	4	7	+
Reoviridae	DS	15	7	+

[a] SS = single-stranded, DS = double-stranded.

[b] Molecular weight in millions of daltons; range indicates extremes within group.

[c] Approximate number of different virus-coded polypeptides in virion.

property, and since morphology represents the integrated result of the operation of many viral genes, morphology is a very useful method of allocating viruses to the major taxonomic groups (see Plate 1-1 and Fig. 1-3).

These "groups" may be considered as the approximate equivalent of the families of botanical and zoological classification and there is now wide agreement on Latinized names for the major families and genera (Table 1-2). Individual viruses within each genus could be considered as equivalent to species, but virology is not yet ready for complete adoption of a system of Latinized binomial names, and they are known by their vernacular names, e.g., measles virus. Tables 1-2 and 1-3 list the properties of the dozen or so major families of vertebrate viruses under the following headings:

1. *Morphology of the virion*
   (a) Shape and size
   (b) Presence or absence of an envelope
   (c) Symmetry of the nucleocapsid

2. *Chemistry of the nucleic acid*
   (a) Type
   (b) Number of strands and number of separate molecules
   (c) Molecular weight
   (d) Polarity and presence or absence of a transcriptase

Some of the names used in the tables have yet to receive universal acceptance, but there is general agreement about the reality of these taxonomic groups. A brief description of each major family is given later in this chapter; those families containing viruses of medical importance are described in greater detail in Chapters 15 to 22.

Classification Based on Epidemiological Criteria

Animal viruses can be transmitted by ingestion, by inhalation, by injection, or by contact. Hence, in discussing the epidemiology and pathogenesis of viral infection we will frequently make use of the following terminology as a useful subsidiary way of classifying viruses.

(a) *Enteric viruses* enter the body through, and multiply primarily in, the gastrointestinal tract. The term embraces the acid- and bile-resistant enterovirus subgroup of the picornaviruses, together with the adenoviruses, reoviruses and hepatitis viruses.

(b) *Respiratory viruses* enter the body through, and multiply in, the respiratory tract. Often the term is restricted to those viruses that remain localized to this part of the body. The group includes the orthomyxoviruses, paramyxoviruses, coronaviruses, and the rhinovirus subgroup of the picornaviruses, as well as adenoviruses, reoviruses, and some enteroviruses, which are also capable of growth in the enteric tract.

(c) *Arboviruses* (*arthropod-bo*rne viruses) infect arthropods which ingest vertebrate blood; they multiply in the arthropod's tissues and can then be transmitted by bite to susceptible vertebrates. Viruses belonging to four families are included: Togaviridae, Bunyaviridae, Rhabdoviridae, and *Orbivirus* (a genus of the family Reoviridae).

The Major Groups of DNA Viruses

Parvoviridae. Many laboratory stocks of adenoviruses are contaminated with very small particles which have been called adeno-associated viruses (AAV). They are about 20 nm in diameter and have icosahedral symmetry and have now been classified as a genus, *Adenosatellovirus*, of the family Parvoviridae (*parvo* = small). Four serotypes have been recognized; none is related serologically to any known adenovirus, nor does their DNA hybridize with adenovirus

DNA. The DNA of AAV is a single-stranded molecule of either polarity (+ or −) in any given virion. Adeno-associated virus is defective in that it is incapable of replication except in the presence of a "helper" adenovirus; it then replicates freely in the nucleus of the cell. Only adenoviruses, but any adenovirus, can function as a helper. One wonders how AAV's have survived the rigorous selective pressure of evolution.

Papovaviridae. The papovaviruses (sigla: Pa = papilloma; po = polyoma; va = vacuolating agent, SV40) are small nonenveloped icosahedral viruses which replicate in the nucleus of vertebrate cells. In the virion their nucleic acid occurs as a cyclic double-stranded molecule.

Apart from the human wart virus and the recently discovered virus of progressive multifocal leukoencephalopathy, the main interest of the papovaviruses to students of medicine is that polyoma virus and simian virus 40 (SV40) have been important tools for unraveling the molecular and cellular events in neoplasia (see Chapter 9).

Adenoviridae. The adenoviruses (*adeno* = gland) are nonenveloped icosahedral viruses which multiply in the nuclei of infected cells. Adenoviruses occur in many different species of animal; 32 serologically distinct types of human adenovirus have been recognized, all of which share the adenovirus group antigen with adenovirus serotypes infecting other mammals.

Human adenoviruses are usually associated with infection of the respiratory tract, and occasionally the eye. Many are characterized by prolonged latency. They also multiply in the intestinal tract and are frequently recovered in feces. Some of the adenoviruses of man and other animals cause malignant tumors when inoculated into newborn hamsters and have been used extensively in experimental studies on oncogenesis (see Chapter 9).

Herpetoviridae. The herpetoviruses (*herpes* = creeping) are icosahedral DNA viruses which multiply in the nuclei of infected cells. They mature by budding through the nuclear membrane, acquiring an envelope in the process.

Among the important herpetovirus diseases of man are herpes simplex, varicella, and zoster, all of which are characterized by vesicular lesions on the skin or mucous membranes. Cytomegalovirus occasionally produces a serious congenital infection, and infectious mononucleosis is caused by the recently discovered EB virus. A feature of all herpetovirus infections is prolonged persistence of the virus in the body, often in latent form. Episodes of recurrent clinical disease of endogenous origin may occur years after the initial infection.

There has been much recent interest in the findings that EB virus and herpes simplex type 2 are regularly associated with certain human cancers (see Chapter 9).

Poxviridae. The poxviruses (*pock* = pustule) are the largest and most complex viruses of vertebrates and contain the largest DNA molecule. The virion consists of a brick-shaped DNA-containing core surrounded by a complex series of membranes of viral origin. Unlike other DNA viruses of mammals, poxviruses multiply in the cytoplasm.

Viruses of several different poxvirus genera may infect man; smallpox, alastrim, and molluscum contagiosum are specifically human diseases, whereas man is only occasionally infected with the viruses of orf, cowpox and milker's nodes, which normally affect livestock. All produce skin lesions, and smallpox and alastrim are serious systemic diseases.

The Major Groups of RNA Viruses

Picornaviridae. The picornaviruses (sigla: *pico* = small; *rna* = ribonucleic acid) include a very large number of viruses of vertebrates. They are small nonenveloped icosahedral viruses which contain a single molecule of single-stranded RNA, and multiply in the cytoplasm.

The subdivision of this family into genera is currently under discussion; from the point of view of human medicine we can concentrate upon the two major genera: *Enterovirus* and *Rhinovirus*. The enteroviruses in turn contain 3 polioviruses, 33 echoviruses, and 30 coxsackieviruses, all of which usually produce inapparent enteric infections. However, the polioviruses may also cause paralysis, while the other enteroviruses are sometimes associated with meningitis, rashes, myocarditis, myositis, or mild upper respiratory tract infections. The rhinoviruses are the major cause of the common cold.

Togaviridae. There are currently about 300 known arboviruses, distributed between about 30 antigenic groups. It is now recognized that, although the arboviruses comprise a logical epidemiological group, they are in fact made up of very different viruses that belong to several different families. The majority are bunyaviruses or togaviruses. The family Togaviridae (*toga* = cloak) is divided into two genera: *Alphavirus* (previously known as "group A" arboviruses) and *Flavivirus* (previously known as "group B" arboviruses).

The togaviruses can be defined on physicochemical data without reference to their capacity to multiply in arthropods (see Tables 1-2 and 1-3). They are spherical enveloped viruses containing single-stranded RNA enclosed within a core of cubic symmetry. They multiply in the cytoplasm and mature by budding from cytoplasmic membranes.

In nature the togaviruses usually produce inapparent viremic infections of birds, mammals, or reptiles, being transmitted by arthropod vectors, in which they also multiply. When man or his domestic animals are bitten an inapparent infection is the usual consequence, but severe generalized disease can result. These include some of the most lethal of human viral diseases, such as hemorrhagic fever, yellow fever, and encephalitis. The virus of rubella, although not arthropod-transmitted, is in other respects a typical togavirus.

Bunyaviridae. Over 100 bunyaviruses (Bunyamwera = a locality in Africa) comprise the largest single group of arboviruses. They are enveloped RNA viruses of helical symmetry. Several can infect man but most cause relatively minor fevers.

Arenaviridae. Arenaviruses (*arena* = sand) acquired their name because of the presence of cellular ribosomes (resembling "grains of sand") incorporated within the rather pleomorphic enveloped virion. All cause natural inapparent infections of rodents. Man may occasionally contract a serious generalized disease, e.g., Lassa fever and lymphocytic choriomeningitis, from a domestic (laboratory) or wild rodent.

Coronaviridae. The coronaviruses (*corona* = crown) are medium-sized somewhat pleomorphic viruses, with widely spaced, club-shaped peplomers in their lipoprotein envelope, which encloses an RNA-containing core of undetermined symmetry. The group includes several agents that cause common colds in man.

Retroviridae. This name (sigla: *re* = reverse; *tr* = transcriptase) has been approved as a family name instead of *"Leukovirus"* for viruses that are characterized by the fact that their genome is RNA, the virion contains a reverse transcriptase that transcribes DNA from RNA, and their multiplication involves this process. The DNA copy of the viral genome thus transcribed is integrated into the cellular DNA before multiplication occurs, and the DNA copy of the genome of some of them (subfamily Oncovirinae) is permanently carried in the DNA of "normal" mouse and chick cells.

The family Retroviridae is subdivided into three subfamilies: Oncovirinae, which include the mammary tumor virus of mice, the leukosis and sarcoma viruses of birds and the leukemia and sarcoma viruses of mice and other mammals; Spumavirinae (*spuma* = foam) the "foamy agents" found in some cultured cells; and Lentivirinae (*lenti* =

slow), which includes several viruses that cause "slow" (progressive) diseases in sheep.

Most infections with oncoviruses remain latent for prolonged periods, but they may eventually cause fatal disease, usually a neoplasia of the lymphoid or hemopoietic systems. A few strains cause the rapid production of solid tumors, notably Rous sarcoma virus in birds and Moloney and Harvey sarcoma viruses in rodents. Much recent cancer research has focused on the possible significance of putative oncovirus particles recovered from human tumors.

Orthomyxoviridae. The orthomyxoviruses (*myxo* = mucus) occur as spherical or as filamentous RNA viruses with a tubular nucleocapsid enclosed within an envelope acquired as they bud from the cytoplasmic membrane. The envelope is studded with spikes (peplomers) which are of two kinds, one being the hemagglutinin of the virus, the other the neuraminidase.

The family contains one genus, *Influenzavirus*, which consists of two species: influenza virus A, strains of which have been recovered from birds, horses, and swine, as well as man; and influenza virus B, which is a specifically human pathogen.

Paramyxoviridae. Like the orthomyxoviruses, with which they were originally grouped, the paramyxoviruses occur as roughly spherical pleomorphic enveloped viruses or as filaments containing a tubular nucleocapsid.

Some of the paramyxoviruses (the parainfluenza and respiratory syncytial viruses) cause localized infections of the respiratory tract of man. Others produce common generalized diseases of childhood (mumps, measles).

Rhabdoviridae. The rhabdoviruses (*rhabdo* = rod) are large, bullet-shaped, enveloped RNA viruses with single-stranded RNA, which is associated with a helical capsid enclosed within a shell to which is closely applied an envelope with peplomers. The virion matures at the cytoplasmic membrane. The best studied rhabdovirus is vesicular stomatitis virus, an arbovirus that infects horses. Morphologically similar viruses have been discovered in fish, insects, and even plants. The main one infecting man is that of rabies, a fatal disease of the brain, transmitted to man by the bite of an infected animal.

Reoviridae. The family name represents a sigla, Respiratory Enteric Orphan virus, reflecting the fact that members of the genus *Reovirus* are found in both the respiratory and enteric tract of man and most animals, but have yet to be associated with any disease. The distinctive feature of the family is that the virions contain double-stranded

RNA, which occurs as 10 separate pieces, each comprising a single gene. There are two genera, *Reovirus* and *Orbivirus* (*orbis* = ring), which differ slightly in morphology. All orbiviruses are arthropod-transmitted. Reoviridae are nonenveloped icosahedral viruses with two concentric shells of capsomers. They multiply in the cytoplasm.

Recently a new genus has been discovered within this family whose members cause diarrhea in animals, and some members of which are a major cause of infantile diarrhea in man.

Unclassified Viruses

During the last few years most previously unclassified viruses have been allocated to one or other of the families defined above. There remain, apart from viruses yet to be discovered, two important groups of known viruses, the virions of which have not yet been described in sufficient detail to enable them to be allocated with certainty to new or existing families.

The hepatitis viruses seem, at the time of writing, to fall into at least two different groups of small icosahedral DNA viruses, one of which (hepatitis A) probably belongs to the family Parvoviridae (see Chapter 23). Also emerging is a new group of "viruses," of which the proto-type is the agent of scrapie in sheep, causing various rare degenerative diseases of the central nervous system in man and animals (see also Chapter 23). The agents identified with the subacute spongiform en-cephalopathies are very small, relatively radiation- and heat-resistant infectious entities, and have been variously described as naked nucleic acid molecules (*viroids*) or nucleic acid in association with membranes, but are frankly still an enigma that holds out a great challenge to virologists.

SUMMARY

Viruses differ from microorganisms in several fundamental proper-ties which make them completely dependent on the host cell for the replication of their nucleic acid and the synthesis of their proteins.

In size, viruses vary from about 200 nm down to about 20 nm in diameter and, hence, can only be visualized by electron microscopy. They can be purified from cellular contaminants by physicochemical procedures such as rate zonal or equilibrium gradient centrifugation, filtration, electrophoresis, and column chromatography. Individual viral protein and nucleic acid molecules may be separated by dissocia-ting the purified virions with sodium dodecyl sulfate and then sub-jecting the product to acrylamide gel electrophoresis.

Viral infectivity can be destroyed by the action of physical or chemical agents. In general, viruses are heat-labile, especially the enveloped viruses, but infectivity can usually be preserved by storage at very low temperatures (−70°). The infectivity of enveloped viruses is destroyed by lipid solvents and all viruses are susceptible to agents which damage their nucleic acid.

Viruses contain either DNA or RNA, but not both. The nucleic acid may be single- or double-stranded, linear or circular, and may exist as one or as several molecules; in the latter case, each molecule may represent a single gene. Single-stranded viral RNA may have "positive" or "negative" polarity; in the former case, the viral RNA is messenger RNA and is itself infectious; in the latter case, a transcriptase carried in the virion is necessary to transcribe mRNA.

Most animal viruses have a basically symmetrical structure in which repeating units of protein known as capsomers are packed around the viral nucleic acid molecule to form either an icosahedral or a helical nucleocapsid. Helical nucleocapsids are always and icosahedral nucleocapsids sometimes enclosed within an envelope derived from host cell membranes; these envelopes contain cellular lipids and two types of virus-specified proteins: a layer of "membrane protein" lining the inside of the envelope and glycoprotein "peplomers" or spikes protruding from the surface of the virion.

Animal viruses can be classified according to criteria based on the mode of transmission or on their degree of genetic relatedness. The latter classification can be approached by studies of the viral nucleic acids. Since such work is still at a rudimentary stage the viruses of vertebrates are currently classified according to a scheme based on the chemistry of the nucleic acid and the morphology of the virion. The basic properties of the fifteen families currently recognized are described and tabulated.

FURTHER READING

Structure and Chemistry of Viruses

Caspar, D. L. D. (1965). Design principles in virus particle construction. *In* "Viral and Rickettsial Infections of Man" (F. L. Horsfall, Jr. and I. Tamm, eds.), 4th ed., pp. 51–93. Lippincott, Philadelphia, Pennsylvania.

Dalton, A. J., and Haguenau, F., eds. (1973). "Ultrastructure of Animal Viruses and Bacteriophages: An Atlas." Academic Press, New York.

Fenner, F., McAuslan, B. R., Mims, C. A., Sambrook, J. F., and White, D. O. (1974). "The Biology of Animal Viruses," 2nd ed., pp. 72–154. Academic Press, New York.

Horne, R. W. (1974). "Virus Structure." Academic Press, New York.

Knight, C. A. (1974). "Molecular Virology." McGraw-Hill, New York.

Rifkin, D. B., and Quigley, J. P. (1974). Virus-induced modification of cellular membranes related to viral structure. *Annu. Rev. Microbiol.* **28**, 325.

Inactivation of Viruses

Bachrach, H. L. (1966). Reactivity of viruses *in vitro. Progr. Med. Virol.* **8,** 214.
Ginoza, W. (1968). Inactivation of viruses by ionizing radiation and by heat. *In* "Methods in Virology" (K. Maramorosch and H. Koprowski, eds.), Vol. 4, pp. 139–209. Academic Press, New York.

Classification of Animal Viruses

Andrewes, C. H., and Pereira, H. G. (1972). "Viruses of Vertebrates," 3rd ed. Baillière, London.
Diener, T. O. (1974). Viroids: The smallest known agents of infectious disease. *Annu. Rev. Microbiol.* **28,** 23.
Fenner, F. (1974). The classification of viruses; why, when, and how. *Aust. J. Exp. Biol. Med. Sci.* **52,** 223.
Fenner, F. (1976). Classification and nomenclature of viruses: Second report of the International Committee on Taxonomy of Viruses. *Monogr. Virol.* (in preparation).
Wildy, P. (1971). Classification and nomenclature of viruses: First report of the International Committee on Nomenclature of Viruses. *Monogr. Virol.* **5,** 1.

Viruses of Other Forms of Life

Fraenkel-Conrat, H. (1974). "Descriptive catalogue of viruses," "Comprehensive Virology" (H. Fraenkel-Conrat and R. R. Wagner, eds.), Vol. I. Plenum, New York.
Gibbs, A. J., ed. (1973). "Viruses and Invertebrates." North-Holland Publ., Amsterdam.
Hayes, W. (1968). "The Genetics of Bacteria and their Viruses: Studies in Basic Genetics and Molecular Biology," 2nd ed. Blackwell, Oxford.
Lemke, P. A., and Nash, C. H. (1974). Fungal viruses. *Bacteriol. Rev.* **38,** 29.
Maramorosch, K., and Kurstak, E., eds. (1971). "Comparative Virology." Academic Press, New York.
Matthews, R. E. F. (1970). "Principles of Plant Virology." Academic Press, New York.
Padan, E., and Shilo, M. (1973). Cyanophages—viruses attacking blue-green algae. *Bacteriol. Rev.* **37,** 343.

CHAPTER 2

Cultivation and Assay of Viruses

INTRODUCTION

Viruses are obligatory intracellular parasites and cannot replicate in any cell-free medium, no matter how complex. Some viruses are fastidious about the sorts of cells that they infect; for instance, some known human viruses have not yet been cultivated under laboratory conditions. Fortunately, however, most viruses can be grown in cultured cells, embryonated eggs, or laboratory animals; indeed, the cultivation of viruses in experimental animals, or better still in cultured cells, is an essential prerequisite for their detailed study.

CELL CULTURE

Although over 60 years have elapsed since human and animal cells were first grown *in vitro*, it is only since the advent of antibiotics that cell culture has become a matter of simple routine. Aseptic precautions are still essential, but the problems of contamination with bacteria, mycoplasmas, fungi, and yeasts are no longer insurmountable, and many kinds of animal cells can be cultivated *in vitro* for at least a few generations. Since 1949, when Enders, Weller, and Robbins reported that poliovirus could be grown in cultured nonneural cells with the production of recognizable histological changes, a large number of animal viruses have been grown in cultured cells, and hundreds of previously unknown viruses have been isolated and identified. The discovery of the adenoviruses, echoviruses, rhinoviruses, and coronaviruses is directly attributable to the use of cultured cells, as is the revolution in the diagnosis of viral disease, the development of poliomyelitis, measles and rubella vaccines, and the recent dramatic advances in knowledge of the molecular biology of viruses of vertebrates.

Methods of Cell Culture

Cells may be grown *in vitro* in several ways, listed below.

Organ Culture. Slices of organs, if carefully handled, maintain their original architecture and functions for several days or sometimes weeks *in vitro*. Such organ cultures (strictly tissue cultures) of respiratory epithelium have been used to study the histopathogenesis of infection by respiratory viruses; indeed, some respiratory viruses can only be grown outside their natural host by using organ cultures. Organ cultures of fetal intestinal epithelium are being used to study certain enteric viruses.

Tissue Culture. This was originally applied to the cultivation *in vitro* of fragments of minced tissue in suspension or "explants" of tissues embedded in clotted plasma. Subsequently, the term came to be associated with the *in vitro* culture of cells in general. "Tissue culture" in its original sense is now obsolete, and there is no logic in perpetuating the general use of the term; "cell culture" will be used throughout the text that follows.

Cell Culture. Tissue is dissociated into a suspension of single cells or small clumps by mechanical mincing followed by treatment with proteolytic enzymes. After the cells are washed and counted they are diluted in medium and are permitted to settle on to the flat surface of a specially treated glass or plastic container. Most types of cells adhere quickly, and under optimal conditions they divide about once a day until the surface is covered with a confluent monolayer of cells.

Media

Cell culture has been greatly aided by the development of chemically defined media containing almost all the nutrients required for cell growth. The best known of these media, developed by Eagle, is an isotonic solution of simple salts, glucose, vitamins, coenzymes, and amino acids, buffered to pH 7.4, and containing antibiotics to inhibit the growth of bacteria. Serum must be added to Eagle's medium to supply the cells with an additional factor(s), the nature of which is still undefined, but without which most cells will not grow.

Types of Cultured Cells

Some types of cells are capable of undergoing only a few divisions *in vitro* before dying out, others will survive for up to a hundred cell generations, and some can be propagated indefinitely. These differences, the nature of which is not fully understood, give us three main types of cultured cells.

Primary Cell Cultures. When cells are taken freshly from animals and placed in culture, the cultures consist of a variety of cell types, most of which are capable of very limited growth *in vitro* — perhaps five or ten divisions at most. However, they support the replication of a wide range of viruses, and primary cultures derived from monkey kidney, human embryonic kidney or amnion, and chicken and mouse embryos are commonly used for this purpose, both for laboratory experiments and vaccine production.

Diploid Cell Strains. These are cells of a single type that are capable of undergoing up to about 100 divisions *in vitro* before dying. They retain their original diploid chromosome number throughout. Diploid strains of fibroblasts established from human embryos are widely used in diagnostic virology and vaccine production and have some use in experimental studies. It should be kept in mind that certain aspects of the expression of the viral genome may require the use of cells that retain their normal state of differentiation.

Continuous Cell Lines. These are cells of a single type that are capable of indefinite propagation *in vitro*. Such immortal lines usually originate from cancers, or by "transformation" occurring in a diploid cell strain. Often they no longer bear any close resemblance to their cell of origin, as they have doubtless undergone many sequential mutations during their long history in culture. The most usual indication of these changes is that the cells are "dedifferentiated," i.e., they have lost the specialized morphology and biochemical abilities that they possessed as differentiated cells *in vivo*. For example, it is no longer possible to distinguish microscopically between the various epithelial cell lines arising from cells of ectodermal or endodermal origin, or between the "fibroblastic" cell lines arising from cells of mesodermal origin (Plate 2-1). Cells of continuous cell lines are often aneuploid in chromosome number, and may be malignant.

Continuous cell lines such as HEp-2, HeLa, and KB, all derived from human carcinomas, support the growth of a number of viruses. These lines, and others derived from mice (L929, 3T3) and hamsters (BHK-21), are widely used in experimental virology; there are now available a range of continuous cell lines derived from a variety of animals. The American Type Culture Collection holds deep-frozen samples of many cell lines, tested for freedom from extraneous viruses and mycoplasmas.

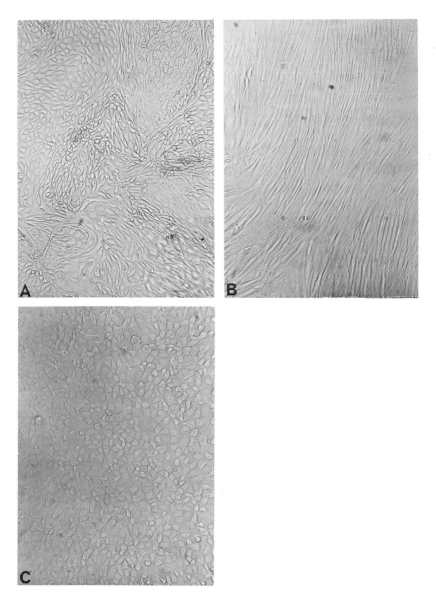

PLATE 2-1. *Cultured cells. Confluent monolayers of the three main types of cultured cell are shown as they normally appear in the living state, viewed by conventional low-power light microscopy through the wall of the tissue culture bottle ×60 (A) Primary monkey kidney epithelium. (B) Diploid strain of human fetal fibroblasts. (C) Continuous line of epithelial cells. (Courtesy I. Jack.)*

The great advantage of continuous cell lines over primary cell cultures is that they can be propagated indefinitely by subculturing the cells at regular intervals. Furthermore, they retain viability for many years when suspended in glycerol or dimethyl sulfoxide and stored at low temperatures [−70° (solid carbon dioxide) or −196° (liquid nitrogen)].

Some continuous cell lines have been adapted to grow in suspension culture, i.e., as a suspension of single cells continuously stirred by a spinning magnet. Such "spinner cultures" are particularly useful for biochemical studies of viral multiplication and for the commercial production of some vaccines.

Cell Type and Viral Growth

Of paramount importance for virologists is the selection of cell lines that will support the optimal growth of the virus under study. Some viruses will multiply in almost any cell line; some cell lines are favorable for supporting the replication of many different types of viruses. On the other hand, many viruses are quite restricted in the kinds of cell in which they will multiply, although repeated blind passage may lead to adaptation.

Cultured cells can serve three main purposes: (a) primary isolation of viruses from clinical specimens, in which emphasis is placed on high sensitivity and readily recognized cytopathic effects (CPE), (b) vaccine production, where the emphasis is placed on high viral yield, and (c) basic biochemical research for which continuous cell lines, preferably growing as suspension cultures, are usually chosen.

Recognition of Viral Growth in Cell Culture

The growth of many viruses in cell culture can be monitored by a number of biochemical procedures indicative of the intracellular increase in viral macromolecules and virions, as described in Chapter 3. In addition, there are simpler methods that are commonly used for diagnostic work, some of which also have an important place in research laboratories.

Cytopathic Effects (CPE). Many, but by no means all, viruses kill the cells in which they multiply, so that infected cell monolayers gradually develop histological evidence of cell damage, as newly formed virions spread to involve more and more cells in the culture. These changes are known as *cytopathic effects;* the responsible virus is said to be *cytopathogenic.* Most CPE can be readily observed in unfixed, unstained cell cultures under low power of the light microscope with

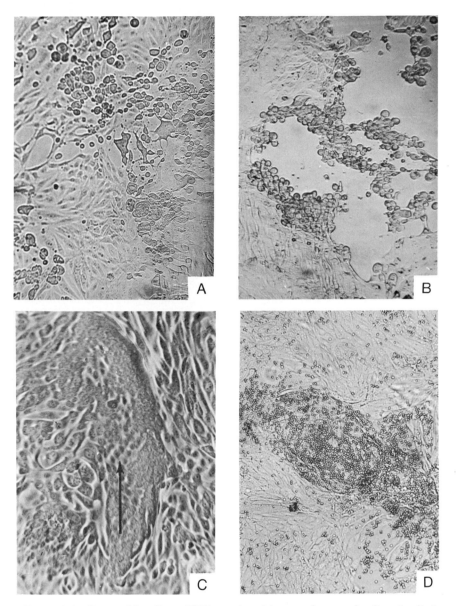

PLATE 2-2. *Cytopathic effects (CPE) produced in monolayers of cultured cells by different viruses. The cultures are shown as they would normally be viewed in the laboratory, unfixed and unstained.* ×60. *(A) Enterovirus—rapid, complete cell destruction. (B) Herpetovirus—focal areas of swollen rounded cells. (C) Paramyxovirus—focal areas of fused cells (syncytia). (D) Hemadsorption Erythrocytes adhere to those cells in the monolayer that are infected. The technique is applicable to any virus that causes a hemagglutinin to be incorporated into the cellular membrane. Most of the enveloped viruses that mature by budding from cytoplasmic membranes produce hemadsorption. (Courtesy I. Jack.)*

TABLE 2-1
Cytopathic Effects of Viruses in Cell Culture

CYTOPATHIC EFFECT	VIRUS
Pyknosis, shrinkage, cell destruction	Enteroviruses, poxviruses, reoviruses,[a] togaviruses,[a] rhinoviruses[a]
Aggregation	Adenoviruses
Cell fusion to give syncytia	Paramyxoviruses, herpetoviruses
Minimal	Orthomyxoviruses, rabies virus, coronaviruses, retroviruses, arenaviruses

[a] Often produces incomplete cytopathic effect.

the condenser racked down and the iris diaphragm partly closed to obtain the contrast required in looking at translucent cells. A trained observer can distinguish several types of CPE in living cultures (Plate 2-2 and Table 2-1), but fixation and staining of the cell monolayer is necessary in order to see details such as inclusion bodies and syncytia. Fluorescent antibody staining, described in Chapter 14, is widely used to recognize viral antigens in such cultured cells.

Observation of CPE is an important tool for the diagnostic virologist, who is concerned with isolating viruses from patients. Some viruses multiply readily in cell culture on primary isolation; the time at which cytopathic changes first become detectable depends to some extent on the number of virions that the specimen contained, but, far more important, on the growth rate of the virus in question. Enteroviruses and herpes simplex virus, for example, which have a short latent period and a high yield, often show detectable CPE after 24 to 48 hours, destroying the monolayer completely within about 3 days. On the other hand, cytomegaloviruses, rubella, respiratory syncytial virus, and some of the more slowly growing adenoviruses may not produce detectable CPE for weeks. Since the cell cultures may have undergone nonspecific degeneration during this period it may be necessary to subinoculate the cells and supernatant fluid from the infected culture on to fresh monolayers. CPE often appears soon after such "blind passage," either because this enhances the titer or selects variants adapted to grow better in the cultured cells.

Hemadsorption. Cultured cells infected with orthomyxoviruses, paramyxoviruses and togaviruses, all of which bud from cytoplasmic membranes, acquire the ability to adsorb erythrocytes. The phenomenon, known as hemadsorption, is due to the incorporation into the plasma membrane of newly synthesized viral protein that has an affinity for red blood cells (Plate 2-2D). Hemadsorption can be used to recognize infection with noncytopathogenic viruses, as well as the early stages of infection with cytocidal viruses.

Interference. The multiplication of one virus in a cell usually inhibits the multiplication of another virus entering subsequently (see Chapter 4). Rubella virus was first discovered by showing that infected monkey kidney cell cultures, showing no CPE, were nevertheless resistant to challenge with an unrelated echovirus. The phenomenon was also exploited for a time for the isolation of rhinoviruses. It is no longer widely used for either, because cell lines have become available in which these viruses produce CPE, but interference is a useful technique to use when searching for new noncytopathogenic viruses.

EMBRYONATED EGGS

Prior to the 1950's, when cell culture really began to make an impact on virology, the standard host for the cultivation of many viruses was the embryonated hen's egg (developing chick embryo). The technique was devised by Goodpasture and extensively developed by Burnet over the ensuing years. Nearly all of the viruses that were known at that time could be grown in the cells of one or another of the embryonic membranes, namely the amnion, allantois, chorion, or yolk sac (Fig. 2-1).

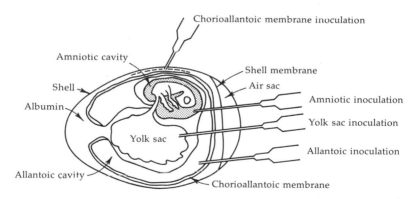

FIG. 2-1. *Routes of inoculation of the embryonated egg. Yolk sac inoculation is usually carried out with 5-day-old embryos; amniotic and allantoic inoculation with 10-day-old embryos; chorioallantoic inoculation with 11- or 12-day-old embryos. (Modified from B. D. Davis, R. Dulbecco, H. N. Eisen, H. S. Ginsberg, and W. B. Wood, "Microbiology." Harper, New York, 1967.)*

TABLE 2-2
Growth of Viruses in Embryonated Eggs

MEMBRANE	VIRUSES	SIGNS OF GROWTH
Yolk sac	Herpes simplex	Death
Chorion	Herpes simplex	
	Poxviruses	Pocks
	Rous sarcoma	
Allantois	Influenza	Hemagglutination
Amnion	Mumps	Death

Eggs are inoculated 5–14 days after fertilization, depending on the state of development of the membrane it is proposed to infect. A hole is drilled in the shell, and virus is injected into the fluid bathing the appropriate membrane. Following incubation for an additional 2–5 days, viral growth can be recognized by one or more of the criteria listed in Table 2-2.

Embryonated eggs are rarely employed now for viral isolation, but the allantois produces such high yields of certain viruses like the influenza viruses that this system is used both by research laboratories and for vaccine production.

LABORATORY ANIMALS

Like embryonated eggs, laboratory animals have almost disappeared now from diagnostic laboratories, since cell cultures are so much simpler to handle and much more versatile, although infant mice are still used in the isolation of some coxsackieviruses and in many arbovirus laboratories.

However, laboratory animals are still essential for many kinds of virological research. Primates are used to study a few human viruses, like the subacute spongiform encephalopathy agents and hepatitis viruses, that will not grow in other laboratory animals or in cultured cells. Hamsters are widely used in tumor virology, because they are highly susceptible to cancer production by oncogenic viruses and can yield valuable antisera. Experiments on pathogenic mechanisms and the role of the immune response can only be carried out with suitable laboratory animals, usually primates, hamsters, rabbits, or mice. Finally, since serology looms large in much virological research, laboratory animals, usually rabbits, are extensively used for producing antisera.

ASSAY OF VIRAL INFECTIVITY

All scientific research depends upon reliable methods of measurement, and with viruses the property we are most obviously concerned with measuring is infectivity. The content of infectious virions in a given suspension can be "titrated" by infecting cell cultures, chick embryos, or laboratory animals with dilutions of viral suspensions and then watching over the next few days for evidence of viral multiplication. Two types of infectivity assay should be distinguished: quantitative and quantal.

Quantitative Assays

A familiar example of this type of assay is the bacterial colony count on an agar plate. Each viable organism multiplies to produce a discrete clone; the colony count therefore represents a direct estimate of the number of organisms originally plated. The parallel systems in virology are the counting of pocks on the chorion of the chick embryo or plaques on monolayers of cultured cells.

Plaque Assays. Dulbecco introduced to animal virology a modification of the bacteriophage plaque assay that is now used very widely for the quantitation of animal viruses. A viral suspension is added to a monolayer of cultured cells for an hour or so to allow the virions to attach to the cells, then the liquid medium is replaced with a solid gel, which ensures that the spread of progeny particles is restricted to the immediate vicinity of the originally infected cell. Hence, each infective particle gives rise to a localized focus of infected cells that becomes, after a few days, large enough to see with the naked eye (Plate 2-3).

Various materials have been used to form the gel with which the cell monolayer is overlaid shortly after inoculation. They include agar, methylcellulose, tragacanth, and starch gel. Agar has been most commonly employed, but suffers from the disadvantage that it contains sulfated polysaccharides that inhibit the growth of some viruses; however, this inhibitory substance can be neutralized by the addition of DEAE-dextran to the agar. Of course, the gel must also incorporate the usual nutrient medium required to maintain the cells in a viable condition.

After an incubation period of 2 days to 4 weeks, depending on the virus under study, the cell monolayers are stained with a vital dye, such as neutral red or tetrazolium. The living cells take up the stain and the plaques appear as clear areas against a red background. Noncytocidal viruses can be titrated in a similar fashion, plaques being

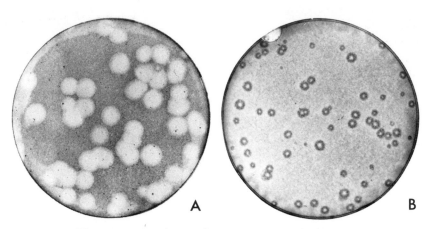

PLATE 2-3. *Plaques produced by influenza virus in monolayers of a continuous cell line derived from human conjunctival cells (Chang). Neglecting possible overlap of plaques or clumping of virions, each plaque is initiated by a single virion and yields a clone. (A) Normal plaques, seen as clear areas in monolayer stained with neutral red. (B) "Red" plaques, characteristic of certain strains of influenza virus, and some other viruses; demonstrated by using half strength neutral red in overlay medium [From E. D. Kilbourne, Bull. W. H. O.* **41**, *643 (1969); courtesy Dr. E. D. Kilbourne].*

recognized by such techniques as hemadsorption, interference, or fluorescent antibody staining. Some viruses, e.g., herpesviruses and poxviruses, will produce plaques even in cell monolayers grown in liquid medium, because most of the newly formed virus remains cell-associated, so that plaques form by direct spread to adjacent cells through intercellular bridges.

There are a number of technical variations of Dulbecco's plaquing method. For instance, virus may be allowed to attach to cells in suspension. The cells can then be permitted either to adhere to the glass and be overlaid with agar medium, or, alternatively, the cell suspension with virus attached may be embedded in agar medium directly. The latter method is suited only to viruses small enough to diffuse from cell to cell through the agar; the plaque that eventually results is then spherical.

Ordinarily, infection with a single virus particle is sufficient to form a plaque, so that the infectivity titer of the original viral suspension can be expressed in terms of "plaque-forming units" (pfu) per milliliter. The error will be minimal if the titer is determined from plates inoculated with a dilution of virus producing about 20 to 100 plaques per plate, depending on plaque size—enough to minimize errors attributable to the Poisson distribution (which governs the distribution of very small numbers of discrete particles), while avoiding overlap of adjacent plaques.

Transformation Assays. Methods other than the plaque assay have been devised to quantitate the infectivity of those viruses that do not cause cell death. We have already referred to the use of hemadsorption or interference to assay some noncytocidal viruses; in addition, there are some oncogenic viruses that interact with some cultured cells in a cytocidal fashion, but "transform" other types of cells. Such cells are not killed, but their social behavior is changed so that they take on many of the properties of malignant cells. Compared with noninfected cells, transformed cells show little contact inhibition, so that they grow in an unrestrained fashion to produce a heaped-up "microtumor" that stands out conspicuously against the background of normal cells in the monolayer. Like malignant cells excised from a tumor, the transformed cells also assume the ability to grow in sloppy suspension of agar or methocel. For certain oncogenic viruses, like Rous sarcoma virus, transformation is the basic method for assaying viral infectivity; for others, such as the DNA tumor viruses, transformation is a relatively inefficient process and these viruses are usually assayed by conventional procedures.

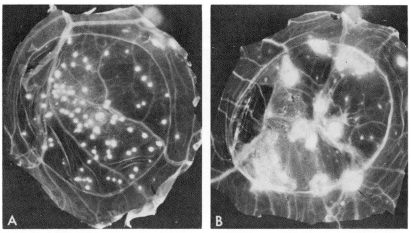

PLATE 2-4. *Pocks produced on the chorioallantoic membrane of embryonated eggs 3 days after inoculation with variola virus (A) and vaccinia virus (B). Small secondary pocks of vaccinia are visible. (Courtesy Professor A. W. Downie.)*

Pock Assays. A much older assay, still occasionally used for the pox-viruses, is the titration of viruses on the chorion of the chick embryo. Newly synthesized virus escaping from infected cells spreads mainly to adjacent cells, so that each infecting particle eventually gives rise to a localized lesion, known as a "pock" (Plate 2-4). The nature of the pock is often highly characteristic of a particular group of viruses or even a particular mutant.

Quantal Assays

The second type of infectivity assay is not quantitative but quantal, i.e., it does not register the number of infectious virus particles in the inoculum, but only whether there are any at all. Being an all-or-none assay, it is not nearly as precise as a quantitative assay; accordingly, it is only used for viruses that do not form plaques. Serial dilutions of virus are inoculated into several replicate cell cultures, eggs, or animals. Adequate time is allowed for virus to multiply and spread to destroy the whole cell culture, or kill the animal, as the case may be. Hence, each host yields only a single piece of information, namely, whether or not it was infected by that particular dilution of virus. A more economical procedure can be used with viruses such as vaccinia that produce localized skin lesions in an animal like the rabbit. Twenty or thirty skin sites can be separately inoculated, each giving an all-or-none answer equivalent to the death or survival of a cell culture of a mouse.

Statistical Considerations. With most viruses one particle is sufficient to initiate infection ("one-hit kinetics"). The evidence for this is the Poissonian distribution of "takes" (infected mice or eggs, lesions in rabbit skin, etc.) when closely spaced dilutions near the end point are tested, or more conveniently, when the linear relationship between dose and plaque count is observed. However, a number of situations have now been recognized in which the infectivity assay (usually a plaque count) follows two-hit rather than one-hit kinetics, indicating that two different types of virus particle must infect a single cell in order that one of them (or occasionally both) may replicate. The adeno-associated viruses, some of the SV40-adenovirus hybrids, and the murine sarcoma viruses cannot multiply except in cells coinfected with a "helper" virus. Assays of such defective viruses provide examples of two-hit kinetics.

The results of typical quantal and quantitative infectivity titrations are given in Table 2-3. It will be seen that at high dilutions of the inoculum all the hosts remained uninfected because they failed to

TABLE 2-3

Comparison of Quantitative and Quantal Infectivity Titrations

VIRUS DILUTION	QUANTITATIVE ASSAY (PLAQUE COUNT)[a]	QUANTAL ASSAY (CPE)
10^{-2}	C, C, C, C, C	++++++++++
10^{-3}	50, 42, 54, 59, 45	++++++++++
10^{-4}	5, 7, 3, 6, 4	++−+++++++
10^{-5}	0, 0, 1, 0, 1	−++−+−−+−+
10^{-6}	0, 0, 0, 0, 0	−−−−−−+−−−
10^{-7}	0, 0, 0, 0, 0	−−−−−−−−−−

[a] C, confluent (uncountable).

receive even a single infectious unit. The "endpoint" of a quantal titration is taken to be that dilution of virus which infects 50% of the
inoculated hosts; the infectivity titer of the original virus suspension is
then expressed in terms of "50% infectious doses" (ID_{50}) per milliliter.
At first sight, it may be thought that Table 2-3 shows some anomalous
results, in that some hosts have become infected following inoculation
with higher dilutions of virus than those that failed to infect others.
This sort of finding is quite normal and explicable in terms of the
Poisson distribution. At any given multiplicity (m) of virus there is a
finite probability (e^{-m}) that a given aliquot of the inoculum will contain no virus particle. In the example shown, each 0.1 ml sample of the
10^{-5} dilution contained an average of one ID_{50}, i.e., 0.5 of an infectious
unit. Hence, each 0.1 sample of the 10^{-6} dilution contained an average
of 0.05 infectious units, i.e., about one host in twenty received one infectious particle; whereas, each 0.1 ml sample of the 10^{-4} dilution contained an average of five infectious units, i.e., e^{-5} (about one in a
hundred) of these hosts failed to receive an infectious particle at all. In
practice, quantal assays rarely produce such a nicely balanced result as
the example presented in Table 2-3, and statistical procedures must be
used to calculate the endpoint of the titration.

Infectious Nucleic Acids

One dramatic demonstration of the basic differences between
viruses and cellular microorganisms is that purified nucleic acid extracted from some viruses is infectious, i.e., such viruses are able to
reproduce from their genome alone. Positive results are regularly obtained only with viruses whose genomes consist of a single molecule
of nucleic acid and whose virions do not contain a transcriptase (Table
2-4).

TABLE 2-4

Viruses Yielding Infectious Nucleic Acids

DNA viruses	
Parvoviridae	Positive results with genus *Adenosatellovirus,* in cells preinfected with helper adenovirus
Papovaviridae	Readily demonstrable; single-strand scission does not destroy infectivity
Adenoviridae	Positive results claimed with certain simian and human adenoviruses
RNA viruses	
Picornaviridae	Readily demonstrable; both single-stranded viral RNA and double-stranded replicative form are infectious
Togaviridae	Readily demonstrable; infectious RNA is one of the distinguishing characteristics of the family

The demonstration that some viral nucleic acids are infectious is of considerable theoretical and practical importance. More clearly than any other observation, it showed that the viral nucleic acid molecule was the repository of all viral genetic information. In the laboratory, it has extended the host range of viruses; for example, poliovirus RNA will infect many cells and animals that are resistant to infection with poliovirions due to the absence of cellular receptors for the viral capsid. This demonstrated clearly that such "resistant" cells are in fact capable of supporting viral growth if a mechanism can be found for facilitating the entry of undegraded viral RNA into the cell. With a few viruses, notably SV40, the use of infectious DNA has been of great assistance in dissection of the functions of the genome. Persistent failure to demonstrate infectivity of the nucleic acids of some viruses led to the successful search for polymerases in their virions. Finally, infectious viral nucleic acids provide a convenient model for studying the more general problem of introducing functional nucleic acids into vertebrate cells.

ASSAY OF OTHER PROPERTIES OF VIRIONS

Hemagglutination

Many viruses contain in their outer coat virus-coded proteins capable of binding to erythrocytes (Table 2-5). Such viruses can, therefore, bridge red blood cells to form a lattice. The phenomenon, known as hemagglutination, was first described in 1941 by Hirst, who then

TABLE 2-5
Hemagglutination by Viruses

VIRUS		
FAMILY	GENERA OR SPECIES	RBC
Parvoviridae	Adeno-associated virus type 4	Human, guinea pig, 4°
Papovaviridae	Polyoma virus	Guinea pig, 4°
Adenoviridae	Most types	Monkey, rat, 37°
Herpetoviridae		Nil
Iridoviridae		Nil
Poxviridae	Smallpox, vaccinia	Fowl (some birds only), 37°
Picornaviridae	Coxsackievirus (some serotypes)	Human
	Echovirus (some serotypes)	Human
	Rhinovirus (some serotypes)	Sheep, 4°
Togaviridae	*Alphavirus, Flavivirus,* rubella	Goose, pigeon; pH and temperature critical
Orthomyxoviridae	Influenza types A and B	Fowl, human, guinea pig, 4°
Paramyxoviridae	Parainfluenza, mumps	Fowl, human, guinea pig, 4°
	Measles	Monkey, 37°
Rhabdoviridae	Rabies	Goose, 4°
Coronaviridae	Human strains	Rat, mouse, fowl, 37°
Arenaviridae		Nil
Bunyaviridae		Goose, pH critical
Reoviridae	*Reovirus*	Human

went on to analyze the mechanism of hemagglutination by influenza virus. In this case, the hemagglutinating protein (hemagglutinin) on the virion is a glycoprotein, which occurs in the form of many short projections (see Plate 20-1). The virus will attach to any species of erythrocyte carrying complementary receptors, which are glycoproteins of a different sort. Hemagglutination by influenza and the paramyxoviruses, but not other viruses, is complicated by the fact that the virion also carries an enzyme, neuraminidase, which destroys the glycoprotein receptors on the erythrocyte surface and allows the virus to elute, unless the test is carried out at a temperature too low for the enzyme to act (4°). About 10^7 influenza virions are required to cause macroscopic agglutination of a convenient number of chick erythrocytes (conventionally 0.25 ml of a 1% suspension of red cells). Thus, hemagglutination is not a sensitive indicator of the presence of small numbers of virions, but because of its simplicity it provides a very convenient assay if large amounts of virus are available. The viral suspension is diluted serially (usually in twofold steps) in a plastic tray using a calibrated wire loop, and erythrocytes are added to each well.

Unclumped red blood cells settle to form a "button," whereas agglutinated cells form a "shield" (see Plate 14-6).

Direct Particle Counts

Negative staining, e.g., with potassium phosphotungstate, is such a simple technique that with many viruses the number of particles in relatively crude suspensions can be counted directly in the electron microscope. The virus can be mixed with a known concentration of latex particles to provide an easily recognizable marker. Alternatively, the virions contained in a known volume can be sedimented in an ultracentrifuge directly onto an electron microscope grid.

Comparison of Different Assays

If a given preparation of virus particles were to be assayed by all of the methods described above, the "titer" would be different in every case. For example, an influenza virus suspension may provide the data tabulated below:

METHOD	AMOUNT (PER MILLILITER)
1. Direct electron microscope count	10^{10} EM particles
2. Quantal infectivity assay in eggs	10^9 egg ID_{50}
3. Quantitative infectivity assay by plaque formation	10^8 pfu (plaque-forming units)
4. Hemagglutination (HA) assay	10^3 HA units

The difference between the EM(1) and HA(4) titers reflects merely the difference in sensitivity between the two assays. On the other hand, the ratio between the infectivity titer (2 or 3) and the EM particle count (1), known as the "plating efficiency," requires deeper analysis. Superficially it might suggest that most of the virus particles visible by electron microscopy are noninfectious, i.e., perhaps inactivated by heat or maltreatment, or "incomplete" in that they lack a complete genome. This may indeed be true, but the situation is complicated by another factor, namely the susceptibility of the cell system in which viral infectivity is assayed. Even a fully infectious virus particle has only a certain chance of successfully negotiating all the hazards it may encounter in the course of infecting a cell. The susceptibility of one cell system (3) may be much lower than that of another (2), hence a given preparation of virus may have a lower efficiency of plating in one system than in the other.

SUMMARY

Animal viruses can be satisfactorily studied in the laboratory only when they can be grown in an experimental system. Laboratory animals (including primates) are important for studies on pathogenesis, and in the past laboratory animals like infant mice and embryonated eggs have been used for isolating and assaying viruses. Nowadays, however, viral isolation and assay, as well as biochemical studies of viral multiplication, are almost always carried out with cultured cells. These are classified as primary cultures, diploid cell strains, or heteroploid cell lines. The growth of viruses in cultured cells can be conveniently monitored by the development of cytopathic effects (CPE), hemadsorption, immunofluorescence, or interference.

Viral infectivity can be titrated by quantitative tests like the plaque or pock assays, or by quantal tests which involve an all-or-none response of the experimental animal or cultured cell system. Many human viruses can be assayed by hemagglutination titrations, and the total number of particles in a viral suspension can be counted by electron microscopy. The quantitative relationship between the infectivity titers of a given viral suspension obtained using two different types of assay is a complex function involving the proportion of infectious, inactivated, and incomplete virions present, as well as the degree of susceptibility or resistance of the host cell or animal systems used.

FURTHER READING

Cooper, P. D. (1967). The plaque assay of animal viruses. In "Methods in Virology" (K. Maramorosch and H. Koprowski, eds.), Vol. 3, p. 243. Academic Press, New York.

Fenner, F. J., McAuslan, B. R., Mims, C. A., Sambrook, J. F., and White, D. O. (1974). "The Biology of Animal Viruses," 2nd ed., pp. 35–71. Academic Press, New York.

Habel, K., and Salzman, N. P., eds. (1969). "Fundamental Techniques in Virology," Vol. 1. Academic Press, New York.

Hayflick, L. (1965). The limited in vitro lifetime of human diploid cell strains. Exp. Cell Res. 37, 614.

Hoorn, B., and Tyrrell, D. A. J. (1969). Organ cultures in virology. Progr. Med. Virol. 11, 408.

Howe, C., and Lee, L. T. (1972). Virus-erythrocyte interactions. Advan. Virus Res. 17, 1.

Kruse, P. F., Jr., and Patterson, M. K., Jr., eds. (1973). "Tissue Culture: Methods and Applications." Academic Press, New York.

Lennette, E. H., and Schmidt, N. J., eds. (1969). "Diagnostic Procedures for Viral and Rickettsial Infections," 4th ed. Amer. Pub. Health Ass., New York.

Maramorosch, K., and Koprowski, H., eds. (1967–1971). "Methods in Virology," Vols. 1–5. Academic Press, New York.

Pagano, J. S. (1970). Biologic activity of isolated viral nucleic acids. Progr. Med. Virol. 12, 1.

Paul, J. (1970). "Cell and Tissue Culture," 4th ed. Livingstone, Edinburgh.
Rose, G. G. (1970). "Vertebrate Cells in Tissue Culture." Academic Press, New York.
Rosen, L. (1969). Hemagglutination with animal viruses. *In* "Fundamental Techniques in Virology" (K. Habel and N. P. Salzman, eds.), Vol. 1, p. 276. Academic Press, New York.
Schmidt, N. J. (1969). Tissue culture technics for diagnostic virology. *In* "Diagnostic Procedures for Viral and Rickettsial Infections" (E. H. Lennette and N. J. Schmidt, eds.), 4th ed., p. 79. Amer. Pub. Health Ass., New York.

CHAPTER 3

Viral Multiplication

INTRODUCTION

Viral multiplication, in all its complexity, is the central focus of most experimental virology. It will be impossible to explore it in depth in this short text, but for the student of medicine some knowledge of the mechanism of viral multiplication is needed as a basis for understanding how viruses get into and out of cells, how they kill cells or transform them to the malignant state, and what stages are likely to be susceptible to chemotherapeutic agents.

All the genetic information in a virus is carried in its nucleic acid molecule(s). Since the nucleic acid is surrounded by one or more layers of protein, it cannot begin to multiply until the particle has been taken into a susceptible cell and these protein coats have been removed. Having reached an appropriate intracellular site, the viral nucleic acid molecule(s) replicates many times, and specifies the proteins needed for its coat. These new viral components are then assembled into virions, which are ultimately released from the cell to initiate another cycle of infection.

METHODS OF INVESTIGATION OF VIRAL BIOSYNTHESIS

Following the pattern established in experiments with bacteriophages, studies of the multiplication of animal viruses are based on *one-step (single-cycle) growth experiments* (Fig. 3-1). In such experiments events in individual cells are synchronized by infecting all cells in a culture simultaneously. Excess input virus is removed or neutralized, and the subsequent increase in numbers of infective virus particles can be followed in the cells and the surrounding medium.

Shortly after infection substantially less infectious virus can be recovered than was inoculated. This "eclipse" of infectious viral par-

47

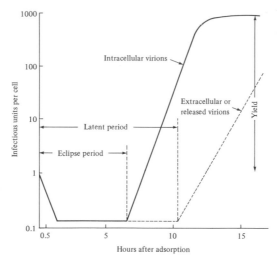

FIG. 3-1. *Generalized diagram of the multiplication cycle of an animal virus that is released from the cell late and incompletely. In other cases, virions mature as they are released and the curves for intracellular and extracellular infectivity correspond.*

ticles continues until progeny become detectable some hours later. A distinction is ordinarily made between the first appearance of new virions intracellularly (end of *eclipse period*) and the appearance of progeny virions in the medium (end of *latent period*). These two periods coincide in the case of viruses that mature by budding from the plasma membrane.

Early studies, relying on quantitative electron microscopy and assay of infectious virions, provided a limited amount of information about the very early and the late events in the multiplication cycle but could not tell us anything about the all-important events that occurred during the eclipse period. Investigation of replication of the viral nucleic acid, the production of new proteins, and to a lesser extent their assembly into virions became possible only with the development of several biochemical techniques some of which are listed below. All utilize radioactive labeling.

Viral nucleic acids can be separated from those of the cell by such techniques as acrylamide gel electrophoresis, column chromatography, and rate-zonal or equilibrium gradient centrifugation. They are identified on the basis of their infectivity, unique base composition, or ability to hybridize with the nucleic acid extracted from virions. Double-stranded RNA, such as that of the *replicative intermediate* formed in the course of viral RNA replication, can be distinguished from single-stranded viral RNA by its relative resistance to ribonuclease, lower sedimentation coefficient, lower buoyant density, etc.

Viral proteins can also be separated by acrylamide gel electrophoresis. Enzymes may be assayed directly, while structural proteins are identified serologically. For example, radioactively labeled capsid proteins can be precipitated from infected cell lysates using antisera against whole or disrupted virions. Individual antigens can be identified in immunodiffusion tests, and localized in the cell by immunofluorescence, or electron microscopy using ferritin- or enzyme-conjugated antibodies.

Chemical inhibitors of DNA synthesis (e.g., cytosine arabinoside or fluorodeoxyuridine), of transcription (e.g., actinomycin D), and of protein synthesis (e.g., puromycin or cycloheximide) have been used extensively in the study of viral biosynthesis. The fact that some RNA viruses (e.g., poliovirus) multiply satisfactorily in cells whose RNA and protein synthesis has been inhibited by actinomycin D has proved a most important aid to the study of the early events in viral multiplication.

THE ESSENTIAL STEPS IN MULTIPLICATION

It is possible to dissect the viral multiplication cycle into several more or less sequential steps (Figure 3-2), although one process grades into the next and after the first few hours several of the processes which follow are going on simultaneously: (1) attachment; (2) penetration, leading to or coincidental with (3) uncoating. This leads to eclipse (the disappearance of infectious virions) during which a complex series of biosynthetic events occurs, viz., (4) transcription of mRNA from specific sequences of the parental viral DNA or RNA, (5) translation of this mRNA into virus-coded enzymes and other "early" proteins, (6) replication of the viral DNA or RNA, (7) transcription of further mRNA from progeny as well as parental DNA or RNA, and (8) translation of these "late" mRNA's into structural and other virus-coded proteins some of which are involved in regulatory functions. Finally the eclipse phase ends with (9) assembly and (10) release of new virions.

Attachment

Attachment of virions to the plasma membrane is electrostatic and follows the establishment of contact by collision between virions and cells. It is independent of temperature except insofar as this affects

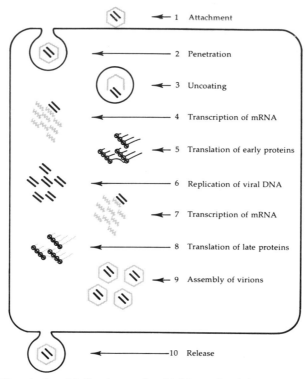

FIG. 3-2. *The viral multiplication cycle — highly stylized diagram, using an icosahedral DNA virus as a model and ignoring questions of nuclear or cytoplasmic localization.*

Brownian movement of virions and cells and thus the likelihood of their collision.

Firm attachment will occur only if there is a certain affinity between the cell surface and the virions; areas of the cell membrane showing this affinity are called *viral receptors.* Lack of receptor sites is the cause of the resistance of some cells and some organisms to infection by certain viruses. The specificity of attachment has been demonstrated most clearly in experiments with poliovirus and influenza virus. Poliovirions will attach to primate but not to rodent cells because the former possess, and the latter lack poliovirus-specific receptors. The importance of glycoprotein receptors for the attachment of influenza virus has been demonstrated in experiments in which they have been destroyed by bacterial neuraminidase ("receptor-destroying enzyme"). The virus cannot attach to the enzyme-treated cells.

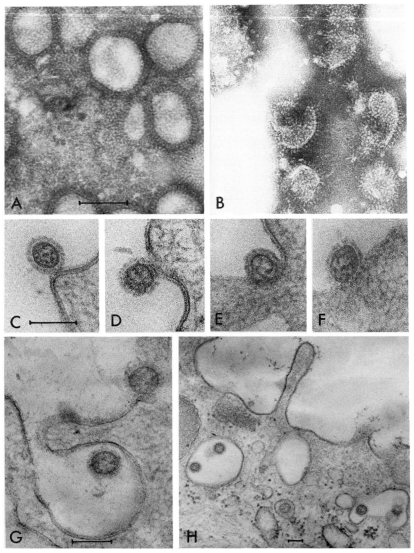

PLATE 3-1. *Entry of influenza virus; fusion or engulfment? Electron micrographs of chorioallontoic membrane at intervals after infection, suggesting either fusion of viral envelope and plasma membrane (A–F) or engulfment of intact virions and subsequent intracellular uncoating (G and H) (bars = 100 nm). A and B, negatively stained preparations of virions attached to the surface of endodermal cells at 4° after 1 minute (A) and 10 minutes (B) at 37°. C to F, thin sections showing successive stages suggesting fusion of envelope with plasma membrane. G and H, thin sections showing engulfment (G) and intact intravesicular virions seen 20 minutes after attachment (H). [A–F, from C. Morgan and H. M. Rose, J. Virol.* **2**, *925 (1968); courtesy Dr. C. Morgan; G and H, from S. Dales and P. W. Choppin, Virology* **18**, *489 (1962), courtesy of Dr. S. Dales.]*

Penetration

With bacterial viruses the coat of the virion remains outside the cell; only the nucleic acid penetrates. This is probably an exceptional event with animal viruses. Opinions differ as to whether disruption of the virion occurs at the plasma membrane, or only after the engulfment of intact virions (Plate 3-1); there are probably important differences between viruses in this regard. Nonenveloped viruses are probably taken in by phagocytosis ("viropexis"); on the other hand the envelope of enveloped viruses may fuse with the plasma membrane of the cell, releasing the core or helical nucleocapsid directly into the cytoplasm. Unlike attachment, engulfment is a temperature-dependent step, and can be inhibited by metabolic poisons that inhibit phagocytosis.

Uncoating

This term implies release of the infectious nucleic acid from the viral coat. Again, a variety of situations exists, ranging from release of the viral nucleic acid at the cell membrane (as appears to happen with enteroviruses) to the complex intracellular uncoating of poxviruses. Some viruses, for example, reoviruses, may never undergo complete uncoating; the virion-associated transcriptase functions within viral cores that continue to produce viral mRNA throughout the multiplication cycle (Plate 3-2).

THE MULTIPLICATION OF DNA VIRUSES

After uncoating, DNA viruses and RNA viruses diverge in important details of their mechanisms of multiplication, and shall be described separately.

Transcription

Most DNA viruses multiply in the nucleus and depend on cellular RNA polymerases. Poxviruses, which multiply in the cytoplasm, carry their own transcriptase. In both cases, transcription is regulated by mechanisms that are unknown. "Early" genes may be defined as those segments of parental DNA which are transcribed when cytosine arabinoside is present to prevent DNA replication. "Late" genes are those whose transcription rate becomes significant only after DNA starts to replicate.

Early transcripts usually represent only a minority of the viral genes.

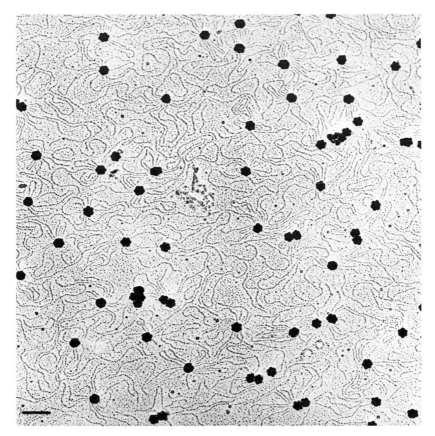

PLATE 3-2. *Reovirus messenger RNA (bar = 200 nm). Reovirus "reaction cores" that have synthesized mRNA for 8 minutes at 37° were prepared for electron microscopy by the Kleinschmidt technique, stained with uranyl acetate and shadowed at a low angle with platinum–palladium. The results of acrylamide gel analysis of similar mRNA molecules are illustrated in Fig. 3-5.* [From N. M. Bartlett, S. C. Gillies, S. Bullivant, and A. R. Bellamy, J. Virol. **14,** 315 (1974), courtesy Dr. A. R. Bellamy.]

In the main, they code for "early" enzymes and other proteins of unknown function, whereas late mRNA's code for the structural proteins of the virion, but the distinction is not always clear-cut. In general, the transcripts, whether early or late, are large polycistronic molecules which are subsequently cleaved into shorter, monocistronic messengers. Polyadenylic acid [poly(A)] (in stretches of up to 150 adenine residues) is then added to the 3' end of each mRNA molecule before it is transported out of the nucleus into the cytoplasm.

Not all mRNA is transcribed from the same strand of double-stranded viral DNA; early and late mRNA's may differ in this regard, and it doubtless represents a control mechanism we know nothing of. Furthermore, in the case of papovaviruses and adenoviruses, whose DNA may become integrated with that of the host cell, long sequences of cellular and viral DNA may be transcribed to give a continuous "hybrid" RNA molecule which is subsequently cleaved.

Translation

Monocistronic mRNA's are translated into proteins on cytoplasmic polyribosomes. The "early" proteins, synthesized even in the presence of inhibitors of DNA synthesis, include the T antigens detectable in both cytocidal and transforming infections by papovaviruses or adenoviruses. They also include many enzymes concerned in DNA replication. Some of these enzymes are virus-specified, but in other cases (e.g., in Papovaviridae) they represent cellular enzymes which have been derepressed by infection.

Arginine-rich proteins are present in the core of most DNA viruses. These histone-like proteins may be closely associated with the viral DNA during its replication and/or transcription. Under conditions of arginine starvation assembly of new viral cores is inhibited.

Table 3-1 summarizes the types of protein that may be specified by viruses.

Replication of Viral DNA

Once an adequate concentration of the necessary enzymes has developed, viral DNA begins to replicate by standard Watson–Crick base-pairing. Cellular DNA polymerases may be involved but the larger DNA viruses may code for their own.

Evidence is accumulating that arginine-rich proteins may remain at-

TABLE 3-1
Products of the Viral Genome

1. Structural polypeptides of the virion (may also be enzymes or act as regulatory proteins)
2. Enzymes of the virion (for use in the next cell infected)
3. Enzymes, nonstructural, involved in DNA transcription or synthesis
4. T antigens and other proteins detectable serologically inside the infected cell or in plasma membrane
5. Regulatory proteins suppressing cellular transcription or translation
6. Regulatory proteins suppressing expression of "early" viral genes

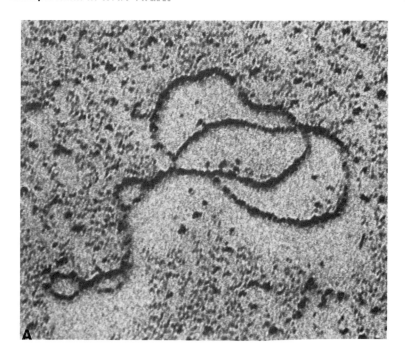

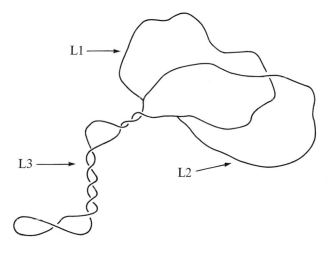

FIG. 3-3. *Replication of viral DNA. (A) Electron micrograph of a twisted SV40 DNA replicating molecule. In some regions the individual strands which comprise the superhelical branch can be seen. Magnification. 1.5 × 10⁵. In an interpretive drawing of the molecule (B) the two replicating branches (not superhelical) are designated L1 and L2. The superhelical unreplicated section is designated L3. [From E. D. Sebring et al., J. Virol. **8,** 478 (1971).]*

tached to viral DNA throughout replication. Association with membranes may also be crucial at certain stages.

Special problems attend the replication of the circular DNA of the papovaviruses. To overcome insuperable physical problems in twisting the superhelix, an endonuclease is induced which "nicks" one strand while a short region is replicated, then a ligase repairs the break again (Fig. 3-3).

Assembly and Release

In the case of the poxviruses and iridoviruses, both DNA replication and structural protein synthesis occur in a common cytoplasmic "factory," hence, there are minimal topological constraints on automatic assembly of virions when the concentration of the two "reactants" reaches an adequate level. On the other hand, herpesviruses, adenoviruses, and papovaviruses are assembled in the nucleus, after structural proteins, synthesized in the cytoplasm, have migrated back to the nucleus, where DNA replication and transcription have been occurring. Little is known about the factors that control these movements of molecules.

All DNA viruses except the smallest (Parvoviridae and Papovaviridae) have a complex structure consisting of several concentric layers of protein. These are laid down stagewise, the basic "core" proteins being the first to associate with viral DNA and the capsomers of the outer capsid being added last. The herpetoviruses, like many of the RNA viruses, acquire a lipoprotein envelope by budding through cellular membrane that has been modified by the incorporation of viral proteins (mainly glycoproteins).

The release of the first virion into the supernatant fluid technically marks the end of the latent period. From each cell, several thousand virions are released one (or a few) at a time over many hours, but they often remain cell-associated for long periods, sometimes in crystalline aggregates. Spread of progeny virions to adjacent or distant cells initiates a second cycle of infection.

Regulation of Viral and Cellular Macromolecular Synthesis

Reference has been made to the fact that both transcription and translation of the viral genome are subject to certain controls. Only a minority of viral cistrons are transcribed before DNA replicates, and even after the whole genome becomes available for transcription the various "late" genes are transcribed at different rates. Moreover, the half-life of these mRNA's varies, so affecting translation rates. More subtle mechanisms also regulate translation. For example, structural viral proteins may act as repressors of transcription or translation by binding to particular segments of viral DNA or mRNA. Little is known about these mechanisms but they are bound to attract increasing attention from virologists over the next few years.

Cellular macromolecular synthesis is also affected by viral infection. DNA, RNA, and protein synthesis may all be "shut down" to varying degrees by different viruses. The advantages of this to the virus are clear enough; if the virus carries all the genetic information it requires, there is a selective advantage in blocking the supply of cellular mRNA's to ribosomes, thus freeing them for the exclusive use of viral messengers. Moreover, interferon will be one of the cellular proteins whose synthesis is turned off.

Virus-coded proteins are responsible for the shutdown. Under artificial laboratory conditions of very high multiplicity, infection may cause a rapid suppression of synthesis of cellular macromolecules, even if the viral genome has been inactivated by irradiation; clearly, structural proteins of the virion are responsible. At low multiplicity, however, viral proteins must be synthesized in order to bring about these effects. Cellular transcription and translation seem to be separately suppressed; DNA synthesis is blocked as a consequence of the inhibition of either of these steps.

Cell shutdown is rapid and almost complete within a few hours of infection by poxviruses and herpesviruses, slow in the case of adenoviruses, and does not occur at all with papovaviruses. In fact, papovaviruses actually stimulate the synthesis of cellular DNA; the viral DNA becomes integrated with that of the cell in productive cytocidal infections as well as in nonproductive transformation.

DNA VIRUSES: A COMPARATIVE SURVEY

The major groups of DNA viruses exhibit great diversity in structural complexity, and their DNA content spans a fiftyfold range (see Tables 1-2 and 1-3); it is not surprising therefore that they should also reveal marked differences in their mode of multiplication (Table 3-2). For example, the DNA of viruses of most families (Parvoviridae, Papovaviridae, Adenoviridae, and Herpetoviridae) is replicated in the nucleus, where the virions are finally assembled, while with others (Poxviridae and Iridoviridae) the entire multiplication cycle occurs in the cytoplasm. The timing of events in the multiplication cycle also varies characteristically with different families; Fig. 3-4 shows approximate growth curves for the fast-growing laboratory mutants commonly used for studying multiplication of the major families.

TABLE 3-2

Major Features of the Multiplication of DNA Viruses

	POXVIRIDAE	HERPETOVIRIDAE	ADENOVIRIDAE	PAPOVAVIRIDAE
Inhibition of host cell				
RNA and protein	Early	Early	Late	No effect
DNA	Early	Early	Late	Stimulated
Site of viral DNA synthesis and assembly of virions	Cytoplasm	Nucleus	Nucleus	Nucleus
Viral release	Minimal, late	Minimal, late	Minimal, late	Minimal, late

Two main types of association of DNA virus and cell can be recognized. Most commonly, the genome of the virus replicates independently of the cellular DNA, and new virions are produced—the infection is said to be *productive*. If the virus is *cytocidal*, cellular functions are arrested and infected cells are destroyed. Some viruses affect cellular synthesis less severely, the extreme case being *steady-state infections* in which both viral and cell multiplication proceed. In the second type of virus–cell interaction the genome of the virus may become integrated into that of the host cell and replicate in perpetuity as part of

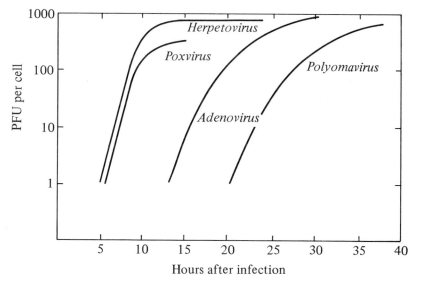

FIG. 3-4. *Idealized multiplication curves (total infectious virus per cell at intervals after infection at high multiplicity) of representatives of the major families of DNA viruses.*

the cellular DNA. In such cases, the outcome may be either (a) a *cryptic infection*, if expression of the viral genome is completely repressed, or (b) the expression of a limited number of viral genes, with morphological (and sometimes malignant) *transformation* of the cell.

THE MULTIPLICATION OF RNA VIRUSES

The multiplication of RNA viruses differs from that of the DNA viruses we have just considered in several important respects. Unique among genetic systems, their genome is RNA, which is transcribed, translated, and replicated quite differently from DNA; indeed, the RNA viruses themselves vary widely in the mechanisms whereby these processes are accomplished. Compared with most DNA viruses, all are extremely limited in the amount of genetic information they carry (see Table 4-1), hence they code for relatively few enzymes, although all require at least one RNA-dependent RNA polymerase. Finally, most RNA viruses mature by budding from cytoplasmic membranes.

Transcription and Translation of Viral RNA

Table 3-3 summarizes the various ways in which transcription, translation, and replication of RNA genomes occur in RNA viruses of different families. It will be noted that the RNA extracted from a virus is infectious only if (a) the genome does not occur in pieces, and (b) there is no virion transcriptase. The latter is a reflection of the polarity of the viral RNA. A transcriptase is only necessary when the viral RNA is "antimessage," so that a complementary RNA must be transcribed to serve as a message. The reverse transcriptase of Retroviridae constitutes a special case.

The alternative mechanisms whereby the genetic information encoded in viral RNA (vRNA) can be translated into protein (Table 3-3) can be summarized as follows:

1. The single-stranded vRNA molecule of picornaviruses and togaviruses can be viewed as a giant messenger RNA (mRNA) molecule containing meaningful genetic information ("sense") directly translatable by ribosomes into protein. The huge polypeptide that results is subsequently enzymatically cleaved into progressively shorter polypeptides.

2. The single-stranded RNA molecule of the paramyxoviruses and rhabdoviruses, on the other hand, is "antimessage"; RNA of complementary nucleotide sequence (cRNA) must first be transcribed, and

TABLE 3-3
Expression of Viral RNA

FAMILY	STRANDEDNESS	MOLECULES	MESSENGER FUNCTION	INFECTIVITY	VIRION TRANSCRIPTASE	POSTTRANSLATIONAL CLEAVAGE OF PROTEINS
Picornaviridae ⎱ Togaviridae ⎰	SS	1	+	+	−	+
Paramyxoviridae ⎱ Coronaviridae ⎰ Rhabdoviridae	SS	1	−	−	+	−[a]
Orthomyxoviridae	SS	7	−	−	+	−[a]
Retroviridae	SS	4	+	−	+(Reverse)	−[a]
Reoviridae	DS	10	−	−	+	−[a]

[a] Minor cleavages do occur.

only this is recognized by tRNA's and ribosomes as mRNA that can be translated into proteins. A virus-coded RNA-dependent RNA polymerase ("transcriptase"), carried in the helical ribonucleoprotein nucleocapsid, transcribes cRNA from the vRNA template. Some at least of these transcripts are full length; it is not certain whether or not any are shorter. However, whether by direct transcription or by cleavage of full length cRNA, short lengths of "monocistronic" cRNA arise, each of which has poly(A) added to its 3' end and is separately translated into a separate polypeptide.

3. The seven (or more) separate single-stranded RNA molecules of the orthomyxoviruses are also "antimessage." A separate cRNA molecule, representing a monocistronic mRNA molecule, is transcribed from each one by virion-associated transcriptase.

4. The single-stranded RNA of the retroviruses is transcribed by a virion-associated RNA-dependent DNA polymerase ("reverse transcriptase") into an RNA:DNA hybrid, which in turn serves as a template for the synthesis of double-stranded DNA. Such virus-specific DNA can be integrated into cellular DNA leading either to a cryptic carrier state or to transformation of the cell. It may, in turn, serve as a template for transcription of single-stranded mRNA and vRNA, which are identical in sense but differ in size.

5. The ten double-stranded vRNA molecules of the reoviruses are transcribed into ten distinct single-stranded mRNA molecules by a transcriptase present in the core of the virion; indeed, virions uncoated only to the core stage can produce mRNA for prolonged periods both *in vivo* and *in vitro* (see Plate 3-2 and Fig. 3-5). After addition of poly(A) to the 3' end, each mRNA is translated on ribosomes into a single identifiable protein.

The RNA viruses have evolved along very different paths to reach equally effective answers to the common problem of how to produce a number of separate proteins from a single set of genetic information, for unlike its bacterial counterpart, the mammalian protein-synthesizing apparatus does not seem to have the capacity to initiate translation of individual cistrons at multiple points along a polycistronic mRNA.

Replication of Viral RNA

The replication of RNA is, as far as we know, something unique to RNA viruses; if cells had the capacity to generate RNA from RNA it would greatly complicate the control of the orderly flow of information from DNA → RNA → protein. The precise mechanism of viral RNA replication is therefore a matter of considerable interest. Though de-

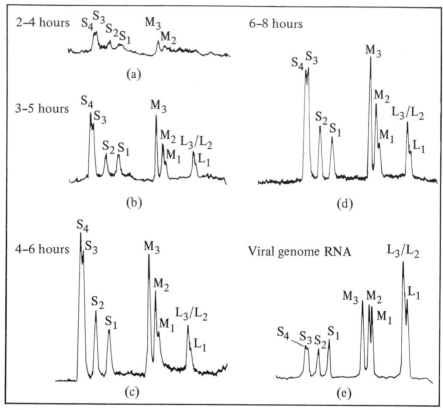

FIG. 3-5. *Transcription of mRNA from the double-stranded viral RNA of reovirus by a viral transcriptase. Tracings from acrylamide gel autoradiograms showing relative rates of formation of reovirus mRNA species during the infection cycle (a–d), as determined by hybridizing labeled RNA from the cytoplasm of infected cells to an excess of genome RNA; compared with the tracing of genome RNA derived from virions (e).* [From H. J. Zweerink and W. K. Joklik, Virology **41,** 501 (1970).]

bated for several years, it is now agreed that viruses with a single-stranded genome replicate by a semiconservative process in which the parental (vRNA) molecule serves as a single-stranded template for the simultaneous transcription of several complementary (cRNA) strands. The "replicative intermediate" (RI) is, therefore, a partially double-stranded structure with single-stranded "tails." The single-stranded cRNA progeny molecules peel off and in turn serve as templates for the synthesis of new vRNA (Fig. 3-6).

Virus-coded RNA-dependent polymerases ("replicases") are re-

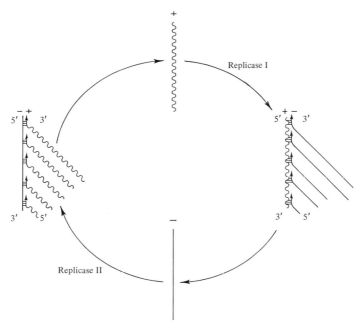

FIG. 3-6. *Diagram illustrating the probable mechanism of replication of single-stranded RNA. Several "−" molecules may be transcribed from the viral "+" RNA simultaneously by separate molecules of replicase I. Released "−" RNA molecules then serve as templates for the simultaneous transcription of several "+" (viral) RNA molecules, perhaps by a second replicase (replicase II). [After S. Spiegelman, N. R. Pace, D. R. Mills, R. Levisohn, T. S. Eikhorn, M. M. Taylor, R. L. Peterson, and D. H. L. Bishop, Cold Spring Harbor Symp. Quant. Biol. **33**, 101 (1968).]*

quired to catalyze this process. These enzymes tend to be specific for the RNA of their own viral species. In some cases, different polymerases are involved in the production of "−" and "+" strands, respectively, but the generality of this situation has yet to be established.

Membranes and Budding

Cellular membranes play a dominant role throughout the multiplication of RNA viruses. The plasma membrane is involved in attachment and penetration, translation of viral proteins occurs mainly on polyribosomes associated with membranes of the rough endoplasmic reticulum, glycosylation of viral glycopolypeptides takes place in various smooth cytoplasmic membranes, and RNA replication occurs in membrane-bound "replication complexes." Moreover, since in most genera the virions of RNA viruses are enveloped, cytoplasmic membranes (usually the plasma membrane but sometimes internal cytoplasmic membranes) play an important role in the assembly and release of virus particles by a process of *budding*. In the budding process the sequence of events is as follows:

1. Viral glycoproteins enter the plasma membrane, displacing all of the existing cellular proteins.

2. Viral "membrane protein," a low molecular weight nonglycoprotein, forms a layer on the inside of the modified plasma membrane.

3. Viral nucleocapsid attaches to the membrane protein and the viral glycoproteins develop into visible projections (peplomers) on the outside of the plasma membrane.

4. Budding occurs; this is a sort of "reverse cytosis" in which there is an evagination of the altered area of plasma membrane, and the new virion, enclosing its nucleocapsid, is nipped off (Plate 3-3).

Regulation of Viral Gene Expression

Very little is yet known of the mechanisms whereby viral RNA transcription, translation, and replication are controlled. In the Picornaviridae, the single vRNA molecule is translated directly into a single polypeptide, and there is no control of transcription or translation of individual "genes," only of the vRNA molecule as a whole. However, with the transcriptase-carrying viruses, in all of which monocistronic cRNA's are synthesized, there is evidence that individual vRNA genes are transcribed at different rates and that the resulting cRNA molecules are translated at different rates. The various viral proteins are not made in equimolar proportions, nor in numbers that are inversely proportional to their molecular weights. Moreover, the relative rates of synthesis of the several viral proteins fluctuate in a systematic fashion as the multiplication cycle proceeds, nonstructural proteins and sometimes "core" proteins tending to be most conspicuous early and capsid proteins later. In addition, there is good evidence that some regulatory mechanism must determine whether at any given phase of the multiplication cycle, mRNA molecules are utilized for translation into protein, or for RNA replication, or, in the case of vmRNA, for assembly into virions. We know very little about how such regulation is accomplished in animal cells.

Regulation of Cellular Macromolecular Synthesis

Cellular macromolecular synthesis is abruptly suppressed by some RNA viruses and totally unaffected by others. At one end of the spec-

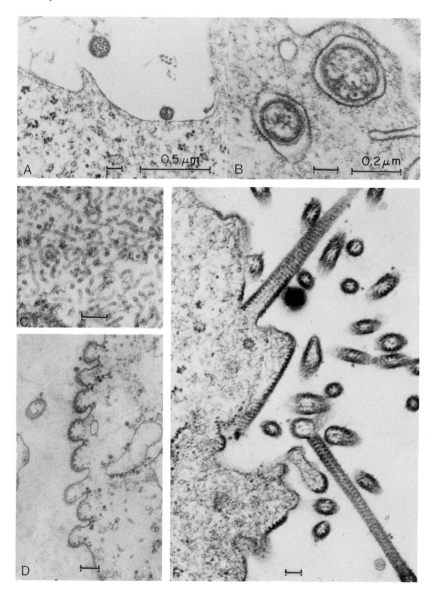

PLATE 3-3. *Many viruses mature by budding from the plasma membrane. Electron micrographs of different stages of multiplication of paramyxovirus SV5 (unlabeled bars = 100 nm). (A) Attachment. (B) Viropexis. (C) Accumulation of nucleocapsids. (D and E) Budding from the plasma membrane, with some filamentous forms. From R. W. Compans et al., Virology* **30,** *411 (1966). (Courtesy Dr. P. W. Choppin.)*

trum, picornaviruses code for proteins that separately shut down the transcription of cellular RNA and its translation into protein; DNA synthesis stops as a consequence. Many of the paramyxoviruses, on the other hand, have virtually no effect on cellular metabolism and bud from cells for prolonged periods without affecting their continued growth and division; such persistent noncytocidal infections, or "carrier cultures" are discussed in Chapter 8. Retroviruses are not only noncytocidal, but they may transform infected cells to malignancy following integration of virus-coded DNA into cellular chromosomes (Chapter 9). There is some evidence that the orthomyxoviruses may require the transcription of RNA from cellular DNA during the first part of their multiplication cycle; this contrasts sharply with the paramyxoviruses and picornaviruses which multiply quite normally in the presence of actinomycin D or even in enucleated cells.

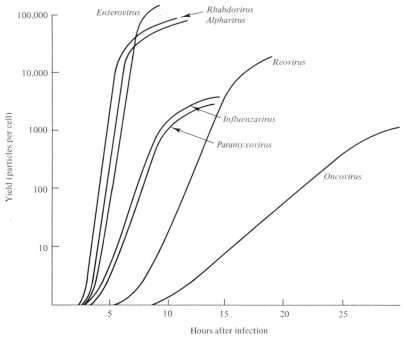

FIG. 3-7. *Idealized multiplication curves (total infectious virus particles per cell at intervals after infection at high multiplicity) of representatives of the major families of RNA viruses. Latent periods, multiplication rates, and final yields are all affected by species and strain of virus and by cell strain and the kind of culture; latent period and multiplication rate, but not final yield are affected by the multiplicity of infection.*

RNA VIRUSES: A COMPARATIVE SURVEY

The basic multiplication cycle of different RNA viruses varies substantially in both the length of the latent period and the viral yield. Figure 3-7 illustrates representative examples of the multiplication curves of model viruses of the best studied families.

SUMMARY

Viral replication is a complicated process, the details of which have only recently been elucidated. The multiplication cycle may be regarded as an overlapping sequence of discrete steps (Fig. 3-2).

1. Attachment (adsorption)
2. Penetration
3. Uncoating of viral nucleic acid
4. Transcription of "early" mRNA
5. Translation of "early" proteins
6. Replication of viral nucleic acid
7. Transcription of "late" mRNA
8. Translation of "late" proteins
9. Assembly (maturation) of virions
10. Release of virions

Quite complex control mechanisms, under the direction of virus-coded proteins, regulate the transcription and translation of both cellular and viral genetic information.

Virions attach to specific viral receptor sites on the plasma membrane of susceptible cells; cells that lack these specific receptors cannot adsorb virus particles, but may nevertheless be infectible, though inefficiently, with purified viral nucleic acid. After attachment, the virus penetrates the cell by phagocytosis, or by fusion of viral envelope with plasma membrane, leading to liberation of nucleocapsid directly into the cytoplasm.

Viruses reproduce from their nucleic acid, which directs not only its own replication but also, via messenger RNA, the synthesis of virus-specific proteins. Some of these are structural components of the virion, while others are proteins leading to the selective shutdown of synthesis of cellular macromolecules, or enzymes involved in the synthesis of viral components.

The viral nucleic acids replicate by standard base-pairing; in the case of single-stranded RNA viruses this involves the synthesis of a "replicative intermediate." The RNA polymerases involved are usually virus-specified enzymes.

New viral nucleic acid and protein molecules are assembled into virions (or nucleocapsids) within nuclear or cytoplasmic factories. Enveloped viruses mature by "budding" through cell membranes that have been modified by the replacement of cellular proteins by viral proteins.

The multiplication of viruses representing the principal families is compared in Table 3-2 and Figs. 3-4 and 3-7. Major variations result from differences in the type and amount of genetic information, the intracellular site of replication, the requirement for continuing cellular function, and the relevance of the cytoplasmic membrane in the morphogenesis of enveloped viruses.

RNA viruses have evolved several alternative routes to protein synthesis (Table 3-3), depending on whether:

(a) The viral RNA (vRNA) has positive or negative polarity, i.e., is equivalent to messenger RNA (mRNA) or to "anti-message" RNA.

(b) The viral nucleocapsid carries a transcriptase which transcribes complementary RNA (cRNA), equivalent to mRNA, from viral RNA of negative polarity.

(c) The viral RNA is segmented into several molecules representing individual genes from which monocistronic mRNA is transcribed, or vRNA is in the form of a single polycistronic molecule and the mRNA transcribed from it is either cleaved into monocistronic messengers or is translated into a giant polypeptide that is subsequently cleaved into the several functional viral proteins.

FURTHER READING

Allison, A. C. (1971). The role of membranes in the replication of animal viruses. *Int. Rev. Exp. Pathol.* **10**, 182.

Baltimore, D. (1971a). Polio is not dead. *Perspect. Virol.* **7**, 1.

Baltimore, D. (1971b). Expression of animal virus genomes. *Bacteriol. Rev.* **35**, 235.

Barry, R. D., and Mahy, B. W. J., eds. (1975). "Negative Strand Viruses." Academic Press, London.

Burke, D. C., and Russell, W. C., eds. (1975). "Control Processes in Virus Multiplication," 25th Symp. Soc. Gen. Microbiol. Cambridge Univ. Press, London and New York.

Choppin, P. W., Klenk, H.-D., Compans, R. W., and Caliguiri, L. A. (1971). The parainfluenza virus SV5 and its relationship to the cell membrane. *Perspect. Virol.* **7**, 127.

Dales, S. (1973). Early events in cell-animal virus interactions. *Bacteriol. Rev.* **37**, 103.

Fenner, F., McAuslan, B. R., Mims, C. A., Sambrook, J. F., and White, D. O. (1974). "The Biology of Animal Viruses," 2nd ed., pp. 155–273. Academic Press, New York.

Fox, C. F., and Robinson, W. S., eds. (1973). "Virus Research." Academic Press, New York.

Fraenkel-Conrat, H., and Wagner, R. R., eds. (1974–75). "Comprehensive Virology: Reproduction of Viruses." Vols. 2–5. Plenum Press, New York.

Hershey, A. D., ed. (1971). "The Bacteriophage Lambda." Cold Spring Harbor Lab., Cold Spring Harbor, New York.

Huang, A. S. (1973). Defective interfering viruses. *Annu. Rev. Microbiol.* **27**, 101.

Joklik, W. K., and Zweerink, H. J. (1971). The morphogenesis of animal viruses. *Annu. Rev. Genet.* **5**, 297.

Lonberg-Holm, K., and Philipson, L. (1974). Early interactions between animal viruses and cells. *Monog. Virol.* **9**, 1.

Maramorosch, K., and Kurstak, E., eds. (1971). "Comparative Virology." Academic Press, New York.

Raspé, G., ed. (1974). "Workshop on Virus-Cell Interactions." *Advan. Biosci.* No. 11. Pergamon, Oxford.

Shatkin, A. J., (1974). Animal RNA viruses: Genome structure and function. *Annu. Rev. Biochem.* **43**, 643.

Spiegelman, S., Pace, N. R., Mills, D. R., Levisohn, R., Eikhom, T. S., Taylor, M. M., Peterson, R. L., and Bishop, D. H. L. (1968). The mechanism of RNA replication. *Cold Spring Harbor Symp. Quant. Biol.* **33**, 101.

Stent, G. S. (1971). "Molecular Genetics: An Introductory Narrative." Freeman, San Francisco, California.

Sugiyama, T., Korant, B. D., and Lonberg-Holm, K. K. (1972). RNA virus gene expression and its control. *Annu. Rev. Microbiol.* **26**, 467.

Watson, J. D. (1970). "Molecular Biology of the Gene," 2nd ed. Benjamin, New York.

Wolstenholme, G. E. W., and O'Connor, M., eds. (1971). "Strategy of the Viral Genome," Ciba Found. Symp., Churchill, London.

CHAPTER 4

Viral Genetics

INTRODUCTION

Originally, animal virology developed largely from the need to control viral infections in man and his domesticated animals, although during the last decade it has been greatly influenced by the belief that viruses might cause human cancers. Since direct experimentation with the natural hosts is often expensive or impossible, much effort has been devoted to two ends: (a) adaption of viruses of medical and economic importance to small laboratory animals and more recently to cell culture, and (b) attenuation of viruses by serial passage in an unnatural host to obtain a live vaccine. In most cases these processes of adaptation and attenuation operate via spontaneous mutation and selection, although, as is usual in science, useful practical results were obtained long before the molecular or, indeed, the biological mechanisms involved were understood. This is well illustrated by the development of an effective rabies vaccine by Pasteur as long ago as 1885.

More recently, stemming from the spectacular advances in molecular genetics that have emerged from the study of bacterial viruses, and stimulated by the tumor virus program, efforts have been made to elucidate the properties and functions of animal viruses by genetic analysis. This involved the selection of appropriate mutants, the construction of genetic maps, and study of the functions of the products of the genes in which mutations occur. With a few viruses, like SV40 and adenoviruses, the techniques of selective cleavage of the genome with "restriction endonucleases," nucleic acid hybridization, and electron microscopy have led to dramatic advances in genetic mapping. So far, genetic analysis has provided little information of practical importance, but its potential is great and, combined with biochemical studies, it provides the only way of fully describing viral multiplication.

Much current experimental work in viral genetics is of interest only to laboratory workers, but almost all viral properties, including those of practical importance, like antigenic properties and virulence, are affected directly by changes in the genetic material, hence the need for this chapter.

NUMBERS OF GENES IN VIRUSES

Most animal viruses contain a single molecule of either DNA or RNA as their genetic material, although a few have fragmented genomes. By making certain assumptions the numbers of genes (cistrons) in different viruses (hence the number of proteins for which they code) can be estimated. Table 4-1 presents calculations made for representative viruses of each of the major families. It is clear that viruses cover a wide range of genetic complexity; polyoma virus contains less than half a dozen genes while the poxviruses carry almost

TABLE 4-1

Comparison of the Numbers of Genes in Animal Viruses[a]

| | | NUCLEIC ACID | | | |
| | | MOLECULAR WEIGHT $(\times 10^6)$ | TYPE[b] | NUMBER OF GENES | POLY-PEPTIDES IN VIRION[c] |
FAMILY	VIRUS				
Parvoviridae	Minute virus of mice	1.8	SS-D	3	3
Papovaviridae	Polyoma	3.0	DS-D	3	3
Adenoviridae	Adenovirus type 2	23	DS-D	23	5–9
Herpetoviridae	Herpes simplex	100	DS-D	100	12–24
Poxviridae	Vaccinia	160	DS-D	160	30
Picornaviridae	Poliovirus	2.6	SS-R	5	4
Togaviridae	Sindbis	4	SS-R	8	3
Orthomyxoviridae	Influenza A	5	SS-R	10(7)[d]	7
Paramyxoviridae	Newcastle disease	7	SS-R	14	6
Retroviridae	Rous sarcoma	12	SS-R	24	10
Rhabdoviridae	Vesicular stomatitis	4	SS-R	8	5
Reoviridae	Reovirus 3	15	DS-R	15(10)[d]	7

[a] Numbers calculated from the estimated molecular weights of their nucleic acids, on the assumption of a nonoverlapping triplet code with genes containing about 1500 nucleotides or nucleotide pairs.

[b] DS, double-stranded; SS, single-stranded; D, DNA; R, RNA.

[c] Data from Chapter 1; does not include virion-associated enzymes.

[d] Figure in parentheses indicates number of separate molecules in genome.

half as much genetic information as some free-living mycoplasmas. Most viral genes code for proteins, and the table indicates the number of different structural proteins that have been recognized in the virions. Even allowing for the fact that gel electrophoresis fails to resolve all the polypeptides of the large DNA virions, it is clear that structural polypeptides account for only a minority of their genes. The remainder presumably code for enzymes involved in DNA replication, regulating proteins, and in the case of the poxviruses, perhaps even enzymes concerned with the synthesis of viral polysaccharides and lipids. At least half the genome of most RNA viruses appears to specify structural proteins.

In viruses with fragmented genomes (Orthomyxoviridae and Reoviridae), each molecule of RNA represents a single gene. There is a reasonably close agreement between the number of genome fragments and the calculated number of genes.

In the case of the Picornaviridae, in which a single molecule of vRNA is translated into a giant polypeptide, which is only later cleaved to yield smaller functional proteins, there is an interesting semantic problem: should such viruses be said to have only one "gene," or should the final cleavage products each be considered to be the product of a separate gene? We have adopted the latter convention, which makes more sense when discussing gene function.

MUTATION

Changes in the hereditary material, i.e., in the information encoded in the nucleic acid, are called mutations. The frequency of spontaneous mutations in viruses varies somewhat but is usually of the order of 10^{-5} to 10^{-6} each time a viral nucleic acid molecule replicates, i.e., it is of the same order of frequency as in organisms. Since in every infection of a vertebrate animal, one or a small number of virus particles multiplies several million times, mutations occur during the course of every natural infection; this is also true every time a population of virions is grown in the laboratory. Whether such mutations emerge to the point of being of practical importance depends upon two factors: (a) whether they afford the mutant virus some selective advantage, and (b) whether they happen to occur early or late in the infection. As we shall see in Chapter 11, mutation and selection (usually during the stage of transmission from one animal to another) are of major importance in the evolution of viruses.

In the laboratory, reasonable genetic constancy of viral stocks (e.g., those used for making viral vaccines) is achieved by (a) isolating a "pure" clone, i.e., a population of virus particles originating from a single particle, usually by growth from a single plaque in cell culture, (b) growing seed stock from the pure clone under carefully controlled conditions, and (c) as far as practicable avoiding serial passage of the virus. If passage is unavoidable a single plaque is again selected to grow up a new seed stock.

Mutagenesis

Mutations can be induced in animal viruses, or their isolated infectious nucleic acids, by treatment of viral suspensions with physical agents like UV- or X-irradiation or with chemicals like nitrous acid, hydroxylamine, ethylmethyl sulfonate, or nitrosoguanidine. Other chemicals, like 5-fluorouracil of 5-bromodeoxyuridine, are mutagenic when virus is grown in their presence because they are incorporated into the viral nucleic acid. Chemical mutagenesis of viruses is an important tool for laboratory studies in viral genetics for it greatly increases the mutation rate and makes it possible to select mutants that would otherwise occur very rarely. It is not yet of much applied medical importance, either to the doctor or for the preparation of viral vaccines.

Kinds of Mutants of Animal Viruses

Recognition of a mutation requires that there be some change in the phenotype of the mutant virus which can be differentiated from that of the wild-type virus. Thus there are plaque- and pock-type mutants, mutants which are resistant to or dependent upon certain chemicals, mutants that differ in the antigenic properties of their coat proteins, and many others.

Conditional Lethal Mutants. A class of mutants different from those just described has become very important in laboratory studies of viruses, namely those that are referred to as *conditional lethal mutants.* These are mutations which so affect a virus that it cannot grow under certain conditions, determined by the experimenter, but can replicate and yield mutant progeny under "normal" conditions. The great importance of these mutations is that a single selective test can be used to obtain mutants in which the lesion may be present in any one of many different genes.

Host-dependent conditional lethal mutants of bacteriophages are those in which a nucleotide triplet originally coding for an amino acid has been altered by mutation to one of the codons responsible for the termination of polypeptide synthesis. As a result they cannot grow in normal bacteria but are able to grow in bacterial mutants that have tRNA's capable of translating such chain-terminating codons into an amino acid. Although host-dependent animal viruses have been found, their defects are different from those of the bacteriophages, and, so far, ways of using such mutants for genetic analysis have not been developed. The only class of conditional lethal mutants used by animal virologists are those showing temperature sensitivity of viral development, the so-called *temperature-sensitive mutants*, in which the selective condition used is a high temperature of incubation of the infected cells. The defects in temperature-sensitive mutants lie in the primary structure of the proteins specified by the affected genes, which ultimately cause an altered secondary or tertiary structure. Thus a change in a single nucleotide of the nucleic acid leads to a single amino acid substitution in a polypeptide which may so weaken the internal bonding of the protein that its structure (and effective function) may be radically changed by a rise in temperature of only a few degrees. Temperature-sensitive mutants are of great importance for the biochemical analysis of viral functions, and, since they are less virulent than the wild-type virus from which they are derived, such mutants are now being used to select attenuated viruses for use as vaccines.

Cold-Adapted Mutants. Many of the attenuated strains of virus currently used for human vaccination have been obtained by growing the wild-type virus in a novel host (usually an embryonated egg or cultured cells) at a low temperature of incubation, for it has been observed empirically that this leads to rapid attenuation of virulence. Sometimes these "cold-adapted" mutants are in fact temperature-sensitive mutants, but there is not a strict correlation between capacity to grow well at a reduced temperature (cold adaptation) and inability to grow at high temperature (temperature sensitivity).

Pleiotropism or Covariation. Mutants selected for a certain phenotypic alteration are often found to be changed in other properties, reflecting the relevance of a single protein to several properties of the virus. For example, a single change in one of the coat proteins of influenza virus could affect *in vitro* properties like hemagglutination, action on glycoprotein inhibitors, antigenicity, and ability to grow in certain hosts.

Pleiotropism is useful in certain applications of viral genetics, e.g., the selection of an attenuated variant for a vaccine. Initial selection can be based on *in vitro* tests, reserving the more cumbersome animal testing for final characterization of the selected strain. For instance, an attenuated poliovirus mutant could be selected on the basis of its ability to grow better at low than at high temperatures before being subjected to expensive tests for lack of neurovirulence in primates.

Adaptation to New Hosts

The essential prerequisite for the experimental investigation of an animal virus or the disease it causes is the production in a laboratory host of recognizable signs of infection associated specifically with the virus in question. These may be produced the first time that the virus is inoculated into an experimental host. In other cases minimal signs of infection are observed initially, yet after serial passage, sometimes prolonged, a lethal infection is regularly produced, as, for example, in the "adaptation" of poliovirus and dengue virus to rodents. Newly isolated viruses may also fail to grow initially in certain kinds of cultured cells; most modern virological research is performed with strains of virus adapted to produce plaques or cytopathic changes in continuous cell lines. A frequent by-product of such adaptation to a new experimental host is the coincident attenuation of the virus for its original host.

ABORTIVE INFECTIONS AND DEFECTIVE VIRUSES

Viral infection is not always *productive*, i.e., does not always lead to the synthesis of infectious progeny; some or even all viral components may be synthesized but not assembled properly. This is called *abortive infection*. It may result from the treatment of the cell with chemicals like proflavine, or may be due to the fact that certain types of cells are *nonpermissive*, i.e., lack some enzyme or tRNA essential for the multiplication of a particular virus, whether it be the wild type or a host-dependent conditional lethal mutant.

A special class of abortive infections of considerable theoretical interest and practical importance are those due to genetically *defective viruses*. For example, conditional lethal mutants are defective under restrictive conditions (e.g., in nonpermissive cells or at high temperature), but can be *rescued*, i.e., helped to yield infectious progeny, by co-infection of the cell with a *helper virus*, which is usually, but not necessarily, a related virus.

Two important examples of defectiveness involve adenoviruses. The small "adeno-associated viruses" (see Chapter 1) appear to be absolutely defective, for they fail to multiply in any cultured cells tested so far, unless an adenovirus is also multiplying in the same cells. Human adenoviruses themselves are defective in monkey cells (but, of course, not in human cells); they will multiply if the restrictive monkey cells are co-infected with the unrelated papovavirus, SV40, or with a simian adenovirus, either of which can function as a helper.

The extreme state of defectiveness is seen in cells transformed by certain DNA viruses (see Chapter 9). Here part or all of the viral genome is integrated with the cellular genome and replicates with it. The genetic information in part of the viral genome is expressed, for virus-specified proteins are synthesized, but ordinarily no infectious virus is produced. If heterokaryons are artificially produced by fusing the transformed cell with a cell that supports productive infection with the virus in question, the integrated viral genome may be rescued (in the case of SV40-transformed cells) and infective virus is then produced.

GENETIC INTERACTIONS BETWEEN VIRUSES

The uses of mutants for fundamental studies are twofold: (a) as the raw material for genetic mapping, by recombination and complementation experiments, and (b) as tools for the study of the functions of viral genes, since, in viruses as in man, normal function can often be explored most effectively by the study of disordered function. When two different virions simultaneously infect the same cell, a variety of types of genetic interaction may occur between the newly synthesized nucleic acid molecules.

Genetic Recombination

Genetic recombination is the name given to the exchange of segments of nucleic acid between different but related viruses, in such a way that some of the progeny contain a novel combination of genes.

Two varieties can be distinguished among animal viruses: *intramolecular recombination* (Fig. 4-1A; Table 4-2), which involves the rearrangement of sequences within a single nucleic acid molecule, and *genetic reassortment* (Fig. 4-1B; Table 4-2), in which separate molecules of the nucleic acid of viruses that have fragmented genomes (e.g., influenza virus and reovirus) are exchanged. Intramolecular recombination is readily demonstrated with the larger DNA viruses (poxviruses, herpetoviruses, and adenoviruses), and it also occurs between viral and cellular DNA's in "transforming" infections by papovaviruses, adenoviruses, and the RNA tumor viruses (as "provirus" DNA). Rarely (with SV40 and adenoviruses) it may occur between

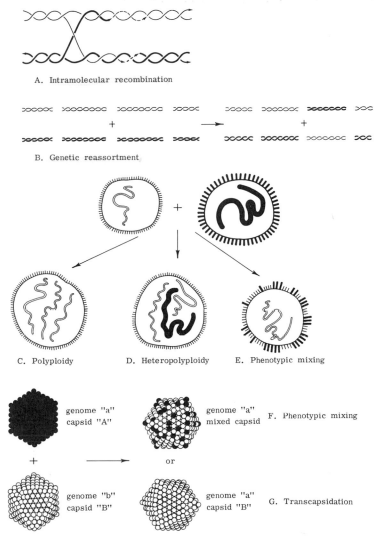

A. Intramolecular recombination

B. Genetic reassortment

C. Polyploidy D. Heteropolyploidy E. Phenotypic mixing

genome "a"
capsid "A" F. Phenotypic mixing

genome "a"
mixed capsid

genome "b"
capsid "B" G. Transcapsidation

genome "a"
capsid "B"

FIG. 4-1. *Genetic recombination, polyploidy, phenotypic mixing and trans-capsidation. (A and B) Genetic recombination: (A) Intramolecular recombina-tion, according to the model of R. D. Hotchkiss, Advan. Genet. **16,** 325 (1971). (B) Reassortment of genome fragments, as in Reoviridae and Orthomyxoviridae. (C and D) Polyploidy: (C) polyploidy, as seen in unmixed infections with Para-myxoviridae; (D) heteropolyploidy, as may occur in mixed infections with Para-myxoviridae. (E–G) Phenotypic mixing: (E) with enveloped viruses; (F) viruses with icosahedral capsids; (G) extreme case of transcapsidation or genomic masking.*

TABLE 4-2

Nucleic Acid Interactions: Genetic Recombination and Reactivation[a]

PHENOMENON	PARENT 1	PARENT 2	PROGENY	COMMENT
Intramolecular recombination	ABC	ABC	ABC, **ABC**,	With mutants of DNA viruses and Picornaviridae
	ABC	A**ST**	AB**T**, A**S**C	With different strains of vaccinia virus
	ABCD	XYZ	ABCZ	Loss of adenovirus genes (D), addition of SV40 genes (Z)
	ABC	123	12ABC3	Oncogenic viruses; integration of viral genes (ABC) into genome of cell (123)
Gene reassortment	A/**B**/C	A/**B**/C	A/**B**/C	With Reoviridae and Orthomyxoviridae
	A/**B**/C	A/**S**/**T**	A/**B**/**T**, A/**S**/C	
Cross-reactivation				
Between UV-inactivated virus and virus of a different but related strain	A̸B̸C	A**ST**	A**S**C	Rescue of gene C from inactivated parent, by intramolecular recombination or gene reassortment
	A̸/B̸/C	A/**S**/**T**	A/**S**/C	
Multiplicity reactivation				
Between virions of same virus inactivated in different genes	ABC̸	AB̸C	ABC	Recombination or reassortment of genes B and C to yield viable virus
	A/B/C̸	A/B̸/C	A/B/C	

[a] A, etc., active viral genes; 1, etc., active cellular genes; **B**, etc., mutant genes; A̸, etc., inactivated genes; ABC, continuous linear genome; A/B/C, segmented genome.

unrelated viruses. With RNA viruses intramolecular recombination has only been conclusively demonstrated with poliovirus, but genetic reassortment is a very common occurrence in mixed infections of cells with influenza viruses or reoviruses.

It is possible, though not yet conclusively proved, that genetic reassortment may be important in the natural history of influenza virus. In the laboratory it has been easy to demonstrate reassortment between influenza viruses of man and of animals (birds, horses, swine) concurrently infecting cultured cells or chick embryos. Similar results have recently been obtained in birds and mammals. These experiments support the hypothesis that pandemic strains of influenza virus may originate by genetic reassortment (see Chapter 11).

Genetic Reactivation

Genetic reactivation may be of two types. *Cross-reactivation* refers to genetic recombination between an active virus and an inactivated virus of a related but distinguishable genotype. By the same process of exchange of nucleic acid described above (molecular recombination or genetic reassortment), some virions are produced which contain active genes from the inactivated virus (Table 4-2). It is of some value in laboratory experiments, e.g., for obtaining recombinants of influenza viruses, but is of no practical importance. *Multiplicity reactivation* is the term applied to the production of infective virus by a cell infected with two or more virions of the same strain, each of which has suffered a lethal mutation in a different gene, e.g., after exposure to UV-irradiation (Table 4-2). The phenomenon has been demonstrated with viruses of several groups: poxvirus, influenza virus, and reovirus. Multiplicity reactivation could theoretically lead to the production of living virus if humans were to be inoculated with UV-irradiated vaccines; accordingly this method of inactivation is not currently used for vaccine production.

Complementation

Complementation is the term used to describe all cases in which interaction between viral gene products in multiply or mixedly infected cells results in an increased yield of one or both agents (see Table 4-3). Unlike genetic recombination, complementation does not involve any exchange of viral nucleic acid, but reflects the fact that one virus provides a gene product which the other cannot make, thus enabling the latter to multiply in the mixedly infected cell. The gene as a unit of function can be defined by complementation tests, that determine whether two mutants, which may exhibit similar phenotypes, are defective in the same function. In experimental virology complementation between related viral mutants is useful as a preliminary step to genetic mapping, to sort out mutants into groups. Temperature-sensitive mutants, for example, can usually be complemented by other mutants of the same virus which are not temperature-sensitive, and different temperature-sensitive mutants can sometimes be allocated to functional groups by testing which pairs of *ts* mutants can complement one another.

The rescue of the defective adeno-associated viruses by the multiplication of an adenovirus in the same cell is another example of complementation. The human adenoviruses themselves are defective in simian cells, but they can be rescued by SV40. SV40 is noncytocidal

TABLE 4-3

Gene Product (Protein) Interactions: Complementation and Phenotypic Mixing[a]

PHENOMENON	PARENT 1	PARENT 2	PROGENY	COMMENT
Complementation				
(a) Between conditional lethal mutants of the same virus (under restrictive conditions)	A **B** C → → → a b c	A B **C** → → → a b c	A**B**C, AB**C**, (ABC)	Reciprocal; both mutants rescued; sometimes recombination also
(b) Between defective virus and unrelated helper virus	A**B́** C → → → a c	B Y Z → → → b y z	AB́C and BYZ	Defective virus is rescued by gene product "b" of helper BYZ
Phenotypic mixing				
(a) Enveloped viruses	$\dfrac{ABC}{a}$	$\dfrac{XYZ}{x}$	$\dfrac{ABC\ XYZ}{ax}$ $\dfrac{ax}{ABC\ XYZ}$	Mixed peplomers in envelopes, genomes unaltered; also parental phenotypes
(b) Nonenveloped viruses	$\dfrac{ABC}{a}$	$\dfrac{XYZ}{x}$	$\dfrac{ABC\ XYZ}{ax}$ $\dfrac{ABC\ XYZ}{a}$	Mixed capsomers in capsids Genome of one parent with capsid of the other (transcapsidation). Not always reciprocal

[a] A, etc., active viral genes; **B**, etc., mutant genes; B́, defective gene B; a, etc., product of gene A, etc.; **b**, etc., product of mutant gene **B**; ABC, genome; ax, proteins in envelope (or capsid).

for rhesus monkey cells but it grows within them, thereby complementing adenovirus growth in these cells. Further, during the course of serial passage of adenoviruses in monkey kidney cells the DNA's of SV40 and of some adenoviruses form hybrid molecules, which become enclosed within adenovirus capsids (see Table 4-2). These hybrid particles, like SV40 itself, can complement adenovirus growth in monkey kidney cells, although they themselves are often defective, since they may lack the genes for the adenovirus capsid protein(s). A comprehensive set of adenovirus-SV40 hybrids has now been recovered which offers the potential for studying the chemistry and function of particular SV40 genes in isolation and thus of determining the role of each in cellular transformation and carcinogenesis.

The discovery of hybrids has wide implications in many areas of virology: (a) the host range of a virus may be extended by the acquisition of a novel capsid from another virus, (b) serological identification of such a virus gives a fallacious picture of its genetic potential, (c) vaccine production could be complicated by such "genetic contamination," (d) being conditionally defective, such hybrids may be prone to induce cancer, and (e) the possibility of genetic interaction at the molecular level between viruses of different groups could be important in the evolution of new groups of viruses.

PHENOTYPIC MIXING

Following mixed infection by two viruses which share certain common features, some of the progeny may acquire phenotypic characteristics from both parents, although their genotype remains unchanged. For example, when cells are infected with antigenically different strains of influenza virus, or with influenza virus and a paramyxovirus, the envelopes of some of the progeny particles contain viral antigens characteristic of each of the parental viruses. However, each virion contains the nucleic acid of only one of the parents, so that on passage it will produce only virions resembling that parent (Fig. 4-1E; Table 4-3).

With nonenveloped viruses phenotypic mixing often seems to involve what has been called *transcapsidation* (Fig. 4-1G); there is very extensive or even complete exchange of capsids. For example, poliovirus nucleic acid may be enclosed within a coxsackievirus capsid, or adenovirus type 7 genome may be enclosed within an adenovirus type 2 capsid.

POLYPLOIDY

Among viruses that mature by budding from cellular membranes it is very commonly found that several nucleocapsids (and thus genomes) are enclosed within a single envelope (Fig. 4-1C). If cells are co-infected with recognizable different strains of such viruses (for example, most paramyxoviruses), almost all the progeny particles are heteropolyploid (Fig. 4-1D) as well as phenotypically mixed in their envelope antigens.

INTERFERENCE

The reader has probably by now gained the impression that most mixed infections lead to some type of genetic or nongenetic interaction between the nucleic acids or the proteins of the progeny of two parents with the result that the virions yielded by that cell are hybrids of one sort or another. In fact, the more common result may be that the two parents, if insufficiently alike, multiply independently of one another, or else they *interfere* with the other's multiplication.

Interference is more likely to occur when one virus preempts the other, in time or in numbers. Preinfection of a cell with one type of virus tends to prevent the multiplication of a second virus with which that cell is subsequently challenged. There are several different mechanisms whereby interference may be established:

(a) The interfering virus may occupy or destroy cellular receptors, so preventing attachment of the challenging virus. While clearly demonstrable under conditions of high multiplicity infection, this is probably something of a laboratory artifact.

(b) The interfering virus may compete for some intracellular site or enzyme, e.g., for RNA polymerase. The best example is the type of homologous interference, known as *auto-interference,* that occurs when large numbers of *defective* (*"incomplete"*) *virions* produced by serial passage of influenza virus at high multiplicity, effectively interfere in the multiplication of infectious virions, leading to a progressive diminution of viable progeny. "Incomplete" influenza virions lack the largest of the seven viral RNA molecules, possibly because this molecule takes longest to replicate and is thus at a survival disadvantage in competing for RNA polymerase, when the cell is over-supplied with templates.

(c) *Interferon,* a virus-induced, cell-coded glycoprotein with the capacity to interfere with the multiplication of virtually all viruses, is produced as a byproduct of viral–cell interaction. This "natural antiviral antibiotic" is probably the most important mediator of interference. Detailed discussion of its mode of action is deferred until the next chapter.

SUMMARY

Since viruses are essentially "infective heredity," many aspects of their behavior can be regarded as facets of viral genetics, and genetic changes in viruses greatly affect their behavior at the levels of cell, organism, and community.

Spontaneous mutations occur among animal viruses with about the same frequency as in other living things, and the mutation rate can be greatly increased by the use of chemical or physical mutagens. The two most useful types of mutation for genetic studies with animal viruses are plaque-type mutants, because they can be rapidly screened, and temperature-sensitive conditional lethal mutants, which can be used to provide insight into the functions of many viral genes. Adaptation of viruses to new hosts and attenuation of viruses for vaccine production are practical ends achieved by making use of mutation and the selection imposed by novel conditions for viral growth.

Multiple infection of cells, with viruses of the same or different kinds, active or inactive, may lead to a variety of interactions either between the viruses themselves or between one virus and the cell, in a way that affects the growth of the second virus. Such interactions fall into five main groups.

1. *Recombination* occurs when two parental viral nucleic acid molecules interact. Exchange occurs between sequences within single molecules from two different parents (intramolecular recombination), or between separate molecules of two parents with segmented genomes (genetic reassortment), so that the resulting hybrid contains genes from both parents. Such hybrids, if viable, breed true thereafter. Variations on the theme of recombination include:

   (a) Genetic recombination (the prototype described above).
   (b) Cross-reactivation, in which inactivated genes may be "rescued by co-infection of the cell with a related virus.
   (c) Multiplicity reactivation, in which two nucleic acid molecules with different genes inactivated may combine to yield viable progeny.
   (d) Hybridization between unrelated viruses, such as SV40 and adenovirus, in which whole or incomplete nucleic acid molecules from the two parents may become covalently linked in

various combinations to yield viable or defective hybrid progeny.

(e) Integration into cellular DNA of part or all of the DNA of papovaviruses, adenoviruses, herpetoviruses or retroviruses to produce either "cryptic infection" or malignant transformation of the cell.

2. *Complementation* occurs when one virus codes for a gene product (a protein) essential for the growth of another. Examples include:

(a) Complementation between conditional lethal mutants of the same virus.

(b) Defective viruses which depend on a "helper" virus to code for an enzyme(s) essential to their multiplication, e.g., (i) adeno-associated virus (defective) plus adenovirus (helper); (ii) adenovirus in monkey cells (conditionally defective) plus SV40 or an SV40-adenovirus hybrid (helper).

3. *Phenotypic mixing (transcapsidation)* occurs when the nucleic acid of one virus is enclosed within an envelope or capsid which contains proteins some or all of which are specified by another virus. With enveloped viruses the envelope proteins are usually mixed; with non-enveloped viruses the genome may be enclosed within a totally foreign capsid.

4. *Polyploidy* occurs when more than one genome is included within a single virion. Mixed infections of cells with viruses that mature by budding from the plasma membrane may yield *heteropolyploid* progeny.

5. *Interference* occurs when one virus preempts the other. Mechanisms of interference include:

(a) Occupation or destruction of receptors on the cell surface.

(b) Intracellular competition, e.g., by attachment of defective viral RNA to the replicase enzyme.

(c) Induction of interferon synthesis.

FURTHER READING

Cairns, J., Stent, G. S., and Watson, J. D., eds. (1966). "Phage and the Origins of Molecular Biology." Cold Spring Harbor Lab., Cold Spring Harbor, New York.

Fenner, F. (1969). Conditional lethal mutants of animal viruses. *Curr. Top. Microbiol. Immunol.* **48**, 1.

Fenner, F. (1970). The genetics of animal viruses. *Annu. Rev. Microbiol.* **24**, 297.

Fenner, F. (1972). The possible use of temperature-sensitive conditional lethal mutants for immunization in viral infections. *In* "Immunity in Viral and Rickettsial Diseases" (A. Kohn and M. A. Klingberg, eds.), p. 131. Plenum, New York.

Fenner, F., McAuslan, B. R., Mims, C. A., Sambrook, J. F., and White, D. O. (1974). "The Biology of Animal Viruses," 2nd ed., pp. 274–318. Academic Press, New York.

Hayes, W. (1968). "The Genetics of Bacteria and their Viruses: Studies in Basic Genetics and Molecular Biology," 2nd ed. Blackwell, Oxford.

Huang, A. S. (1973). Defective interfering viruses. *Annu. Rev. Microbiol.* **27,** 101.

Luria, S. E., and Darnell, J. E. (1967). "General Virology," 2nd ed., pp. 221–263. Wiley, New York.

Rapp, F. (1969). Defective DNA animal viruses. *Annu. Rev. Microbiol.* **23,** 293.

Ritchie, D. A. (1973). Genetic analysis of animal viruses. *Brit. Med. Bull.* **29,** 247.

Rowe, W. P. (1967). Some interactions of defective animal viruses. *Perspect. Virol.* **5,** 123.

Sharp, D. G. (1968). Multiplicity reactivation of animal viruses. *Progr. Med. Virol.* **10,** 64.

Simon, E. H. (1972). The distribution and significance of multiploid virus particles. *Progr. Med. Virol.* **14,** 36.

Stent, G. S. (1971). "Molecular Genetics: An Introductory Narrative." Freeman, San Francisco, California.

Watson, J. D. (1970). "The Molecular Biology of the Gene," 2nd ed. Benjamin, New York.

Webster, R. G., and Laver, W. G. (1971). Antigenic variation in influenza virus. Biology and chemistry. *Progr. Med. Virol.* **13,** 271.

CHAPTER 5

Effects of Viruses on Cells

INTRODUCTION

The student of medicine is concerned with viruses because of their impact on bodily functions, which is the end result of changes occurring at the cellular and subcellular levels. In terms of cellular pathology, a complete spectrum of effects may be observed with different combinations of viruses and host cells (Table 5-1). Some highly *cytocidal* (cell-killing) viruses cause early cessation of cellular macromolecular biosynthesis, leading rapidly to the destruction of the cell. At the other extreme, viruses like lymphocytic choriomeningitis virus in congenitally infected mice multiply in cells without any apparent adverse effect on them. These steady-state *noncytocidal* infections are discussed at greater length in Chapter 8. The third important type of cellular response to infection is *transformation* to a neoplastic state (Chapter 9).

All these changes in virus-infected cells are due to the biochemical effects of virus-specified products. The biochemical changes usually lead to functional disturbances, and eventually, in many cases, to histopathological changes. Since the latter are the most obvious, they have received most attention; we shall therefore take these *cytopathic effects* (CPE) as a starting point for discussion.

CELL DAMAGE CAUSED BY CYTOCIDAL VIRUSES

The terminal event of many types of virus-cell interaction is cell death, and the histological appearance of the damage produced in cultured cells by particular cytocidal viruses is often sufficiently characteristic to be used as a diagnostic criterion (see Chapter 14). Despite the impressive expansion in our knowledge of the molecular biology

86

TABLE 5-1
Types of Virus-Cell Interaction

VIRUS	CELL	INTERACTION	EFFECT ON CELL	YIELD OF INFECTIOUS VIRUS
Poliovirus	HeLa	Cytocidal	Destruction	+
	Mouse	Nil	Nil	−
Poliovirus RNA	Mouse	Cytocidal	Destruction	+
Rabbitpox	Chick	Cytocidal	Destruction	+
	PK	Cytocidal	Destruction	-I
Rabbitpox PK⁻ [a]	Chick	Cytocidal	Destruction	+
	PK	Cytocidal	Destruction	−
Lymphocytic choriomeningitis	Mouse	Steady-state	Nil	+
Avian oncovirus	Chick	Steady-state	Nil	+
Avian oncovirus, RSV[b]	Chick	Noncytocidal	Transformation	−
Polyoma	Mouse	Cytocidal (most cells)	Destruction	+
	Mouse	Noncytocidal (few cells)	Transformation	−
	Hamster	Noncytocidal	Transformation	−

[a] PK negative mutant of rabbitpox virus.

[b] RSV, Rous sarcoma virus—a mutant avian oncovirus.

of viral multiplication over the last decade, we still are not at all certain about how viruses kill cells—or indeed, at the level of the organism, how they cause disease.

Gross biochemical changes occur in cells infected with cytocidal viruses. Early virus-coded proteins often shut down host RNA and protein synthesis, which in itself is incompatible with survival of the cell. Furthermore, large numbers of viral macromolecules accumulate late in the infectious cycle; some of these, particularly certain capsid proteins, may be directly toxic. Viral proteins or virions themselves sometimes congregate in large crystalline aggregates or inclusions that visibly distort the cell. Infected cells usually swell substantially; it would be surprising if there were not consequential changes in membrane permeability. Eventually the cell's own lysosomal enzymes leak out and result in autolytic digestion of the cell.

Thus, there are numerous changes in the virus-infected cell which, individually or cumulatively, are lethal. In a sense, only the first lethal change is relevant, even though it may not necessarily be the one that produces the first visible "cytopathic effect," or the final dissolution of the cell. Most of the literature on the subject is concerned with identifying the event responsible for visible cytopathic damage. Such studies have demonstrated that cell damage may still occur when expression of the viral genome is (a) limited as a result of the virus being inactivated (e.g., by UV irradiation) or a conditional lethal mutant, or (b) blocked by chemical inhibitors. The occurrence of severe cytopathic effects in cells infected by polioviruses whose multiplication is blocked with guanidine early in the growth cycle, or by host-dependent mutants of rabbitpox virus that fail to synthesize DNA or more than a few recognizable proteins (Table 5-1), provide good examples. Abortive infections of this type are, of course, not transmissible to other cells, although the products of the damaged cells may be "toxic."

Shutdown of Cellular Macromolecular Syntheses

Most cytocidal viruses code for early proteins which shut down the synthesis of cellular RNA and protein; DNA synthesis is usually affected secondarily. The shutdown is particularly rapid and severe in infections of cultured cells by picornaviruses, some poxviruses, and herpetoviruses, all of which are rapidly cytopathogenic. With viruses of several other families the shutdown is late and gradual, while with noncytocidal viruses such as retroviruses there is, by definition, no shutdown and no cell death.

Cytopathic Effects of Viral Proteins

Viral capsid proteins in high dosage are often toxic for animal cells and may be the principal cause of cytopathic effects. This may follow the accumulation of viral proteins late in the multiplication cycle after infection at low multiplicity or it may be seen quite early as a laboratory artifact resulting from the experimental use of very large inocula. The best studied example is adenovirus, which contains two capsid antigens capable of causing cytopathic effects. Purified adenovirus penton antigen produces reversible cell clumping and detachment from the substrate, whereas the purified fiber antigen depresses cellular synthesis of RNA, DNA, and protein and inhibits the multiplication of adenoviruses and other viruses in the treated cells. This generalized shutdown of all viral and cellular activity is not to be confused with the more specific shutdown of cellular (but not viral) RNA and protein synthesis by virus-coded "early" proteins which is an important feature of the multiplication cycle of many cytocidal viruses.

Release of Enzymes by Lysosomes

Activation of lysosomal enzymes may be an important mechanism in the production of some kinds of cellular damage by viruses. Three stages of activation can be distinguished by staining fixed and unfixed cells for a lysosomal enzyme, acid phosphatase. In first-stage activation the membranes of lysosomes show increased permeability, but the enzymes remain within them; this state is reversible. Second-stage activation is marked by diffusion of the enzymes into the cytoplasm, often with secondary adsorption to the nucleus; the cells become rounded. In third-stage activation, the acid phosphatase marker disappears, due to diffusion or inactivation.

Slight first-stage activation may occur in cells infected with noncytocidal viruses, and the cell detachment produced by the penton antigen of adenoviruses is associated with first-stage activation of lysosomes. In some viral infections lysosomal changes do not progress beyond the first (reversible) stage, and since such lysosomes take up more neutral red than usual, red plaques may be produced (see Plate 2-3). Second-stage activation is found in cells infected with cytocidal viruses; it leads to a failure of the affected cells to take up neutral red and the occurrrence of the familiar unstained plaques. The final dissolution of the dead cell probably results from autolysis brought about by the liberation of enzymes from the cell's own lysosomes.

Nonspecific Histological Changes

In addition to changes due to the specific effects of viral multiplication, most virus-infected cells also show nonspecific lesions, very much like those induced by physical or chemical agents. The most common early and potentially reversible change is what histopathologists call "cloudy swelling," which is associated with changes in the permeability of the plasma membrane. Electron microscopical study of such cells reveals diffuse changes in the nucleus, plasma membrane, endoplasmic reticulum, mitochondria, and polyribosomes, prior to any changes in the lysosomes. Margination of the nuclear chromatin and pyknosis are late nonspecific changes that probably occur after death of the cell.

Inclusion Bodies

The most characteristic morphological change in virus-infected cells, recognized by histologists since before the turn of the century, is the

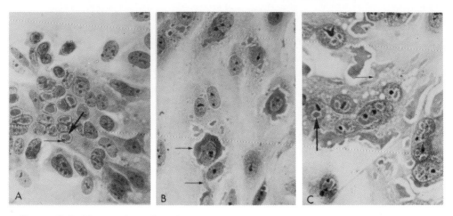

PLATE 5-1. *Types of viral inclusion body (H and E stain)* ×200. *(A) Intranuclear inclusions; cells form syncytium—herpetovirus. Small arrow, nucleolus; large arrow, inclusion body. Note also margination of nuclear chromatin. (B) Intracytoplasmic inclusions—reovirus. Arrows indicate inclusion bodies in perinuclear locations. (C) Intranuclear and intracytoplasmic inclusions; cells form syncytium—measles virus. Small arrow, intracytoplasmic inclusion body; large arrow, intranuclear inclusion body. (Courtesy I. Jack.)*

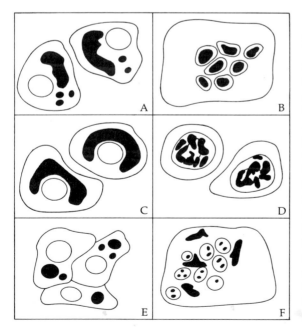

FIG. 5-1. *Inclusion bodies in virus-infected cells. (A) Vaccinia virus—intracytoplasmic acidophilic inclusion (Guarnieri body). (B) Herpes simplex virus—intranuclear acidophilic inclusion (Cowdry type A); cell fusion produces syncytium. (C) Reovirus—perinuclear intracytoplasmic acidophilic inclusion. (D) Adenovirus—intranuclear basophilic inclusion. (E) Rabies virus—intracytoplasmic acidophilic inclusions (Negri bodies). (F) Measles virus—intranuclear and intracytoplasmic acidophilic inclusions; cell fusion produces syncytium.*

"inclusion body"—an area with altered staining behavior. Depending on the causative virus, such inclusions may be single or multiple, large or small, round or irregular in shape, intranuclear or intracytoplasmic, acidophilic or basophilic. They develop progressively as the multiplication cycle proceeds and generally represent "virus factories," i.e., foci in which viral nucleic acid or protein is being synthesized or virions assembled.

The most important viral inclusion bodies are the intracytoplasmic inclusions found in cells infected with poxviruses, paramyxoviruses, reoviruses, and rabies viruses, and the intranuclear inclusion bodies produced by herpetoviruses and adenoviruses (Fig. 5-1; Plate 5-1). Most such inclusions have now been shown, by fluorescent-antibody staining and electron microscopy, to be sites of viral synthesis; some, like the intranuclear inclusion bodies of herpetoviruses, are late degenerative changes which confer altered staining characteristics on the cell. In a few instances (e.g., adenoviruses, reoviruses) the inclusions may represent crystalline aggregates of virions.

Although they are most commonly due to viruses, inclusion bodies are not diagnostic of viral infection, since similar histological appearances may be produced by chemicals or occasionally by bacteria.

ALTERATIONS TO THE CELL MEMBRANE

All cytocidal and many noncytocidal viruses cause alterations in the cytoplasmic membrane, although this may sometimes be secondary to death of the cell from other causes.

Cell Fusion

Live or UV-irradiated paramyxoviruses can cause rapid fusion of cultured cells (or lysis of red cells) if applied to the cells at high multiciplicity. The process is called *early polykaryocytosis*. Cell biologists have used this phenomenon to produce functional heterokaryons by fusing cells from different species of animal.

Most of the viruses that produce cell fusion when artificially large doses are used, also produce late polykaryocytosis, i.e., late in their multiplication cycle they cause changes in the cell membrane which result in the fusion of the infected cell with neighboring uninfected cells. These *syncytia*, i.e., giant cells with many nuclei, are a feature of the pathology of certain viral infections in man (e.g., Warthin's cells in measles), as well as being a conspicuous result of infection of cell monolayers by paramyxoviruses and herpetoviruses.

The mechanism of virus-induced polykaryocytosis is not fully understood. Almost certainly the fusion of cell membranes is caused by enzymes, but it is not certain whether they are of viral or of lysosomal origin.

New Antigens on the Cell Surface

There is abundant evidence that new antigens, specified by the viral genome, may appear in the plasma membrane of virus-infected cells. The most familiar example is the way in which the plasma membranes of cells infected with enveloped RNA viruses (e.g., influenza virus) incorporate viral hemagglutinin, so that red blood cells can be adsorbed to their surface (*hemadsorption*). Such virus-coded antigens on the surface of infected cells constitute a target for the body's specific immune mechanisms, such as cytotoxic antibodies and "T" lymphocytes, which destroy the cell and hasten recovery, while at the same time (in many cases) precipitating symptoms of the "disease" (see Chapter 7).

Another class of new antigens found on the cell membrane are the *transplantation antigens* formed on the surface of cells that are "transformed" by viruses. Malignant transformation induced by physical, chemical, or viral agents ia always accompanied by the appearance of new antigens at the cell surface. With physical and chemical carcinogens these antigens are different in every instance; with the oncogenic viruses, however, the transplantation antigens are characteristic of the virus, being proteins specified by viral genes. These novel transplantation antigens are recognized as "foreign" by the body's immunological surveillance mechanisms, and the resulting immune response may lead to rejection of the tumor; indeed study of these antigens and the organism's response to them constitutes the currently fashionable and important field of tumor immunology.

EFFECTS OF VIRAL INFECTION ON MITOSIS

In steady-state noncytocidal infections, the synthesis of cellular macromolecules and cell division itself may be virtually unaffected by the presence of the virus. In some such cases intercellular transfer of virus is effected by the release of infective virions, in other cases by the partitioning of intracytoplasmic virus between daughter cells at mitosis. At the other extreme, highly cytopathogenic viruses like poliovirus or herpes simplex virus rapidly and irreversibly shut down the synthesis of cellular RNA and protein, and thus prevent the cell from dividing. However, if infection occurs too late in the cell cycle to block mitosis, cellular macromolecular syntheses take precedence and viral multiplication is delayed for the duration of mitosis. In borderline cases mitosis may proceed in an aberrant fashion. To cite two examples: herpes simplex virus infection of mitotic cells causes aberrant cleavage of nuclei, and mitoses are often synchronized within the polykaryocytes formed by infection with measles virus.

CHROMOSOMAL ABERRATIONS

With the development of improved techniques for the demonstration of chromosomes in vertebrate cells, chromosomal changes have been found in cultured cells infected with either cytocidal or noncytocidal viruses. Hamster cells have been particularly favored for this work because they contain such a small number of chromosomes. In some instances, e.g., with herpes simplex virus, particular chromosomes appear to be broken more often than others. However, it is unlikely that such changes have any deep significance. These changes can follow the interaction of virus with cells that do not support viral growth, and in any case the changes are usually qualitatively similar to those produced by ionizing radiation or radiomimetic chemicals.

Chromosomal breaks, translocations, and deletions are also common in virus-induced tumors (as indeed they are in nonviral tumors) but it is difficult to ascribe to them a primary role in carcinogenesis; they are more likely to be a nonspecific concomitant of the disordered cell division found in such cells.

CELL DAMAGE IN THE INFECTED ANIMAL

All of the changes described in virus-infected cells in culture also occur in the whole animal, although infected cells are often more difficult to locate in stained sections of tissue because infection may be limited by the body's defenses or by lack of susceptibility of all but a few cell types in a given organ. Nevertheless, inclusion bodies, multinucleate giant cells, etc., are sufficiently in evidence to be pathognomonic of certain viral infections in animals and man, e.g., Negri bodies in rabies, Warthin's giant cells in measles, and cytomegalic cells in cytomegalovirus infection.

Destruction of infected cells by direct viral action clearly plays an important part in pathogenesis. However, sublethal changes such as cloudy swelling can also have profound physiological effects if they occur in certain cells, such as the endothelial cells of small vessels, or in organs like the brain. Minor functional abnormalities in myocardial cells or neurons may be of critical importance. The central nervous system shows extreme vulnerability to early damage, as evidenced, for example, by the early paraesthesias in poliomyelitis in man.

Cell damage is often the result not of direct viral cytotoxicity, but of immunological responses to the products of viral infection (see Chapter 7). Further, the intact organism differs from cultured cells in having a circulatory system. Circulatory disturbances due to inflammation or to direct or indirect viral damage to blood vessel walls inevitably contribute to pathological changes, with production of edema, anoxia, hemorrhage, or infarction. If the vascular disturbances are severe the tissues supplied by affected vessels undergo necrosis.

TRANSFORMATION

Most viruses capable of inducing tumors in experimental animals (see Chapter 9) can also "transform" cells to the malignant state *in vitro*. Transformation is recognized primarily by a morphological change. Transformed cells are altered in shape and lose the property of contact inhibition, so that they pile up over one another instead of forming a monolayer (Plate 5-2). Some investigators believe that this loss of contact inhibition is the crucial change that confers on the cell the capacity for unregulated growth. Table 5-2 summarizes the respects in which virus-transformed cells differ from normal cells.

INTERFERON

A very common and important response of vertebrate cells to viral infection is the production of the antiviral substance, interferon.

Although it has been investigated intensively for nearly 20 years, it is still impossible to give a simple and satisfactory account of inter-

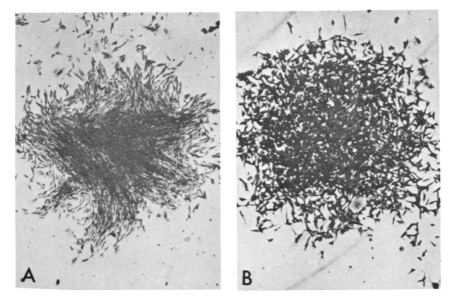

PLATE 5-2. *Transformation of cultured BHK21 cells by polyoma virus. (A) Normal colony, illustrating the regular parallel arrangement of elongated fibroblastic cells. (B) Transformed colony, illustrating the criss-cross random orientation of more rounded cells.* ×25. [*From W. House and M. G. P. Stoker, J. Cell Sci.* **1,** *169 (1966).*]

feron. Indeed the current view is that there is no such single substance, but rather that interferons are a class of cellular glycoproteins, occurring as monomers or polymers of a fundamental unit of molecular weight, 12,000–15,000 daltons. Today we would define interferons as "cell-coded proteins which are (a) induced by foreign nucleic acids, (b) nontoxic for cells, and (c) able to inhibit the multiplication of vertebrate viruses in cells of homologous species." Virtually all viruses are capable of inducing interferon synthesis in the cells they infect, and this interferon is capable of inhibiting the multiplication of that virus or virtually any other virus growing in cells of the same (but not unrelated) animal species. In other words, any particular interferon shows cellular specificity but not viral specificity. Nevertheless, different viruses differ in their sensitivity to interferon and, in general, "virulent" strains are less sensitive than attenuated strains of the same virus.

The role of interferon in the process of recovery from viral infections is discussed in Chapter 7 and its potential for chemotherapy in Chapter 13.

TABLE 5-2

Characteristics of Cells Transformed by Viruses

---

1. Capacity to produce malignant neoplasms when inoculated into isologous animals
2. Greater growth potential *in vitro*
   (a) Formation of three-dimensional colonies of randomly oriented cells in monolayer culture, usually due to loss of contact inhibition
   (b) Capacity to divide indefinitely in serial culture
   (c) Higher efficiency of cloning
   (d) Capacity to grow in suspension or in semi-solid agar or in serum-deficient medium
3. Altered cell morphology
4. Altered cell metabolism
   (a) Greater production of organic acids
   (b) Greater production of acid mucopolysaccharides on the plasma membrane
   (c) Changes in the glycolipid and glycoprotein composition of the plasma membrane
5. Chromosomal abnormalities
6. Novel (virus-specified) antigens
   (a) New surface antigens (transplantation antigens)
   (b) New intracellular antigens (e.g., T antigens)
7. Enhanced agglutinability by lectins

Induction of Interferons

Some viruses are much more effective inducers of interferon than others, and some are effective even after inactivation by ultraviolet light. Viral nucleic acid, but not other viral components, will induce interferon synthesis, and a discovery of considerable potential importance for antiviral chemotherapy is that several synthetic polynucleotides [notably poly(I)·poly(C); see Chapter 13] are powerful inducers of interferon. Double-stranded RNA's and polynucleotides appear to be particularly effective, perhaps because of their resistance to cellular nucleases. Highly cytocidal viruses, which shut down cellular macromolecular synthesis rapidly, induce little interferon. In many cell systems, interferon production is promoted and viral production depressed by temperatures around 40°.

Actinomycin D, which prevents transcription from the cellular genome, and puromycin, which prevents protein synthesis, both inhibit the production of interferon. Using these inhibitors, it has been demonstrated that interferon production has a definite temporal course; little is found within the first hour after viral infection and peak titers are usually found late in infection. Once interferon synthesis has begun, production continues at a linear rate for many hours. In most systems interferon is released from the cell almost as soon as it is formed.

In recent years a variety of infectious agents other than viruses (e.g., certain protozoa and bacteria, and bacterial endotoxin) have been shown to induce cellular proteins that interfere with the growth of viruses and have therefore been called "interferons." These usually have a higher molecular weight (about 90,000, compared with 24,000–30,000 for virus-induced human interferon). In general such inducers are effective only in intact animals and not in cultured cells; the kinetics of production suggests that preformed "interferon" is released rather than newly synthesized. The mode of action of these substances is not known; it is possible that further work will demonstrate that, although they have antiviral activity, the cellular proteins induced by microorganisms and endotoxin act quite differently from classical virus-induced interferon.

Mechanism of Action of Interferons

Interferon is completely without effect on the virion itself; it exerts its effect on the cell. Viral adsorption, penetration and uncoating all occur normally in the presence of interferon. There is experimental evidence, from different cell-virus systems, that the virus-inhibitory effect may be exerted, not by interferon itself, but by a cellular protein induced by the interferon. Some experiments suggest that the block is at the level of translation; others indicate that transcription is inhibited. Indeed it now seems likely that, just as there is a range of proteins that can be classed as "interferons," so these substances may inhibit viral multiplication by effects exerted at several different stages of the cycle.

SUMMARY

Viruses may affect cells in many different ways. Three common types of response are (a) cytocidal infection, in which the cells are killed by the virus, (b) steady-state noncytocidal infection, in which cells produce virus but are metabolically unaffected by it, and (c) cellular transformation, in which the virus does not kill the cell but permanently alters its growth characteristics.

The cytopathic effects of viruses are due to the disruptive actions of products of the viral genome on cell metabolism. Some viruses produce changes in cell membranes. One result may be the fusion of the infected cell with neighboring uninfected cells to form a syncytium. Another important change is the incorporation into the plasma membrane of novel virus-specified antigens; these may be structural proteins of the virion (in the case of enveloped viruses) or transplantation antigens (in virus-transformed cells).

Some viruses produce characteristic "inclusion bodies" in infected cells; these may have diagnostic value, as with the cytoplasmic inclusions of the paramyxoviruses, rhabdoviruses, reoviruses, and poxviruses or the nuclear inclusions of the herpetoviruses and adenoviruses.

One almost universal effect of viral infection is the induction of interferon by the infected cell. This cellular protein, which is derepressed by viral (or other) nucleic acid, is released from the cell and renders other cells in the animal or culture resistant to infection by a variety of different viruses by blocking viral transcription or translation.

FURTHER READING

Allison, A. C. (1967). Lysosomes in virus-infected cells. *Perspect. Virol.* **5,** 29.

Allison, A. C. (1971). The role of membranes in the replication of animal viruses. *Int. Rev. Exp. Pathol.* **10,** 182.

Bablanian, R. (1972). Mechanisms of virus cytopathic effects. *Symp. Soc. Gen. Microbiol.* **22,** 359.

Ciba Foundation. (1971). "Symposium on Interferon." Churchill, London.

Colby, C., and Morgan, M. J. (1971). Interferon induction and action. *Annu. Rev. Microbiol.* **25,** 333.

Fenner, F., McAuslan, B. R., Mims, C. A., Sambrook, J. F., and White, D. O. (1974). "The Biology of Animal Viruses," 2nd ed., pp. 323–346. Academic Press, New York.

Finter, N. B., ed. (1973). "Interferons and Interferon Inducers." North-Holland Publ., Amsterdam.

Harris, H. (1970). "Cell Fusion." Oxford Univ. Press (Clarendon), London and New York.

Macpherson, I. (1970). The characteristics of animal cells transformed *in vitro.* Advan. Cancer Res. **13,** 169.

Nichols, W. W. (1970). Virus-induced chromosome abnormalities. *Annu. Rev. Microbiol.* **24,** 479.

Poste, G. (1972). Mechanisms of virus-induced cell fusion. *Int. Rev. Cytol.* **33,** 157.

Poste, G., and Nicolson, G., eds. (1975). "Virus Infections and the Cell Surface." North-Holland Publ., Amsterdam.

Stich, H. F., and Yohn, D. S. (1970). Viruses and chromosomes. *Progr. Med. Virol.* **12,** 78.

CHAPTER 6

Pathogenesis of Viral Infections

INTRODUCTION

At the molecular and cellular levels, viruses behave quite differently from bacteria and protozoa, but this distinction largely disappears when we consider viruses as infective agents at the levels of organism and of community. In this chapter and the next we consider the ways in which viruses produce disease, i.e., the pathogenesis of viral infections. This involves three types of interaction between viruses and their vertebrate hosts: the way in which viruses spread (or fail to spread) through the body, the immune response and its influence on viral infections, and the many nonimmunological factors which affect virus-host interactions.

It is relevant here to consider two much-abused words, *pathogenecity* and *virulence,* as they apply to viruses. Neither term can be used without reference to the host, for they apply to the interaction between host and parasite. A virus is *pathogenic* for a particular host if it can infect that host and produce signs and symptoms of disease in it. But infection is not synonymous with disease; many infections, even with virulent viruses, may be *subclinical* (*inapparent*). A given strain of virus is said to be more *virulent* than another if it regularly produces more severe disease in a host in which both strains are pathogenic. Live vaccines, for example, are usually attenuated strains, i.e., they are less virulent than the wild-type virus. It is best not to use the word "virulent" to describe the killing of cultured cells by viruses. As explained in Chapter 5, the appropriate term is "cytocidal." Cytocidal viruses are not invariably highly virulent in intact animals; conversely, noncytocidal viruses like rubella or the leukemia viruses may cause severe disease.

Since viruses must get into susceptible cells before they can produce disease it is necessary to consider first the important portals of entry

99

into the body, which are the three major epithelial surfaces: the skin, the respiratory tract, and the alimentary tract. Localized pathological changes may be produced when viruses breach these barriers, although invasion may also occur without the development of a local lesion. After the surface has been breached the infection may remain localized, or it may spread through the organism via the lymphatics, blood vessels, or nerves.

VIRAL INFECTIONS OF THE RESPIRATORY TRACT

Animals are constantly sampling their environment during breathing, so that the respiratory tract is a very important portal of entry for viruses. Some viruses cause silent initial infection of the respiratory tract prior to systemic distribution, others remain localized to the upper or lower respiratory tract and produce disease as a result of their multiplication there. Such respiratory viruses (see Table 24-1), include the rhinoviruses, orthomyxoviruses, paramyxoviruses, and coronaviruses, as well as several adenoviruses and enteroviruses. As an example of the way in which these viruses produce respiratory disease, we describe the pathogenesis of influenza, the most important respiratory viral infection of man.

Human infection can be initiated by influenza virus in the wet or dry state, in small or large aggregates, and when lodged on the upper or lower respiratory passages. Symptoms are produced by multiplication in the cells of the upper and lower respiratory tract; generalization via the bloodstream is a very rare event.

Virus particles alighting on the mucus film that covers the epithelium of the upper respiratory tract may undergo one of several fates. If the individual has previously recovered from an infection with that strain of influenza virus, antibody (mainly IgA) in the mucous secretion may combine with the virus and neutralize it. Mucus also contains glycoprotein inhibitors which combine with virus and prevent it from attaching to the specific receptors on the host cell; eventually these inhibitors are destroyed by the viral neuraminidase.

The successful infection of a few cells by influenza virus, and the passage of newly synthesized virions from these cells into the respiratory mucus and then into other cells may lead to progressive infection, which is aided by the transudation of fluid that follows cellular injury and helps to disperse the virus. On the other hand, the inflammation also results in an increased diffusion of plasma constituents, including antibody and nonspecific inhibitors, which may inactivate the virus and cut short the infection. Interferon, which has been demonstrated in the lungs of mice infected with influenza virus, may also play a role in limiting the spread of infection.

Although the release of virus does not cause obvious cell damage, the end result of infection is necrosis and desquamation of the respiratory epithelium. Usually damage is confined to the epithelial cells of the upper respiratory tract, but in cases of pneumonia, fluorescent antibody staining shows that there are foci of infection in the epithelial cells of the bronchi, bronchioles and alveoli, and in alveolar macrophages but not in the vascular endothelium. Destruction of the respiratory epithelium by influenza virus lowers its resistance to secondary bacterial invaders, especially pneumococci and staphylococci.

VIRAL INFECTIONS OF THE ALIMENTARY TRACT

Only nonenveloped viruses regularly cause infection of the human alimentary tract, because the infectivity of enveloped viruses is destroyed by the bile. Furthermore, the rhinoviruses are too acid-labile to pass the stomach. The main viruses that infect the gut are therefore the nonenveloped, icosahedral enteroviruses (the major genus of the Picornaviridae), reoviruses, adenoviruses, and hepatitis viruses.

Most enteroviruses are parasites of the intestinal tract, but proof of the etiological role of coxsackieviruses and echoviruses in gastroenteritis has been difficult to obtain. Most infections are symptomless. Less frequently, enteroviruses may spread from the alimentary tract to cause generalized infection. Important though these infections are, we know very little of their pathogenesis before they leave the gut. Adenoviruses are primarily pathogens of the respiratory tract but many serotypes can multiply asymptomatically in the alimentary tract and are excreted in the feces. Recent work has shown that a newly discovered group of reoviruses ("rotaviruses") are a major cause of nonbacterial diarrhea in infants, while the "Norwalk agent" and certain coronaviruses have been associated with gastroenteritis in adults. Hepatitis A virus, long known to be transmitted primarily by the alimentary route, has now been tentatively identified as a parvovirus; hepatitis B virus (which causes "serum hepatitis") is morphologically quite dissimilar.

VIRAL INFECTIONS OF THE SKIN

When viruses are introduced into the skin by injection, arthropod bite, or through a breach caused by mechanical trauma, lesions may be produced at the site. The only localized viral infections of the skin in man are human warts, which are a trivial and very common hyperplasia due to a virus of the papilloma subgroup of the papovaviruses, and the proliferative inflammatory lesions produced by poxviruses: vaccinia, cowpox, orf, and molluscum contagiosum (see Chapter 17). More commonly skin lesions occur secondarily to systemic spread, when they constitute a rash, as in the exanthemata of childhood.

THE SYSTEMIC SPREAD OF VIRUSES

Systemic disease usually follows the distribution of virus via the bloodstream; the presence of virus in the bloodstream is known as *viremia*. Such a generalized infection may be inapparent, or it may produce systemic symptoms or symptoms associated with lesions in a particular organ, which differs characteristically with different viruses. The principal "target" organs are the skin and the central nervous system; more rarely the liver, heart, and certain glands may be chiefly affected. Viremia occurs in most viral infections, even though it may not be readily detectable. Rhinovirus infections of the upper respiratory tract and infection of the skin with warts virus may be exceptions.

In the blood, viruses may occur free in the plasma or may be characteristically associated with particular types of leukocytes, with platelets, or with erythrocytes. Leukocyte-associated viremia is a feature of several types of infection, including measles and smallpox, for example. Many viruses multiply in macrophages; others, e.g., EB virus, multiply in lymphocytes. Occasionally virus is adsorbed to erythrocytes as in Rift Valley fever, Colorado tick fever, and lymphocytic choriomeningitis. Often virus circulates free in the plasma; all the togaviruses and the enteroviruses that cause viremia fall into this group. Finally, in some infections the viremia is mixed, i.e., the virus is partly in the plasma and partly cell-associated.

Whether virus circulates free in the plasma or is cell-associated affects its passage from the circulation to extravascular sites. Leukocytes can pass through the walls of small vessels by diapedesis, and infected leukocytes can thus initiate infection in various parts of the body. On the other hand, virus in the plasma may escape from circulation by being ingested by a cell in contact with the blood, either a macrophage or a capillary endothelial cell. Extravascular infection by such viruses may follow growth of the virus through the endothelial cells of the small blood vessels, or the virus may be transferred across the cell without growing in it.

Virus circulating in the blood is continually removed by cells of the reticuloendothelial system. Viremia can therefore be maintained only if there is a continued release of virus into the blood from cells in contact with it, or if the clearance system is grossly impaired. The circulating leukocytes could themselves constitute a source of replicating virus; indeed blood leukocytes maintained in culture (from which polymorphonuclear leukocytes are rapidly lost) support limited replication of many viruses. However, viremia is maintained primarily by organs with extensive sinusoids, like the liver, spleen and bone marrow, the endothelial cells of the blood vessels themselves, and the lymphoid tissues (via the thoracic duct). Cells of the voluntary muscles may be an important site of multiplication of some enteroviruses and togaviruses.

GENERALIZED INFECTIONS WITH RASH

Rashes are more easily seen in human beings than in furred and feathered beasts, hence most of the descriptive data are derived from human infections. The individual lesions in generalized rashes are described as macules, papules, vesicles, or pustules. A lasting local dilation of subpapillary dermal blood vessels produces a macule, which becomes a papule if there is also edema and an infiltration of cells into the area. Primary involvement of the epidermis usually results in vesiculation, ulceration, and scabbing, but prior to ulceration a vesicle may be converted to a pustule if there is a copious cellular exudate. Secondary changes in the epidermis may lead to desquamation. More severe involvement of the dermal vessels may lead to hemorrhagic and petechial rashes, although coagulation defects and thrombocytopenia may also be important in the genesis of such lesions. Information on the occurrence of rashes in human viral infections is set out in Table 24-3.

As a background for the discussion of generalized infections with rash we will briefly describe the sequence of events in smallpox (Fig. 6-1). Epidemiological studies have established clearly that there is an incubation period of 10–12 days before any symptoms of illness appear.

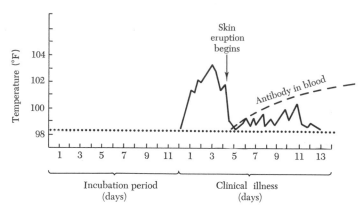

FIG. 6-1. *The stages of infection in human smallpox.* [*From A. W. Downie, Vet. Rec.* **75,** *1125 (1963).*]

These usually begin abruptly with fever, headache, loss of appetite, and nausea and are followed in a few days by signs associated with infection of the skin and the mucous membrane of the oropharynx. The skin lesions progress from macules through papules, vesicles, and pustules which begin to dry up 10–11 days after first appearance of the macules. In milder cases fever abates soon after the rash appears, and antibody can usually be detected in the serum at about this time. Cases are not infectious during the incubation period.

In its essential features smallpox is characteristic of exanthematous infections caused by a variety of viruses. Pertinent questions concerning pathogenesis include the following: What is happening to the virus during the incubation period and why is the patient both symptom-free and noninfectious during this period? What determines the sudden onset of illness? Why does the virus multiply especially in certain of the internal organs and why does it localize in the skin? What determines recovery? What is the mechanism by which viral multiplication causes signs of disease, and in some cases, death of the host? We are still very far from being able to answer most of these questions, but some of them have been solved by studies of model infections in experimental animals.

The Incubation Period

What happens during the long incubation period of generalized viral infections was first studied many years ago with mousepox, a disease that closely resembles human smallpox in many of its essential features. Various tissues and organs were shown to become infected in regular sequence (Fig. 6-2). In mousepox induced by infection of the skin the virus was carried through the lymphatics to the local lymph node, then to the bloodstream (*primary viremia*), and then localized in the liver and spleen. A *secondary viremia* occurred after the virus had multiplied extensively in the parenchymal cells of the liver and in the spleen, and this was followed by focal infection of the skin and mucous membranes producing the rash. Meanwhile, a *primary lesion* developed at the site of implantation of the virus; its appearance marked the end of the incubation period.

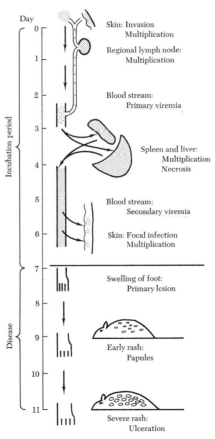

FIG. 6-2. *The probable sequence of events during the incubation period and early stages of illness in mousepox, a model for the acute exanthemata of man.* [*From F. Fenner, Lancet* **2,** *915 (1948).*]

With variations in the "central focus" and "target organ," it is probable that the same pattern underlies all generalized hematogenous infections. In different viral infections the lung, the bone marrow, the lymph nodes, the muscles, or the endothelial cells of capillaries throughout the body may serve as the "central focus." "Target organs" include the skin, brain, liver, heart, and pancreas. It is not yet possible to provide a convincing explanation for the tendency of particular viruses to multiply preferentially in certain organs.

The Production of Symptoms

The incubation period ends with the production of symptoms. In smallpox this appears to occur at about the time of the massive secondary viremia. The severity of the symptoms is related to the level of the viremia, which is an index of the degree of multiplication of the virus in the internal organs and affects the severity of the rash.

The general symptoms that usher in disease are common to a wide variety of infections: viral, protozoal, and bacterial. The usual explanation of these symptoms is that they are due to the absorption of "toxic" or "pyrogenic" products of cells destroyed during viral multiplication. However, this does not explain why the onset is sometimes so sudden, and it is tempting to believe that the abrupt onset of symptoms, like the synchronous development of the skin lesions, may have an immunological basis.

Localization of Virus

The factors that determine the secondary localization of virus in generalized infections are unknown. To some extent secondary localization must be related to the nature of the viremia, i.e., whether the virus is cell-associated or free in the plasma.

In the skin, poxviruses first involve the endothelium of the capillaries and venules in the dermis, and from there spread to the epidermis where typical degenerative changes occur resulting in the formation of vesicles. The focal eruption of the exanthemata also involves the mucous membranes of the mouth and upper respiratory tract (as an enanthem). Since there is no keratinized squamous epithelium here, these lesions break down more rapidly than the skin lesions and are important in determining the infectiousness of the patient in the early stages of disease.

GENERALIZED INFECTIONS INVOLVING
THE CENTRAL NERVOUS SYSTEM

Disease of the central nervous system (CNS) is an exceptional complication rather than the normal consequence of infection, even with the togaviruses and enteroviruses that are the viruses most commonly incriminated (see Table 24-4). Some togaviruses, like Japanese encephalitis virus, involve the CNS but never produce a rash, whereas others, like dengue, do the reverse. Many of the enteroviruses occasionally produce meningitis, and polioviruses characteristically attack the anterior horn cells of the spinal cord in a small proportion of infected human beings. Several of the herpetoviruses affect nerve cells; herpes simplex virus is probably the most common cause of sporadic fatal encephalitis in man, while herpes zoster is due to the activation of chickenpox virus, latent in cells of the posterior ganglia. Encephalitis occurring after measles is thought to be an autoimmune disease, possibly associated with limited viral multiplication in the neurons. The rare disease called subacute sclerosing panencephalitis is a late manifestation of CNS infection with measles virus, and another rare disease called progressive multifocal leukoencephalopathy is caused by a papovavirus.

Hematogenous Spread

Viruses can spread from the blood to the brain cells by several routes (Figure 6-3). Growth through the endothelium of small cerebral vessels has been clearly demonstrated in several systems, and there is suggestive evidence that virions may sometimes be passively transferred across the vascular endothelium. The production of meningitis rather than encephalitis by enteroviruses, and the ease with which these agents are recovered from the cerebrospinal fluid, can be explained by postulating that the virus in the blood either grows or passes through the chorioid plexus.

Neural Spread

Spread from the periphery to the CNS is possible without generalization of viruses through the blood stream, for the peripheral nerves and the nerve fibers of the olfactory bulb offer potential direct pathways. Although earlier workers thought in terms of spread via "con-

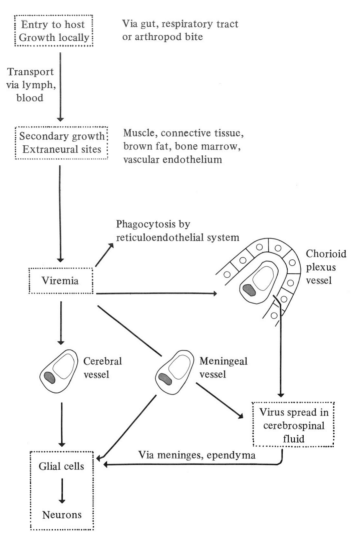

FIG. 6-3. *Steps in the hematogenous spread of virus to the central nervous system.*

duits" — the axon, the lymphatics, and the tissue spaces between nerve fibers — the route of spread along peripheral nerves is usually by growth within endoneural cells, a fact demonstrated experimentally with herpetovirus infections by fluorescent antibody staining methods. However, this cannot be the only method of neural spread, for in experiments with rabies virus, no fluorescent endoneural cells could be found, the dorsal root ganglion cells being the first cells showing specific staining (see Chapter 22).

Apart from rabies and B virus (simian herpes) infections of man after monkey bites, no other examples of neural spread have been established in natural infections. However, transmission in the reverse direction, from the dorsal root ganglia down the corresponding peripheral nerves, appears to be the most likely mode of spread of virus in herpes zoster and recurrent herpes simplex (see Chapters 8 and 16).

Production of Disease

Cytocidal infections of neuronal cells, whether due to poliovirus, a togavirus, or a herpetovirus, are characterized by the three hallmarks of encephalitis: cell necrosis, phagocytosis by glial cells (neuronophagia), and perivascular infiltration of inflammatory cells, which is an expression of cell-mediated immunity (see Chapter 7). The cause of symptoms in other CNS infections is more obscure. Rabies virus is noncytocidal in cultured cells; in infected animals it evokes none of the inflammatory reactions or cell necrosis found in the encephalitides, yet it is highly lethal for most species of animal. With some other viruses, infection of neurons causes no symptoms; for example, the extensive CNS infection of mice congenitally infected with lymphocytic choriomeningitis virus, readily demonstrable by fluorescent antibody staining, has no deleterious effect. Still other changes are produced by some of the viruses that cause slowly progressive diseases of the CNS (see Chapter 8). In scrapie of sheep, for example, there is slow neuronal degeneration and vacuolization; in visna (another chronic disease of sheep), changes in cell membranes lead to demyelinization.

Post infection encephalitis is most commonly seen after smallpox vaccination and measles. The pathological picture is predominantly demyelinization without neuronal degeneration—changes unlike those produced by the direct action of viruses on the CNS. Allied with the failure to recover virus from the brain this has led to the view that post infection encephalitis is probably an autoimmune disease.

CONGENITAL INFECTIONS

Oncovirus infections of chickens and rodents persist in these species by the congenital transfer of the viral genome, which is integrated into the chromosomes of the host's cells as a DNA copy of the viral RNA. In addition, viremia in the mother may be followed by congenital transfer of virions (Figure 6-4) in these as well as in a variety of other infections (Table 6-1). Congenital infection may occur at any stage from the development of the ovum up to birth. With noncytocidal viruses, like lymphocytic choriomenigitis in mice, every cell in the embryo's body may be infected.

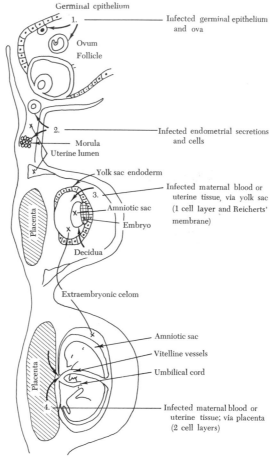

FIG. 6-4. *Schematic drawing of the reproductive system of the mouse to show possible routes of infection of the mammalian embryo by viruses (courtesy Dr. C. A. Mims). In the mouse, infection may be transmitted via the germinal epithelium and ova (1), endometrial secretions and cells (2), maternal blood or uterine tissue, via the yolk sac (3), or the placenta (4). In the human, fetal infection usually occurs via the placenta.*

TABLE 6-1
Congenital Viral Infections

SYNDROME	DISEASE	HOST
Fetal death and abortion	Smallpox	Man
	Bluetongue vaccine	Sheep
Severe neonatal disease	Cytomegalovirus	Man
	Rubella	Man
Congenital defects	Rubella	Man
	Hog cholera vaccine	Pig
Inapparent, with lifelong carrier state	Lymphocytic choriomeningitis	Mouse
Inapparent, with integrated viral genome	Murine leukemia	Mouse
	Avian leukosis	Chicken

In severe acute viral infections (e.g., smallpox), congenital infection may cause fetal death and abortion. More important are the viruses which do not kill the embryo but produce congenital malformations due to interference with the normal development of particular organs or tissues. Some viruses appear to affect particular organs at certain stages of fetal development. By far the most detailed observations on congenital defects are those made on human infants whose mothers were infected with rubella virus early in pregnancy.

Rubella

Since Gregg's initial observations in 1941, it has been recognized that there is an association between certain congenital abnormalities and maternal rubella contracted in the early months of pregnancy. A variety of abnormalities has been recognized of which the most severe are deafness, blindness, and congenital heart and brain defects. These defects may only be recognized after the birth of an apparently healthy baby, or they may be associated with severe neonatal disease—hepatosplenomegaly, purpura, and jaundice—to comprise the "rubella syndrome."

Little is known of the pathogenesis of congenital abnormalities in rubella. Damage occurs mainly in fetuses infected during the first trimester. Immunological tolerance does not develop; children who have contracted rubella *in utero* display high titers of neutralizing antibodies throughout their lives. Usually, in the infected human fetus, as in most types of cultured cells, rubella virus is relatively noncytocidal; few inflammatory or necrotic changes are found. The retarded growth in rubella-infected infants may be due to slowing of cell division leading to the reduced numbers of cells observed in many of their organs. Clones of persistently infected cells would be unaffected by the maternal antibody which develops during the first few weeks after maternal infection, even though such antibody could limit fetal viremia.

Cytomegalovirus

The cytomegaloviruses are members of the family Herpetoviridae which commonly cause subclinical infection in man. Apart from rare activation of latent infections (see Chapter 8), disease due to the cytomegaloviruses usually results from infection acquired congenitally from mothers suffering an inapparent infection during pregnancy. Virus-infected cells are present in the chorionic villi. The important clinical features in neonates include hepatosplenomegaly, thrombocytopenic purpura, hepatitis and jaundice, microcephaly, and mental retardation.

THE INCUBATION PERIOD AND ITS SIGNIFICANCE IN PATHOGENESIS

The incubation period of an infectious disease is the period that elapses between infection and the first manifestation of symptoms. Consideration of the mode of spread of viruses within the infected animal, set out earlier in this chapter, allows us to make some generalizations about the incubation period in natural infections (see Table 10-4). It will tend to be short in diseases in which the symptoms are entirely due to viral multiplication at the portal of entry. Thus, respiratory viruses produce symptoms by virtue of their multiplication in the upper and/or lower respiratory tract and, hence, have short incubation periods (1–3 days). On the other hand, the incubation period will be relatively long (10–20 days) in generalized infections, like the common childhood exanthemata, measles, chickenpox, and rubella, where the virus spreads in stepwise fashion through the body before reaching the target organ in which symptoms are produced.

Other factors also influence the length of the incubation period. The generalized infections produced by togaviruses may have an unexpectedly short incubation period attributable to the direct intravenous injection by an insect of a rapidly multiplying virus. Conversely, the long incubation periods of some localized infections like warts and molluscum contagiosum are presumably due to slow multiplication of the viruses concerned. An extreme but poorly understood case is that of the so-called *slow virus infections* and the viral leukemias, where the virus-cell interaction is initially noncytocidal and symptoms may not appear for many months or even years after infection (see Chapter 8).

INAPPARENT INFECTIONS

During the foregoing discussions we have noted that the majority of viral infections are "inapparent" or "subclinical." Although less dramatic than overt disease, such infections are very common and have great epidemiological importance, for they represent an often unrecognized source of dissemination of virus, and they confer immunity. Strictly speaking, a subclinical or inapparent infection is one in which no symptoms or signs of disease occur. Such an infection can only be recognized retrospectively by serological investigations. However, the term is sometimes used to include slight fever and malaise without target organ involvement. Clearly, the term "subclinical" means one thing to the pampered offspring of a wealthy New England family and another to the tattered urchins in a slum in Rio de Janeiro.

SUMMARY

Virus may multiply at its portal of entry to the body with or without causing an obvious lesion. Localized primary lesions occur characteristically in the skin, with poxviruses and papovaviruses, and in the respiratory tract, with orthomyxoviruses, paramyxoviruses, coronaviruses, and rhinoviruses. The other major portal of entry of viruses is the alimentary tract. The multiplication of enteroviruses in cells of the gut fails to produce symptoms, but other viruses, including the "Norwalk agent" and rotaviruses cause diarrhea.

Systemic infection usually follows dissemination of virus in the blood (viremia). The virus may occur free in the plasma or associated with the cells, especially with macrophages and lymphocytes. In such generalized infections signs and symptoms are usually associated with the multiplication of the virus in particular "target organs," although multiplication almost always occurs in other organs as well. Typical targets are the skin (in the exanthemata), certain glands (as in mumps), the liver (as in hepatitis), and the central nervous system (in encephalitis and meningitis). In generalized infections with rashes, invasion of the body occurs by a stepwise process in which multiplication at the site of entry is followed by multiplication in some "central focus" that is usually reached after passage of the virus through the lymphatic system and the blood. Multiplication in the central focus, often the reticuloendothelial system itself, establishes the viremia, which leads to widespread distribution of the virus in the skin and the production of a rash. Sometimes the rash is due primarily to cellular damage caused by local multiplication of the virus, whereas in other diseases the immune response of the host plays an important part.

The central nervous system (CNS) is an important target organ because damage to it usually has serious effects. Different viruses show well-defined cellular affinities, some causing encephalitis, others, in the meninges, causing meningitis. Viruses may reach the cells of the CNS by growing through small vessels, or they may reach the CNS by growth along the endoneural cells of the peripheral nerve fibers.

The occurrence of viremia leads to the possibility of transfer of virus from mother to embryo. In some infections the placenta is usually a barrier to transfer of virus, in others placental infection leads to abortion. The most serious situations are those in which relatively noncytocidal viruses like rubella irreversibly damage organs that are particularly vulnerable at certain stages of organogenesis. Congenital infection with other noncytocidal viruses, like lymphocytic choriomeningitis virus in mice, leads to no such teratogenic effects nor to any other discernible disease in the newborn, though the offspring carry the virus for life.

Consideration of the pathogenesis of viral infections explains the short incubation period of localized diseases like most of the respiratory infections and the rather long incubation period of the generalized diseases. It is important to recognize that most viral infections are inapparent (subclinical).

FURTHER READING

Allison, A. C. (1965). Genetic factors in resistance against virus infections. *Arch. Gesamte Virusforsch.* **17,** 280.

Blattner, R. J., Williamson, A. P., and Heys, F. M. (1973). Role of viruses in the etiology of congenital malformations. *Progr. Med. Virol.* **15,** 1.

Ciba Foundation. (1973). "Intrauterine Infections." Elsevier, Amsterdam.

Fenner, F., McAuslan, B. R., Mims, C. A., Sambrook, J. F., and White, D. O. (1974). "The Biology of Animal Viruses," 2nd ed., pp. 338–393. Academic Press, New York.

Fuccillo, D. A., and Sever, J. L. (1973). Viral teratology. *Bacteriol. Rev.* **37**, 19.

Johnson, R. T., and Mims, C. A. (1968). Pathogenesis of viral infections of the nervous system. *N. Engl. J. Med.* **278**, 23 and 84.

Mims, C. A (1964). Aspects of the pathogenesis of virus diseases. *Bacteriol. Rev.* **28**, 30.

Mims, C. A. (1966). The pathogenesis of rashes in virus diseases. *Bacteriol. Rev.* **30**, 739.

Mims, C. A. (1968). Pathogenesis of viral infection of the fetus. *Progr. Med. Virol.* **10**, 194.

Robinson, W. S., and Duesberg, P. H., eds. (1974). "Mechanisms of Virus Disease." Academic Press, New York.

Scrimshaw, N. S., Taylor, C. E., and Gordon, J. E. (1968). Interactions of nutrition and infection. *World Health Organ., Monogr. Ser.* **57**.

Smith, H., and Pearce, J. H., eds. (1972). "Microbial Pathogenicity in Man and Animals," 22nd Symp. Soc. Gen. Microbiol. Cambridge Univ. Press, London and New York.

CHAPTER 7

Host Responses to Viral Infections

In Chapter 5 we discussed the effects of viruses on individual cells, and in Chapter 6 we saw how viruses can cause disease in the complex assemblage of highly differentiated cells that constitutes the mammalian organism. Vertebrates have evolved in a world of parasites, and mechanisms for adjusting to parasitic invasion with a minimum of bodily disturbance have evolved in parallel. In this chapter we discuss these responses of the host to viral infections. The best understood is that mounted by the lymphoid system and directed against "not-self," viz., the immune response. In this connection it is fitting to recall that the beginnings of both virology and immunology can be traced to the investigations of Edward Jenner into the protection of man against smallpox by prior inoculation with cowpox virus.

The resistance of vertebrates to viral infection is also affected by many other physiological factors, such as interferon, body temperature, nutrition, and hormones. These factors may interact with each other and with the immune response; they are sometimes called "nonspecific" because they do not exhibit immunological specificity.

Genetic resistance of species, strains, or individuals to viral infections may operate through effects on the immune response, on cellular receptors, or on nonspecific factors; the changes in resistance of genetically susceptible animals as they mature are also due to changes in these physiological responses.

ACTIVE IMMUNITY: THE BASIS OF RESISTANCE TO REINFECTION

Viruses are in general "good" antigens; moreover, in viral infections a minute quantity of antigen, quite insufficient on its own to provoke antibody production, multiplies many millionfold. In the process many different antigens are produced, including structural components of the virion and virus-specific enzymes. The important classes of antibodies that result include not only IgG and IgM, which are the major immunoglobulins of the serum, but also IgA, produced by lymphoid cells in the mucosa of the gastrointestinal and respiratory tracts.

Generalized Infections

The lifelong immunity that follows a single attack of diseases such as smallpox, measles, or yellow fever is proverbial; yet by contrast most adults have suffered from many attacks of the common cold or influenza. A number of factors contribute to this paradoxical situation, the most important being: (a) the greater intensity of the antibody response in generalized viral infections compared with those localized to superficial membranes, and (b) the fact that generalized diseases are usually caused by a single serotype, whereas a multiplicity of antigenic variants cause superficial infections. In generalized infections like smallpox and measles much viral antigen is produced and it gains ready access to the sites of antibody production, namely the spleen and lymph nodes, over a period of time in which the response changes from primary to secondary type. There is a correspondingly large production of circulating IgM and IgG, both of which have marked neutralizing capacity. IgM titers rise rapidly within days of infection but decline after a few weeks, whereas IgG continues to be synthesized for many years, and often confers protection against reinfection for life.

Reinfections with viruses causing generalized infection are usually aborted before the state of symptom production by virtue of the secondary response which occurs during the long incubation period. Repeated unrecognized infections of this sort could play a part in maintaining active immunity at a high level. In measles, for example, the titers of neutralizing antibody may periodically rise in immune individuals who are exposed to reinfection but fail to develop symptoms. Even in the absence of such reexposure, however, immunity may persist for many years. The classic example is Panum's study of measles in the remote Faroe Islands, where successive measles epidemics were separated by intervals of 65 and 29 years. Each epidemic affected virtually all those who had not been infected previously and left untouched all those who had been. In only a few other instances has it been possible to test for antibodies in the absence of possible reexposure over a period of many years, but reports with yellow fever, poliomyelitis among the Eskimos, and Rift Valley fever indicate continuing synthesis of antibody for 75, 40, and 12 years, respectively. The evidence from epidemics in the Faroe Islands also showed that measles virus in 1846 and 1875 was antigenically identical with that of 1781; and experience with immunoprophylaxis since then has confirmed the antigenic stability of measles virus. Similar considerations apply to all the acute exanthemata of man, second attacks being extremely rare.

Superficial Infections

Superficial infections of the respiratory or intestinal tract induce the synthesis of IgA locally, but little circulating IgG. The local production and activity of IgA is responsible for the fact that reinfection is much less common after oral administration of live poliovirus vaccine (Sabin) than after parenteral administration of inactivated vaccine (Salk) even though the latter produces equally high titers of circulating antibody (Fig. 7-1).

Investigations also show that IgA found in human nasal secretions is of major importance in protection against respiratory viruses. In in-

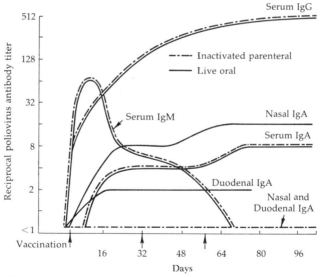

FIG. 7-1. *Comparison of responses to live oral and inactivated parenteral poliovirus vaccines. Serum IgG, IgM, and IgA responses were identical; nasal and duodenal IgA were produced by oral immunization with live vaccine but not by parenteral injection of inactivated vaccine. [Modified from data of P. L. Ogra, D. T. Karzon, F. Righthand, and M. MacGillivray, N. Engl. J. Med.* **279,** *893 (1968).]*

fections with parainfluenza or respiratory syncytial viruses, for example, there is a positive correlation between the titer of IgA in the nasal secretions and protection against reinfection, but not between the serum IgG level and protection. However, IgA titers in mucus do not persist at protective levels for long—less than 5 years for influenza, respiratory syncytial virus and common cold viruses.

Recurrent clinical attacks of common colds or influenza are usually attributable to a succession of antigenically distinct viral serotypes. The continuing change in the antigenic constitution of viruses like influenza was originally called immunological drift, but is now more commonly known as *antigenic drift*. It is an evolutionary change in the envelope antigens of influenza virus imposed by the selective pressure of low levels of IgA in the respiratory mucus of immune individuals, and is discussed at greater length in Chapter 11.

Original Antigenic Sin

The antibody response to serial infections of animals with the same or with antigenically related viruses (or to injections of inactivated vaccines prepared from them) is affected by two processes which at first sight appear to operate in competition: there is a broadening of the specificity of the antibodies produced, but a heightened response to the original virus rather than to that used for challenge.

Early experiments with different strains of influenza A virus in man and experimental animals suggested that vaccination with a series of carefully selected antigenically related viral vaccines might produce a broad immunity which might be effective against all influenza A viruses. However, this aim cannot be achieved because of a feature of immunological memory that has been called "original antigenic sin." A study of the response of various age groups of the human population to vaccination with different strains of influenza A showed that the primary response to the inoculated strain was swamped by an anamnestic (booster) response to the strain first experienced in early life. This dominance of antibody directed against the virus first encountered is characteristic of infections of long-lived animals with any virus of which there are several cross-reacting antigenic types (e.g., togaviruses, paramyxoviruses, and enteroviruses). The phenomenon can be reproduced in laboratory animals either by successive infections with cross-reacting viruses, or successive vaccination with inactive vaccines made from such cross-reacting viruses.

Original antigenic sin is of practical importance in two fields, serological epidemiology (Chapter 10) and vaccination (Chapter 12). In epidemiological studies, efforts may be made to determine the pattern of previous infection by analysis of the antibodies present in a cross-section of the population. Within the limits of the sensitivity of the serological assay, such surveys give unequivocal data about infections with monotypic viruses, like measles, or with the three serotypes of poliovirus. However, in infections with influenza viruses or toga-viruses exact interpretation of the serological results, in terms of whether or not individuals have experienced earlier infection with particular serotypes, is almost impossible. Usually the most that can be said is that a certain percentage of a particular age group in the population gives serological evidence of prior infection with some member of the relevant genus, for example, a type A influenza virus or a flavivirus.

If highly specific antibody is needed to give protection against a particular (new) strain of influenza A, the existence of original antigenic sin imposes barriers to obtaining such a specific response by vaccination in those who have previously been infected (or vaccinated) with another strain of influenza A virus. This can be overcome only by the use of very large doses of antigen, which may be expensive and toxic (see Chapter 12).

Antiviral Functions of Antibody

We have seen that antibodies are of paramount importance in acquired immunity to reinfection, and it will become apparent below that they also play an important role in recovery from some particular viral infections. There are at least two major mechanisms whereby antibodies can assist in the rejection of an invading virus: (a) neutralization of the infectivity of the virion, and (b) activation of complement which lyses virus-infected cells.

Neutralization of Viruses by Antibody. Neutralization of viral infectivity is due mainly to antibodies that are complementary to those antigens on the surface of the virion that are specifically concerned with viral attachment to the cell. Antibodies directed against internal antigens of the virion, or against virus-coded nonstructural antigens synthesized by infected cells, have no relevance to viral neutralization.

Viruses are not irreversibly inactivated by reaction with neutralizing antibody. Indeed, fully infectious virus can regularly be recovered from a noninfectious virus-antibody complex following its artificial dissociation. Dissociation of virus-antibody complexes can be brought about by simple dilution, or in the case of the firmer complexes formed by prolonged incubation of virus and antibody, by any of a variety of chemical or physical treatments, e.g., high or low pH, fluoro-carbon, or sonic vibration.

The precise mechanism of viral neutralization by antibody is still not clearly understood. High concentrations of neutralizing antibody prevent the attachment of virions to cells by steric hindrance. Virions with one or a small number of bound antibody molecules may attach to cells, and even be engulfed by them, but are then rapidly degraded for reasons that are not entirely clear. Antibody attached by both its binding sites to the surface of a virion may distort the configuration of the viral coat sufficiently to affect the normal processes of attachment, engulfment or uncoating, or to admit the entry of nucleases or other enzymes into the particle.

It should be stressed that some types of virus-antibody complexes are infectious; in such cases the antibodies are sometimes said to be "non-neutralizing." The resulting "immune complexes" are not readily eliminated from the body and accumulate in deposits in arteries and the kidney glomeruli to cause "immune complex disease," which is characteristic of several of the chronic infections (see Chapter 8). In some situations viral infectivity may actually be enhanced by antibody which in other circumstances has neutralizing capacity; the biological significance of this phenomenon is unknown.

Intracellular virus is protected from antibody present in the surrounding plasma or tissue fluid. Some viruses, e.g., poliovirus and influenza virus, are released from the surface of the cells in which they multiply and immediately exposed to the action of neutralizing antibody in their environment, which prevents infection of the neighboring cells by interfering with the attachment of the neutralized virus or by modifying its intracellular fate. Other viruses, like vaccinia and herpes simplex, often pass directly from one cell to a contiguous cell, with the result that in cultured cells or in embryonated eggs, which do not develop cell-mediated immunity, expanding lesions can develop even in the presence of a high concentration of neutralizing antibody.

Antibody also plays a key role in the processing of viruses by macrophages. Antibody "opsonizes" bacteria, i.e., increases their rate of uptake by phagocytic cells. With some viruses there is evidence of a similar process; virion-antibody complexes are not only more readily ingested by macrophages but are broken down intracellularly instead of multiplying. Thus some viruses which normally grow in macrophages are effectively neutralized by antibody.

Complement-Mediated Lysis of Virus-Infected Cells. Viral antigens are present in the plasma membrane of cells infected with any budding virus, and many nonbudding viruses also. Specific antibodies recognize and bind to these antigens *in situ*. "Complement," which is abundant in serum, binds to the F_c end of immunoglobulin molecules which have been distorted in shape as a result of the attachment of their F_{ab} ends to antigen. Hence, the several components of complement are activated and the virus-infected "target cell" to which an antibody molecule has bound is lysed by the complement system. In this way an infected cell may be destroyed by "immune cytolysis" long before it would otherwise have been destroyed by the virus itself, hence the yield of virions from that cell may be greatly diminished.

CONGENITAL IMMUNITY

In nature, passive immunity is due to transfer of antibodies from mother to progeny. Different mechanisms operate in different species of animals. IgA in the colostrum or milk may be important in providing local (gut) immunity in newborn animals, and is responsible for the low proportion of "takes" following Sabin oral poliovirus vaccination of breast-fed infants. In humans, IgG, but not IgM or IgA, is readily transferred across the placenta and provides congenital immunity against most generalized viral infections.

Congenital immunity is of considerable epidemiological and evolutionary importance. It protects the very susceptible young animal from infection with the bacteria and viruses that occur commonly in its environment, or it may so modify infection with an otherwise lethal agent that the infected young become actively immunized without serious injury, and are then immune to subsequent reinfection. The role of congenital immunity in viral infections of man has been described in a study of the incidence of herpes simplex infections in an orphanage. Babies living in a heavily contaminated environment failed to become infected until they were about 11 months old, because virtually all of them had passively acquired maternal antibody. Even though this neutralizing antibody had fallen below the level of laboratory detection by the seventh month, immunity against primary infection lasted for an additional 3–4 months, indicating the high efficiency of antibody in the prevention of infection. Herpes simplex is potentially lethal to the newborn, but, in general, the primary herpetic stomatitis seen in young children is a mild disease. There is little doubt that congenital immunity is also important in other more serious viral infections, such as poliomyelitis, smallpox, and measles. Whereas infection with these agents in childhood is less severe than in later life, infection in the newborn infant is severe and even lethal if the infant is serologically unprotected.

RECOVERY FROM VIRAL INFECTION

In the past, the significance of the immune response for the infected host has been largely assessed in terms of acquired immunity to reinfection, in which antibody plays the key role. Viral vaccines are designed to generate such antibodies. Less attention has been paid to the role of the immune response in *recovery* from viral infections. The coincidence in time between recovery and the appearance of circulating antibodies, and the protection conferred by "passive immunization" with antiserum given during the incubation period of measles, originally led immunologists to believe that the humoral response was important in promoting recovery. However, the natural experiment of human dysgammaglobulinemia has cast doubt on the importance of antibodies in recovery from viral infection. Though subjects with such B cell deficiencies suffer from recurrent and intractable bacterial infections, which can be partially controlled by the administration of γ-globulin, they recover from viral infections in a normal fashion in spite of very low levels of γ-globulin and often no detectable production of specific antibodies. By contrast children with congenital or acquired T cell deficiencies are known to be extremely vulnerable to viral infections and often succumb to otherwise trivial diseases such as measles, varicella, or cytomegalovirus.

The respective roles of antibodies, T cells, macrophages, and interferon in recovery from viral infections has been the focus of much recent research in animals and cell culture.

Antibody

Experiments in which the immunological responsiveness of animals is totally suppressed by irradiation or cytotoxic drugs, then selectively restored by administration of antibody, reveal that viruses that produce systemic diseases with a plasma viremia can be controlled by antibody. For example, if adult mice that have been inoculated with coxsackievirus or yellow fever virus are treated with cyclophosphamide, which greatly reduces antibody production, they die from infections that cause no symptoms in untreated animals. Passive immunization with specific antibody as late as 2–4 days after infection greatly reduces the amounts of virus in the blood and in target organs and protects such immunosuppressed mice against death. Neonatal thymectomy, which suppresses T cell-mediated immunity, does not increase the susceptibility of mice to enterovirus infections, suggesting that this immune response plays only a minor role in recovery from these infections. Likewise, children with severe hypogammaglobulinemia, but intact cell-mediated immunity, are more liable to develop paralytic poliomyelitis after exposure to vaccine strains than are normal children.

Restoration experiments with secretory IgA in respiratory virus infections are difficult to carry out. However, since antilymphocyte serum has no detectable effect on the pathogenicity for mice of either influenza or parainfluenza viruses, antibodies available on the epithelial surfaces (i.e., secretory antibody) probably play a role in recovery.

Although attention is usually concentrated on the effects of neutralizing antibodies in relation to recovery, other kinds of antibody may also be important. For example, antibody directed against any viral antigen present in infected tissues could, by forming immune complexes, induce the inflammatory infiltrates that lead to an antiviral effect. Conceivably antibody, cell-mediated immunity, and interferon each play a part in recovery from all viral infections, although their relative importance may vary considerably in different situations.

T Cell-Mediated Immunity

There is ample evidence that viruses elicit "cell-mediated" immunity (CMI). This is demonstrable by delayed hypersensitivity following intradermal injection of viral antigens into a previously infected animal, including man; the capacity to respond in this way is transferable by lymphocytes but not by serum. T cell-mediated immunity is largely abrogated by neonatal thymectomy and treatment with antilymphocyte (or better still, in mice, anti-θ) serum. Such treatments greatly aggravate infections of mice due to herpetoviruses and poxviruses, but have little effect on enterovirus or togavirus infections (Table 7-1).

Restoration experiments confirm the importance of T cell-mediated immunity in recovery from systemic infections in which viremia is cell-associated. For example, mice infected with sublethal doses of ectromelia virus (a poxvirus) died if they were treated with antilymphocyte or anti-θ serum, apparently as a result of the uncontrolled growth of virus in the liver. Antilymphocyte serum suppresses the CMI response, but not the antibody or interferon responses to ectromelia virus. Transfer of immune splenic lymphocytes to mice infected one day earlier with ectromelia virus caused striking inhibition of viral growth and a fall in viral titer in target organs, the liver and spleen. This protective effect was greatly reduced when transfused cells were first treated with anti-θ serum plus complement, which kills T but not B lymphocytes. The recipients of the immune cells did not develop detectable antibody. When mouse hyperimmune anti-ectromelia serum was transfused, antibody reached a high titer in recipients (far higher

TABLE 7-1

The Effects of Depletion of T or B Lymphocytes
on Recovery from Various Viral Infections

ANIMAL	IMMUNODEFICIENCY	LYMPHOCYTES INVOLVED	INFECTIONS AGGRAVATED	INFECTIONS UNAFFECTED
Man	Hypogammaglobulinemia with intact CMI	B	Paralytic poliomyelitis	Smallpox vaccination
Man	Deficient CMI (with or without normal immunoglobulins)	T	Vaccinia Herpes simplex Varicella-zoster Cytomegalovirus Measles	
Mouse	Suppression of antibody by cyclophosphamide	B	Coxsackievirus B	
Mouse	Impairment of CMI by neonatal thymectomy or antilymphocyte (or anti-theta) serum	T	Herpes simplex Mousepox Vaccinia	Influenza Sendai Yellow fever

than that achieved in a normal primary response), and there was some inhibition of viral growth, but no fall in viral titer such as is produced by transfer of immune cells. Passively administered interferon had no effect. Prior irradiation of recipients markedly impaired the antiviral effects of immune cells or hyperimmune serum, probably by inactivating radiosensitive precursors of blood monocytes (macrophages), since in unirradiated recipients monocyte invasion of virus-infected foci in the liver coincided with elimination of the virus. This train of experimental evidence shows that "sensitized" (i.e., immune) T cells are the primary agents of the antiviral effect and that macrophages collaborate with them in the recovery process.

Several mechanisms have been postulated to account for the antiviral effects of sensitized T cells. By liberating lymphokines on exposure to antigens in tissues they induce the migration and activation of macrophages, which, perhaps with the help of opsonizing antibody, phagocytose and digest virions and infected cell debris. Further, sensitized T cells that encounter intact infected cells which bear virus-induced antigens on their surface kill such cells before virus is liberated; such T cell-mediated lysis has been demonstrated *in vitro* with several viruses. Finally, sensitized lymphocytes may liberate interferon on exposure to antigen, and in some infections this could have a significant local antiviral effect.

Thus, T cell-mediated immunity plays a central role in recovery from at least some viral infections, especially those generalized infections in which infected cells display virus-induced antigens on their surfaces.

IMMUNOPATHOLOGY

In certain viral infections it is clear that the immune response itself is a major factor in causing pathological changes and, hence, disease. A good example is lymphocytic choriomeningitis (LCM) virus, which is lethal for untreated adult mice but nonpathogenic in the absence of an immune response. Thus mice infected as adults can be protected by X-irradiation, neonatal thymectomy, antilymphocyte or anti-θ serum, or other immunosuppressive treatments, despite the fact that viral growth is similar in normal and immunosuppressed animals. In normal adult mice the cell-mediated immune response to LCM infection, though beneficial in eliminating virus from organs such as lung, liver, and spleen, is also responsible for lethal inflammatory changes in the brain. Adoptive transfer of immune T cells has been shown to induce lesions in the chorioid plexus and meninges in adult mice infected with LCM virus and immunosuppressed with cyclophosphamide treatment. Recipients of immune T cells do not produce significant amounts of antibody before dying of classic LCM disease, thus suggesting that cell-mediated immunity rather than humoral factors is responsible. Further, T cells cytotoxic for LCM-infected target cells *in vitro* can be isolated from the cerebrospinal fluid of moribund mice, and levels of such T cells in the spleen are maximal at the time of onset of the lethal disease.

In other situations, the humoral immune responses to LCM virus can be pathogenic. Mice infected neonatally or *in utero* show immunological tolerance (see below) with lifelong and widespread infection of tissues, but their tolerance is incomplete. Throughout life, small amounts of antibody are produced which react with virus in the blood to form "immune complexes," and these are precipitated out in kidney glomeruli, eventually causing glomerulonephritis. Antibody can be eluted from glomeruli, and treatments that increase or decrease antibody formation will increase or decrease the severity of glomerulonephritis. Glomerulonephritis (and arteritis) due to such immune complexes is important in several other persistent infections (see Chapter 8).

When certain classes of antibodies react with viral antigens on the surface of infected cells, fixation of complement can lead to cell lysis. This has been shown to occur *in vitro,* not only with LCM but with a number of other noncytopathic and cytopathic viral infections. There is no clear evidence about the part played by such reactions in the infected host, but activation of complement would result in the release of mediators of inflammation and thus cause pathological changes.

The immune response has been invoked as a major cause of pathological changes in many other viral infections, though the evidence is generally less complete than with the LCM model system. To take just one example, there are good reasons for supposing that the T cell-mediated response to measles infection is responsible not only for recovery from the disease but also for the rash which characterizes the disease itself. On infection with measles virus, patients with a defective T cell response due to various lymphoreticular tumors with or without associated immunosuppressive chemotherapy, or thymic aplasia, may show progressive growth of virus in the lungs leading to a fatal "giant cell pneumonia," but no rash. Presumably the measles rash in normal children results from T cell-mediated cytolysis of virus-infected cells in the skin or capillary endothelium, and heralds recovery because those cells are destroyed immunologically before they have had time to yield large numbers of new virions.

Killed measles virus vaccines, now no longer used, were found to elicit the formation of circulating antibodies rather than CMI, and on subsequent infection with live virus there were unusual lesions in the skin and lungs, presumably immunopathological in nature. There was a similar experience with an experimental inactivated respiratory syncytial viral vaccine. It is likely that immunopathological mechanisms contribute to other viral diseases, such as dengue hemorrhagic fever, viral diseases of the central nervous system, and respiratory syncytial viral bronchiolitis in infants inheriting maternal antibody, as well as the several persistent infections described in Chapter 8.

The balance between the humoral and the cell-mediated immune responses to a viral infection (or between the numbers of immune B and T cells respectively) probably plays a key role in both recovery and pathogenesis. Recovery often depends on one rather than the other arm of the immune response. Elucidation of the types of experimental manipulation favoring either type of response may introduce a new era of "immunological engineering" that will make vaccines more effective, and at the same time increase our understanding of viral immunopathology.

IMMUNOLOGICAL TOLERANCE

Under natural conditions it may be expected that viruses would have greater opportunities than other foreign agents to evoke immunological tolerance. However, it is now becoming clear that not all congenital viral infections render tolerant the individual exposed to them in embryonic life. If the virus is cytocidal the embryonic tissues are so highly susceptible, and the embryo is so vulnerable, that fetal death is the more common result.

The classic example of immunological tolerance in a viral infection is lymphocytic choriomeningitis (LCM) in mice. Tolerance is subtotal; small amounts of virus-specific antibody are produced in mice congenitally infected with LCM virus. Disease can be induced in the "tolerant" animal by adoptive immunization, i.e., the injection of a large number of immune lymphocytes. Important natural noncytocidal viral infections which produce tolerance are those due to the onco-viruses of birds and rodents.

IMMUNODEPRESSION BY VIRUSES

It has long been known that when tuberculin-positive individuals suffer from measles they temporarily become tuberculin-negative, and studies of Eskimos in Greenland have shown that measles exacerbates preexisting tuberculosis. *In vitro* studies show that cultured human T lymphocytes will support the growth of measles virus, and such infected cells fail to respond to tuberculin, like lymphocytes taken from individuals rendered temporarily tuberculin-negative by measles.

In the past 10 years, direct tests for immune function have shown that several other viral infections influence the immune responses. Immunodepression is seen, for instance, in mice infected with leukemia viruses, cytomegalovirus, lymphocytic choriomeningitis virus and in chickens with Marek's disease or avian leukosis. The reduced response is to unrelated antigens, and is thus distinct from the immunologically specific effect seen in tolerance. Most reports deal with depressed antibody responses. Cell-mediated immunity is more difficult to measure, but delayed skin graft rejection has been described in mice infected with leukemia and lactic dehydrogenase viruses. Human patients with leukemia, lymphomas, or Hodgkin's disease, diseases which may be caused by viruses (see Chapter 9), usually exhibit immunodepression even in the absence of treatment with cytotoxic drugs, but one must be cautious about too facile an interpretation of this observation.

The mechanism of immunodepression is not understood, but probably results from the multiplication of virus in lymphocytes and/or macrophages. Many viruses are capable of replication in macrophages, and several viruses have now been shown to grow in T cells which have been stimulated by mitogens (concanavalin A or phytohemag-glutinin) to divide *in vitro*, while a few others have been grown in cultured B cells. The immune responses are depressed, not abolished, however, and individual lymphocytes infected with murine leukemia virus, for instance, can at the same time produce antibody to sheep red cells.

Immunodepression by infectious agents is not restricted to viruses; it has been demonstrated in several nonviral infections including leprosy, malaria, and leishmaniasis.

NONIMMUNOLOGICAL RESISTANCE

Many physiological responses other than the immune response affect the resistance of animals to viral infections. Our knowledge of these is sketchy; here we shall do little more than mention the existence of some factors and discuss briefly those on which there is more precise information.

PHAGOCYTOSIS

The phagocytosis of bacteria by polymorphonuclear leukocytes, which is aided by the opsonization of the bacteria by specific antibody, is an important mechanism of defense against bacterial infection. Polymorphs play no part in protection against viral infections, which are usually characterized by leukopenia with a relative lymphocytosis, due largely to a deficit of polymorphs in the blood.

However, macrophages, particularly the "fixed" macrophages of the reticuloendothelial system, are very important in the pathogenesis of viral infections. Several organs (liver, spleen, bone marrow) contain blood sinuses which are partially or completely lined by macrophages and similar phagocytic cells which monitor the lymph, the pleural and peritoneal cavities, the respiratory tract, and the connective tissue throughout the body. Macrophages play an important role in clearing viruses from the bloodstream and preventing the infection of susceptible cells in target organs.

Yet some viruses, far from being digested by the macrophages that ingest them, actually multiply preferentially in these cells. Indeed, circulating macrophages (the monocytes of the blood) may transport replicating viruses around the body (see Chapter 6; cell-associated viremia), so helping to disseminate the infection. On the other hand, infected macrophages produce substantial amounts of interferon and thus protect susceptible cells from infection. They also appear to be important in the process of recovery from infection which is triggered by immune T cells, as described above. Clearly, the role of the macrophage in recovery from viral infections is a paradoxical one, about which we have much to learn. Under some circumstances it seems to play a key role in viral clearance, interferon production and the generation of T cell-mediated immunity; other viruses seem to have evolved the capacity to exploit it to their own advantage.

INTERFERON

In Chapter 5 we describe the molecular basis of the production and action of interferon, the cellular protein which is induced by viruses and renders cells insusceptible to viral infection. Here we are concerned with the role of interferon in recovery from natural infection. The potential of interferon as a chemotherapeutic agent is discussed in Chapter 13.

Interferon produced during the course of a viral infection protects some cells from infection, especially locally but perhaps also in distant target organs. However, it has proved difficult to devise decisive experiments to evaluate the importance of these effects in promoting recovery, since there are no known naturally occurring diseases of man or animals in which there is a specific deficit in interferon production. However, there is much circumstantial evidence that interferon is important in recovery from viral infection. Decreased interferon production caused by altered temperature, chemical inhibitors, or different viral strains has been correlated with impaired recovery, but the situation is often complex with multiple determinants of viral virulence. Virulence of some viruses is associated with a weak interferon response, but in other instances there is no correlation. The responsiveness of infected cells to interferon action may be important. For example, mice of the C3HRV strain are much more resistant than C3H mice to flavivirus infections. Although both produce equal amounts of interferon, the cells of the resistant mice are more susceptible to the action of interferon, i.e., a genetic difference in interferon sensitivity rather than interferon production appears to determine the severity of infection. Several other "nonspecific" factors involved in resistance may operate at least in part by their effects on interferon production; this appears to be the case with body temperature, with some types of stress, and with the effects of some hormones.

Despite the wealth of evidence (a) that interferon is produced as a by-product of most or all viral infections, and (b) that passively administered interferon can protect against certain viruses under limited experimental circumstances, there are some clinical observations which cast doubt on the primacy of the role of interferon in recovery from natural infections and at the same time temper optimism in relation to its possible therapeutic use. Neither natural infections with viruses nor immunization with live attenuated vaccines confers substantial protection against simultaneous or subsequent heterologous viral infections elsewhere in the body. Indeed, certain attenuated viruses can be combined to produce potent multivalent vaccines (see Chapter 12). Perhaps, under natural circumstances, interferon expedites recovery because transient high concentrations occur in the immediate vicinity of infected cells. It may be impossible, in terms of antiviral therapy, to produce enough interferon in the right place at the right time.

BODY TEMPERATURE

Exposure of animals to abnormally low temperatures increases thyroid activity and the stress response, lowers the metabolic rate, and affects such varied physiological responses as the state of the peripheral circulation, the immune response, and inflammation. Exposure to abnormally high temperatures has a similarly wide range of physiological effects.

Studies with experimental animals show that lowered body temperatures enhance, and raised body temperatures diminish the severity of viral infections. Several different mechanisms operate. Many viruses multiply better at temperatures somewhat below 37°, whereas their multiplication may be inhibited at temperatures of 39° or higher. Indeed, there is a correlation between virulence and the ability to multiply at elevated temperatures which holds true for many viruses, notably for the polioviruses. Virulent wild-type polioviruses grow well at 40° but "temperature-sensitive" mutants are relatively avirulent and are suitable for use in vaccines. In the host, both the immune response and interferon production are stimulated at high temperatures, and in different situations increased body temperatures (or the higher temperature of some parts of the body than of others) may suppress viral multiplication directly, by an effect on some temperature-sensitive step in multiplication, or indirectly, by promoting the efficiency of the host response in terms of the production of antibody and/or interferon. Low body temperatures usually have the reverse effect.

Quite apart from the question of environmental temperature, the body itself attains temperatures of 38°–41° as a febrile response to almost all severe viral infections. Fever usually develops at the end of the incubation period, when the distribution of virus throughout the body has been completed. It may be an important protective mechanism in promoting recovery by limiting further viral multiplication.

There is one situation in which elevated body temperature promotes viral multiplication, namely the production of recurrent herpes simplex ("fever blisters") in man. Artifically induced fever precipitates an attack of herpes simplex in about 50% of cases. Moreover, fever blisters are a frequent complication of some febrile diseases (malaria, influenza, streptococcal and pneumococcal infections), but for some unknown reason, are very rarely found in others (tuberculosis, smallpox, typhoid fever).

HORMONES

The corticosteroids have readily demonstrable effects on resistance to viral infections, both in man and in many experimental animals. Adult mice treated with cortisone die following inoculation with coxsackievirus B1, which is normally lethal only in newborn mice. Cortisone greatly increases the susceptibility of hamsters and monkeys to poliovirus infection. Exacerbation of otherwise mild viral infections (e.g., vaccinia and varicella) is a well-recognized complication in human patients receiving corticosteroid therapy, and these drugs have caused blindness when incautiously used in the treatment of herpetic keratoconjunctivitis. There are several possible explanations of these findings. The anti-inflammatory effects of the corticosteroids, including impairment of monocyte recruitment and their capacity to depress the immune response, have been amply documented. Moreover, a number of investigators have shown that hydrocortisone inhibits the production of virus-induced interferon in chick embryos, cultured cells, and intact animals, and that it "stabilizes" lysosomes against breakdown and release of nucleases that might otherwise destroy viral nucleic acid.

Pregnancy affects the pathogenesis and the severity of several viral diseases, probably because of hormonal changes. Smallpox, hepatitis, influenza, and poliomyelitis are all more severe in pregnant women.

NUTRITION

Malnutrition can interfere with any of the mechanisms that act as barriers to the multiplication or progress of viruses through the body. It has been repeatedly demonstrated that almost any severe nutritional deficiency will interfere with the production of antibodies and the activity of phagocytes, while the integrity of the skin and mucous membranes is impaired in many types of nutritional deficiency. The synergistic effects of malnutrition and infectious disease are evident in the data on childhood mortality in underdeveloped countries. Measles, rarely a cause of death in affluent countries, produces a substantial mortality in most underdeveloped countries. In 1959, in various South American countries, for example, the mortality rate of measles was 100 to 400 times higher than in the United States of America, most deaths occurring in children in the lowest socioeconomic groups.

In animal experiments, mice which from the age of 3 weeks received one-quarter the food intake necessary for optimal growth developed marked atrophy of lymphoid tissue and lymphocytopenia. They were then much more susceptible than normal mice to coxsackievirus B3 infection, which produced severe and often fatal lesions in liver and heart, with a marked absence of cell infiltration. There were indications that the effect involved an inadequate host immune response to the infection. A similar depletion of the lymphoid tissue and depression of cell-mediated immunity has been observed in children with protein-calorie malnutrition.

Studies of nitrogen balance in human beings have demonstrated that even mild viral infections lead to substantial nitrogen losses. These are readily sustained in well-nourished individuals, but in those suffering from protein deficiencies relatively mild infections can lead to severe disease. Even smallpox vaccination may lead to lowered serum levels of vitamins A and C, as well as negative nitrogen balance, in individuals suffering from malnutrition.

OTHER NONSPECIFIC FACTORS

Isolated reports have been published of a variety of other nonspecific factors that raise or lower the resistance of animals or human beings to viral infections.

Stress

"Stress" is a very general term that is used to describe almost any type of abnormal or unusual environmental influence on animal behavior; thus an animal may be subjected to nutritional stress, heat stress, psychological stress, the stresses of infection, and so on. These stresses call into play an adaptive host response involving the pituitary-adrenal axis (corticosteroid hormones) and the autonomic nervous system (adrenaline), which functions by combating the harmful effects of the stresses. Various stresses may therefore influence a viral infection by inducing an increased output of corticosteroids, whose adverse effects have been referred to earlier. Infection itself is often an important stress, and can cause changes in lymphoid tissues and liver which are attributable to the action of corticosteroids. Synergism between malnutrition and the stresses of infectious diseases may be of great importance in determining the severity of such infections in man.

Trauma

Most of the evidence concerning the effects of trauma on viral infections relates to poliomyelitis. To affect the severity of the disease, trauma must occur during the early stages of the incubation period, or if the effects of the "trauma" are prolonged, shortly after infection. The effects of the traumatic experience are to increase the likelihood of paralysis, its severity, and sometimes its localization. Physical exertion and nonspecific trauma increase the likelihood of paralysis, injections of pertussis vaccine or diphtheria or tetanus toxoids increase the likelihood of paralysis in the injected limb, and tonsillectomy increases the likelihood of bulbar poliomyelitis. There is also statistical evidence to show that persons whose tonsils have been removed at any time in the past are more likely to suffer from bulbar poliomyelitis than those with intact tonsils. This probably has an immunological rather than a "traumatic" basis, in that removal of the tonsils and adenoids removes the major IgA-synthesizing tissues from the throat.

There is abundant evidence that the skin lesions in exanthematous viral diseases localize in "provoked" skin areas, probably because of the increased vascularity and the inflammatory response associated with the provoking factor.

Concurrent Infections

Since latent or chronic infections with bacteria, protozoa, and viruses are common in both man and animals, concurrent infections are also common. Sometimes the two infections apparently have no effect on each other; with other combinations of agents there are interactions which may either increase or decrease the severity of either infection. Veterinary virologists are particularly conscious of the potentiating effects on viral diseases of coinfection with parasites because of the universal infestation of livestock with protozoa and worms.

Infections with respiratory viruses commonly increase the susceptibility of the respiratory tract to infection with bacteria, e.g., after rhinovirus or influenza virus infections, and after measles. Measles also exacerbates the severity of tuberculosis in man.

In other experimental situations, resistance to several kinds of virus [Western equine encephalitis in chickens, mouse hepatitis virus, and Mengo (encephalomyocarditis) virus in mice] appears to be enhanced by concurrent protozoal infections of the challenged animals, presumably due to activation of macrophages.

AGE

It is a commonplace observation in clinical medicine and in experimental virology that the response to many viruses changes greatly with age. Viral infections tend to be very severe in the perinatal period, moderately severe in infancy, mild during childhood, and severe in the aged, the last effect being due to the steady deterioration of homeostatic mechanisms.

An exception to this pattern was provided by the 1918–1919 epidemic of influenza, which showed a unique mortality curve with peak incidence of deaths from pneumonia in young adults. This is not a general characteristic of influenza, for all previous and subsequent epidemics have the characteristic peaks of mortality in the very young and the very old, and the reason for the "young adult" peak of mortality on this occasion is still obscure.

The high susceptibility of newborn animals to many viral infections has been of considerable importance in laboratory studies of viruses. Thus the coxsackieviruses were discovered by the use of suckling mice, which are also the most sensitive hosts for the recovery of arboviruses. In laboratory animals the first few weeks of life are a period of very rapid physiological change. During this time mice, for example, pass from a stage of immunological nonreactivity (to many antigens) to normal responsiveness. This change profoundly affects their reaction to viruses like lymphocytic choriomeningitis (which induces a tolerant state when inoculated into newborn mice), and the oncogenic viruses. Newborn rabbits and chickens also may suffer from severe generalized disease when inoculated with some "oncogenic" viruses, whereas older animals develop tumors. On the other hand, polyoma virus induces tumors only when inoculated into newborn rodents; older animals are infectible but develop no apparent disease. There is good evidence that in all these cases the high susceptibility of the newborn animal is due mainly to immunological immaturity. The human species is reasonably mature immunologically at the time of birth, but no doubt infant mortality from viral infections would be quite high if most newborns were not protected by the passive transfer of maternal antibodies across the placenta.

GENETIC RESISTANCE TO VIRUSES

There is a good deal of circumstantial evidence that human racial groups differ in their genetic susceptibility to certain viral infections, such as yellow fever, to which tropical Africans appear to be more resistant than Europeans, and measles, which seems to be particularly severe in inhabitants of the Pacific Islands and Africa. Accurate human genetic data on resistance to infection is almost unobtainable, because genetic and environmental differences are generally confounded.

In mice, however, it has been possible to study the genetics of resistance to viral infection in some detail. Two situations have been found in which resistance is associated with the response of macrophages. (a) Susceptibility to certain flaviviruses is under single gene control, with resistance dominant; in this system the yield of virus from cultures of splenic macrophages from the susceptible strain of mice is 100–1000 times greater than in cultures from resistant strains. (b) Susceptibility to mouse hepatitis virus (a coronavirus) which is also associated with capacity of macrophages to support multiplication of the virus, is under single gene control with susceptibility dominant.

In a few instances it has been shown that the susceptibility of animals, or particular organs, is a direct consequence of the presence of the relevant cellular receptors. The best genetic evidence relates to the susceptibility of different strains of chickens to Rous sarcoma virus. Here susceptibility of both cell and organism is due to a single gene that determines the nature of the cellular receptor substance, which differs for different antigenic types of the virus. Susceptibility is dominant to resistance.

Poliovirus provides an example of the importance of cellular receptors at the species level. It ordinarily infects primates; mice and other nonprimates are insusceptible because their cells lack the surface receptors for this virus. However, polioviral nucleic acid, either naked or enclosed within the capsid of a mouse-pathogenic coxsackievirus, will undergo a single cycle of multiplication in mouse brain. Since the progeny particles have the capsids of normal poliovirions, they are unable to initiate a second cycle.

Immunological responsiveness to particular antigens varies greatly from one strain of mouse to another, being under the control of particular I_r (immune response) genes. No doubt individual humans also differ in their capacity to mount an immune response to any given antigen, including viruses, but we know nothing yet of the genetics of the situation.

SUMMARY

The response of vertebrates to viral infection is affected by a great variety of different factors, the most important of which is the immune response.

Antibodies neutralize viral infectivity by inhibiting the normal processes of viral attachment, penetration or uncoating. They may also attach to virus-coded antigens on the surface of infected cells thereby bringing about complement-mediated cytolysis.

Antibodies are of paramount importance in acquired immunity to reinfection. IgG continues to be synthesized in large amounts for years after infection, with most of the viruses causing systemic infection, and protects the individual against reinfection with the same agent. By contrast, the local IgA response that characterizes superficial infections of respiratory and gastrointestinal mucosae provides a relatively short-lived immunity to reinfection, even with the same serotype. Such a situation favors antigenic drift, which has generated a large number of distinct serotypes of respiratory viruses showing no cross-immunity; recurrent attacks of the "same" disease are the result.

The role of antibodies in recovery from infection, once it has been established, seems to depend on the pathogenesis of the infection. Generalized infections in which virions circulate freely in the plasma may be halted by serum IgM and IgG, but such antibodies have little effect on generalized infections in which circulating virus is multiplying in lymphocytes and macrophages.

IgG is also vital in congenital immunity and is responsible for the protection of infants for the first few months of life by antiviral antibodies transmitted across the placenta.

Not all antibodies succeed in neutralizing virus. Some persistent infections are characterized by high titers of circulating virus-antibody complexes which are eventually deposited in kidney glomeruli and arteries causing "immune complex disease."

Immune thymus-derived lymphocytes (T cells) lyse virus-infected target cells by direct interaction. Such "immune cytolysis" facilitates recovery of the patient by reducing the yield of virions but, at the same time, may contribute to the pathology and symptomatology of the disease itself, e.g., by producing a skin rash. Children with congenital or drug-induced deficiencies of T cell function may die of otherwise trivial viral infections, such as measles or varicella, whereas those with hypogammaglobulinemia do not.

The role of macrophages in viral infection is complex. Viruses are readily phagocytosed but, unless neutralized by antibody, are not always destroyed. Indeed, some viruses grow preferentially in macrophages. Nevertheless, "immune" macrophages "activated" by exposure to foreign antigen and perhaps "armed" with antiviral antibody do potentiate immune T cells in expediting recovery and produce large amounts of interferon.

The relative importance of interferon in the recovery process is difficult to gauge. It is certainly synthesized in most or all viral infections and may build up to sufficient concentrations locally to slow the growth of virus in the vicinity of a lesion, but whether it plays a determinative role in bringing systemic infections to a halt is open to doubt.

Many types of physiological change also affect susceptibility to, and/or recovery from, viral infection. These include certain hormones, altered body temperature, malnutrition, concurrent infections, and age. Viral multiplication, the immune response, or the production of interferon may be affected.

Several of the determinants of innate susceptibility or resistance to infection by viruses may be heritable. For example, I_r genes presumably dictate the immunological responsiveness of man, as well as mice, to a particular virus, while other genes code for specific viral receptors on the surface of infectible cells or determine whether macrophages will destroy or be destroyed by a given virus.

FURTHER READING

Allison, A. C. (1974). Interactions of antibodies, complement components and various cell types in immunity against viruses and pyogenic bacteria. *Transplant. Rev.* **19**, 3.

Allison, A. C., and Burns, W. H. (1972). Immunogenicity of animal viruses. *In* "Immunogenicity" (F. Borek, ed.), p. 155. North-Holland Publ., Amsterdam.

Blanden, R. V. (1974). T cell response to viral and bacterial infection. *Transplant. Rev.* **19**, 56.

Burns, W. H., and Allison, A. C. (1975). Virus infections and the immune responses they elicit. *In* "The Antigens" (M. Sela, ed.), Vol. 3, p. 479. Academic Press, New York.

Cowan, K. M. (1973). Antibody response to viral antigens. *Advan. Immunol.* **17**, 195.

Doherty, P. C., and Zinkernagel, R. M. (1974). T-cell mediated immunopathology in viral infections. *Transplant. Rev.* **19**, 89.

Fenner, F., McAuslan, B. R., Mims, C. A., Sambrook, J. F., and White, D. O. (1974). "The Biology of Animal Viruses," 2nd ed., pp. 394–451. Academic Press, New York.

Greaves, M., Owen, J., and Raff, M. (1973). "T and B Lymphocytes." Elsevier, Amsterdam.

Lindenmann, J. (1973). The use of viruses as immunological potentiators. *In* "Immunopotentiation," Ciba Found. Symp. No. 18, p. 197. Elsevier, Amsterdam.

Notkins, A. L., ed. (1975). "Viral Immunology and Immunopathology." Academic Press, New York.

Notkins, A. L., and Lodmell, D. L. (1975). Cellular immune response in viral infections. *Perspect. Virol.* **9,** 115.

Notkins, A. L., Mergenhagen, S. E., and Howard, R. J. (1970) Effect of virus infections on the function of the immune system. *Annu. Rev. Microbiol.* **24,** 525.

Ogra, P. L., and Karzon, D. T. (1971). Formation and function of poliovirus antibody in different tissues. *Progr. Med. Virol.* **13,** 156.

Porter, D. D. (1971). Destruction of virus-infected cells by immunological mechanisms. *Annu. Rev. Microbiol.* **25,** 283.

Roitt, I. (1974). "Essential Immunology," 2nd ed. Blackwell, London.

Rossen, R. D., Kasel, J. A., and Couch, R. B. (1971). The secretory immune system: Its relation to respiratory viral infection. *Progr. Med. Virol.* **13,** 194.

World Health Organization. (1973). Cell-mediated immunity and resistance to infection. *World Health Organ., Tech. Rep. Ser.* **519.**

CHAPTER 8

Persistent Infections

INTRODUCTION

The thinking of clinicians and virologists about viral diseases has long been dominated by acute febrile diseases like smallpox, measles, poliomyelitis, and influenza. In such diseases, as is seen in the preceding chapters, the course of the infection may be summarized as follows: The causative virus enters the body, multiplies in one or more tissues, and spreads either locally or through the bloodstream. When viral multiplication has reached a critical level, after an incubation period of 2 days to 2 or 3 weeks, symptoms of disease appear, associated with localized or widespread tissue damage. Nonspecific and specific host defenses are mobilized during the incubation period and, unless the disease is fatal, the host has usually eliminated the infecting agent within 2 or 3 weeks of the onset of symptoms. Virus can ordinarily be isolated from the blood or secretions only in the short period just before and just after the appearance of symptoms. Some viruses (e.g., measles and smallpox in man) almost always cause acute disease; many others produce acute *infections* in which the pathogenic mechanisms are similar but often there is no clinical *disease,* i.e., the infection is *subclinical* (see Table 10-6).

Quite distinct from the acute infections, however, are those in which virus persists for months or years, i.e., *persistent viral infections.* Persistent infections are associated with a great variety of pathogenic mechanisms and clinical manifestations, and it is difficult to classify them satisfactorily. Because of this we shall have to draw rather heavily on experimental and natural infections in animals, as well as on observations made in human diseases. For convenience, we shall subdivide the persistent infections into three categories, recognizing that there is some overlap.

1. Persistent infections with intermittent acute episodes of disease

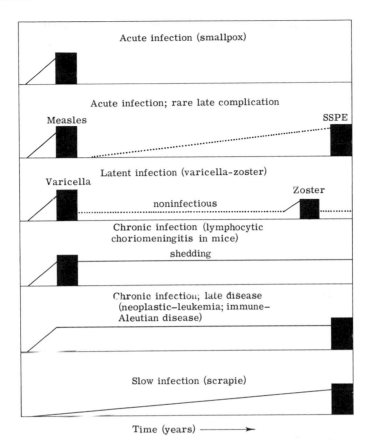

FIG. 8-1. *Diagram illustrating acute infection, and various kinds of persistent infections. Solid line, demonstrable infectious virus; dotted line, virus not readily demonstrable; box, disease episode.*

between which virus is usually not demonstrable: *latent infections* (see Table 8-1).

2. Persistent infections in which virus is always demonstrable and often shed, but disease is either absent, or is associated with immunopathological disturbances: *chronic infections* (see Tables 8-2 and 8-3).

3. Persistent infections with a long incubation period followed by slowly progressive disease that is usually lethal: *slow infections* (see Table 8-4).

The key distinctions between these three groups of persistent infections are illustrated diagrammatically in Figure 8-1. In slow infections, the concentration of virus in the body builds up gradually over a prolonged period until disease finally becomes manifest. Chronic infections, on the other hand, can be regarded as acute infections (clinical or subclinical) following which the host fails to reject the virus; sometimes disease supervenes late in life as a result of an immunopathological or neoplastic complication. The distinction between chronic infections and latent infections, e.g., herpetoviruses between recrudescences of endogenous disease, may be a fundamental one concerned with the state of the virus between attacks, or it may be merely a matter of the ease of demonstration of infectious virus.

Before discussing persistent infections in animals it will be useful to consider examples found in cultured cells.

PERSISTENT INFECTIONS OF CULTURED CELLS

One of the unexpected discoveries that has emerged from the widespread use of cultured animal cells is the common occurrence in such cells of viruses which give no overt sign of their presence. In such latently infected cells the viral nucleic acid may be physically integrated with cellular DNA, as in cells transformed by papovaviruses (see Chapter 9). Alternatively, the viral and cellular genomes may replicate independently, as a *steady-state infection.*

Infective virus may often be recovered by prolonged *in vitro* cultivation of tissue or trypsinized cells from human or animal organs, even though homogenates of the same material fail to yield virus. For example, adenoviruses were originally discovered when the cell monolayers that grew out from explanted fragments of "normal" human adenoids underwent virus-induced degeneration *in vitro.* The same adenoids failed to yield virus when homogenates were inoculated directly onto monolayers of susceptible cells. The nature of the cell–virus association in these tonsils or adenoids remains obscure; probably the virus continues to replicate slowly in a very small proportion of cells in the glands, but is kept in check by antibody *in vivo.*

Another situation in which cultured cells were found unexpectedly to yield virus derives from the extensive use of monkey kidney cells for the isolation of viruses and the manufacture of vaccines. Over fifty different simian viruses have been recovered from "normal" monkey kidney cell cultures. Some of these establish weakly cytocidal or steady-state associations with the cells, but in most cases the simian viruses are probably accidental blood-borne contaminants of the kidneys, kept in check by antibody bathing the tissue *in vivo,* but producing cytopathic effects when the cells are washed and cultured.

In most cases, the relation of these viruses to the cells which carry them is unknown; they have been said to be "occult," a term that "appropriately surrounds our state of ignorance with an aura of mysticism." However, studies of a variety of types of persistent infections in carefully controlled *in vitro* systems are beginning to shed some light on the nature of the virus–cell interactions.

The following information needs to be obtained by experiment before one can determine what is happening in persistently infected cultured cells.

1. Are all cells infected and continuously producing virus, or are only a minority involved?

2. Do infected cells divide, or do they die?

3. Can the culture be "cured" of infection by cloning cells in the presence of antiviral antibody?

4. Can the balance be altered toward cell destruction by washing inhibitory substances from the medium?

Basically, the great variety of virus-cell interactions that have been described as persistent infections of cultured cells can be allocated to one of three categories:

1. The viral genome is integrated with that of all cells in the culture (tumor viruses). These are discussed in Chapter 9.

2. All cells continuously produce a noncytocidal virus (*steady-state infection*).

3. A cytocidal virus is kept in check by inhibitors in the medium or by the presence in the culture of a minority of susceptible and a majority of genetically resistant cells (*carrier culture*).

Steady-State Infections

Steady-state infections are not uncommonly found in cells infected with RNA viruses that mature by budding from the plasma membrane. In such cells large amounts of infectious virus may be continuously released from cells whose metabolism and multiplication are scarcely affected. Such latently infected cells can be superinfected with other viruses without noticeably affecting the growth of either virus. Alternatively, the yield of either the superinfecting virus or the noncytocidal virus may be enhanced (complementation) or reduced (interference) (see Chapter 4).

A typical example is the paramyxovirus SV5, a common contaminant of primary cultures of rhesus monkey kidney cells, in which it multiplies to high titer with little cytopathic effect. The infected cells survive and multiply, and produce large amounts of infective virus for many days. Cellular DNA, RNA, and protein synthesis are hardly affected by the virus; viral RNA synthesis amounts to less than 1% of the cellular RNA synthesis. Electron microscopic observations reveal little cellular damage despite the fact that large numbers of virions are maturing at the plasma membrane at all times. Infection with SV5 does not interfere with the growth of other viruses.

Similar steady-state infections have been described with most para-myxoviruses and with several togaviruses, rhabdoviruses, arena-viruses, and retroviruses, which are relatively noncytocidal viruses that do not shut down the metabolism of the cell and are released by budding from the plasma membrane. Most of the situations described in the literature share the following common features: (a) most or all cells in the culture are infected, (b) virus is released continuously but at a slow rate, and (c) the cultures cannot be "cured" by antibody.

Carrier Cultures

Persistent viral infections commonly arise in cells maintained by serial culture in the laboratory, after they have been deliberately in-fected with any of a variety of viruses. These *carrier cultures* differ from the steady-state infections that we have just described in that virus-free cells can always be recovered by cloning from a carrier culture, whereas this is impossible in steady-state infections. This enduring state of "peaceful coexistence" of virus and cells usually results from the presence in the culture of virus-inhibitory substances, which limit the spread of virus from cell to cell. The medium may contain antibody or other antiviral substances, or infected cells may be synthesizing enough interferon to keep viral multiplication and thus cell destruc-tion in check. Sometimes the situation is complicated by the presence of some cells that are genetically resistant to infection. Often more than one mechanism of cell protection operates and sometimes the system undergoes an evolving series of virus-cell relationships.

Although the investigation of carrier cultures has demonstrated that there are a number of ways whereby an infecting virus can persist in a population of cells and remain relatively inconspicuous, these re-sults have not yet been effectively applied to understanding the latent infections of man.

With several different viruses in cell culture and in several animal systems (foot and mouth disease, vesicular stomatitis, and parainflu-enza virus infections), carrier cultures or persistent infections *in vivo* appear to be associated with temperature sensitive mutants of the viruses involved.

PERSISTENT INFECTIONS IN ANIMALS

Viruses that produce persistent infections in cultured cells may also produce persistent infections in the intact animal. As in cultured cells, such infections may be accompanied by the production of infectious virus or they may be attributable to the integration of viral nucleic acid into the cellular DNA. In addition, there are a few viruses (e.g., herpes simplex virus) that produce cytocidal infections in most cultured cell systems but persistent infections *in vivo*.

As outlined in the Introduction, we have grouped persistent infections into three categories which for convenience can be called *latent*, *chronic*, and *slow* infections. With the help of appropriate tables we shall first describe illustrative examples of infections thus classified, and then consider the pathogenic mechanisms that allow such viruses to persist in the infected host.

LATENT INFECTIONS

We use this expression for a small but important group of herpetovirus infections, of which the best understood are herpes simplex and varicella-zoster in man (Table 8-1). These are characterized by the apparent disappearance of the virus following the acute primary infection. Yet years later acute disease recurs, perhaps more than once, and infectious virus is then readily demonstrable (Figure 8-1). There is ample evidence that the recurrent episodes of acute disease result from endogenous exacerbation of infections that have lain dormant for many years (although exogenous reinfection does sometimes occur).

Herpes Simplex

First infections with herpes simplex virus are usually manifest as an acute stomatitis or more commonly a subclinical infection, contracted in early childhood. Typically, at intervals of months or years after recovery from the primary infection, blisters from which the virus can be recovered appear, usually around the lips and nose. A variety of stimuli can trigger this recurrent viral activity, e.g., exposure to ultraviolet light, fever, menstruation, nerve injury, or emotional disturbances. Patients who have completely recovered from stomatitis may excrete virus intermittently in their saliva for several weeks, and recovery of virus from the saliva has been reported in 2.5% of asymptomatic adults at any given time, suggesting low-grade chronic multiplication and periodic release of infectious virus. Intermittent secretion of virus from the conjunctiva, lacrimal, and salivary glands in patients subject to recurrent herpetic keratitis has also been reported. There is no relationship between antibody levels and the frequency of either shedding or recurrent attacks.

Between attacks, virus cannot be demonstrated directly, either in the

TABLE 8-1

Latent Infections: Virus Occult between Acute Episodes of Disease

		SITES OF INFECTION			
EXAMPLES	SYMPTOMS	BETWEEN DISEASE EPISODES	DURING ACUTE ATTACK	VIRUS SHEDDING	ANTIBODIES
Herpes simplex	1. Primary stomatitis 2. Recurrent fever blisters	Occult, in cerebral or dorsal root ganglion cells	1. Epithelial cells 2. Schwann cells, then cells of corresponding dermatome	Sporadically in saliva between attacks; plentiful in recurrent fever blisters	+
Varicella-zoster	1. Generalized primary infection: varicella 2. Late recurrent skin eruption: herpes zoster	Occult in cerebral or dorsal root ganglion cells	1. Generalized with rash 2. Schwann cells, then cells of corresponding dermatome	1. Varicella: from throat and skin lesions 2. Zoster: from skin lesions. Contacts contract varicella	+

mucous membrane where "fever blisters" habitually recur, or any-where else in the body. However if the trigeminal ganglion (obtained from people dying from other causes) is explanted *in vitro* herpes simplex virus appears in the culture after a few weeks. Electron micro-scopic studies suggest that the virions first appear in neurons them-selves, rather than in satellite cells, but we still know nothing about the state of the viral genome in the neurons during remissions.

Herpes Zoster

This disease, familiarly known as "shingles", is characterized by a rash which is usually limited to an area of skin (and mucous mem-brane) served by a single sensory ganglion. It occurs predominantly in older people, and only in persons who have already had varicella in childhood.

Epidemiological evidence, supported by virological and serological studies, shows that zoster and varicella are two clinical manifestations of the activity of a single virus. There is also very strong evidence for the view that zoster represents a reactivation of virus that has re-mained latent since an attack of varicella, perhaps many years earlier, and that the virus is dormant in the cells of one or more of the dorsal root or cranial nerve (sensory) ganglia.

The incidence of herpes zoster is greatly increased by X-irradiation and the lesions are then often related to the site of the irradiation. Zoster is also a common complication of Hodgkin's disease and other lymphoproliferative disorders.

Pathogenesis of Recurrent Herpes Simplex and Herpes Zoster

Figure 8-2 illustrates the likely pathogenesis of these two latent in-fections. All data are consistent with the view that during a primary attack of herpes simplex or varicella, virus moves to the ganglia along the sensory nerves, probably by spread in Schwann cells of the nerve sheath. In recurrent herpes simplex or zoster the virus moves down the sensory nerves again until it reaches the skin, where it proliferates and produces vesicles.

CHRONIC INFECTIONS

A wide variety of conditions fall within this category (Tables 8-2 and 8-3). We have already referred to the fact that many viruses have been recovered during the cultivation of tissues from "normal" animals.

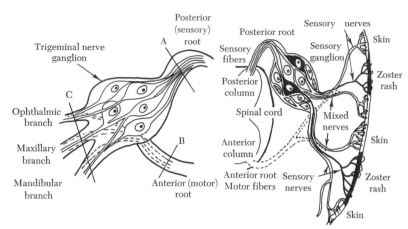

FIG. 8-2. *Diagram illustrating the probable pathogenesis of recurrent herpes simplex (left) and herpes zoster (right). [Latter modified from R. E. Hope-Simpson, Proc. Roy. Soc. Med.* **58,** *9 (1965).] Left, Herpes simplex virus is presumed to be latent in the sensory nerve cells of the trigeminal nerve ganglion. Recurrent viral activity is triggered by fever, ultraviolet light, etc., and also by nerve injury. Section of posterior (sensory) root of trigeminal nerve (A) produces herpes simplex lesions in skin innervated by maxillary and mandibular branches of that nerve. Section of the motor root (B) or the branches (C) has no such effect. Right, varicella virus is presumed to become latent in sensory cells of the dorsal root ganglion. Upon activation the virus grows down the sensory nerve and infects the skin to produce the vesicles of herpes zoster.*

Often organs that yield viruses on explant culture do not yield infectious virus if ground up and inoculated directly into appropriate cultures or experimental animals.

Then there are a number of situations in which the features in common are the absence of disease but persistent or recurrent excretion of infectious virus, which may be shed or transferred by blood transfusion, so that infection of susceptible contacts can occur in the absence of apparent disease in the transmitters. It will be noted that herpes simplex but not varicella-zoster could be correctly classified as a chronic infection as well as being classed as a latent infection, for the reasons already given, just as in some situations infection with EB virus (see below) is correctly described as latent. The two classes overlap.

Cytomegalovirus and EB Virus

Generalized cytomegalovirus infection is occasionally seen in hospitals, and may result from two alternative sources of infection. In

TABLE 8-2

Chronic Infections: Virus Always Demonstrable, Often No Disease

EXAMPLES	HOST	SITES OF INFECTION	VIRUS SHEDDING	ANTIBODIES	DISEASE
Various viruses: (SV40, reovirus, adenovirus, herpetovirus, paramyxovirus)	Monkeys	Kidneys	Variable	+	None recognized; viruses found in explanted cells
Cytomegalovirus	Many species	Salivary glands	Saliva	+	None recognized
Cytomegalovirus	Man	Salivary glands Kidney Circulating leukocytes	Saliva Urine	+	Rarely; activated by immuno-suppression; sometimes transmitted by transfusion of fresh blood
EB virus	Man	Lymphoid tissue Circulating leukocytes	Nil	+	Usually none; association with Burkitt's lymphoma and NPC
Hepatitis B	Man	Liver	Serum	+	Rarely chronic hepatitis; often transmitted by transfer of serum
Rubella	Man	Widespread	Urine Respiratory	+	Rubella syndrome → death or recovery; teratogenic effects

individuals with lymphoreticular disease, or undergoing prolonged immunosuppression, generalized cytomegalovirus infection may represent activation of an endogenous latent infection. More commonly it is an exogenous infection resulting from the transmission of cytomegalovirus during the transfusion of large volumes of fresh blood. Both situations reflect the widespread occurrence of healthy carriers of cytomegalovirus in the general population.

The EB virus, the causative agent of infectious mononucleosis, is another ubiquitous herpetovirus carried asymptomatically by large numbers of normal people. Two types of human cancer, Burkitt lymphoma and a nasopharyngeal carcinoma commonly found in the Chinese, consistently carry several genome equivalents of EB virus DNA and synthesize EBV antigens but no infectious virus. Lymphoblastoid cells from such patients, and indeed from many normal people, carry the viral genome in a repressed state which can be activated by drugs.

Hepatitis B

Perhaps the best known chronic human infection is hepatitis B, which is acquired by the transmission of virus in serum from a chronic carrier (Chapter 23). What is known of the etiological agent was described in Chapter 1, in which attention was drawn to the common occurrence of a large amount of a virus-associated antigen ("Australia" antigen or HB-Ag) in the serum. Following clinical or subclinical infection the antigen disappears quite rapidly from the serum in most cases, but in the remainder it may persist for several years. The concentration of antigen in the serum of carriers may be extremely high (10^{12} to 10^{13} HB-Ag particles per milliliter), and although concentrations of infectious virions are very much lower (about 10^6 infectious doses per milliliter), such sera are still highly infectious for recipients of transfused blood, heroin addicts, and attendants in renal dialysis units.

Sera from some acute or chronically ill individuals, but usually not from chronic asymptomatic carriers, contain antigen–antibody complexes which may play a part in the acute disease and in some forms of chronic hepatic disease, as well as in the glomerulonephritis and polyarteritis nodosa that are occasionally seen in HB-Ag carriers.

Rubella Syndrome

Babies born of mothers infected with rubella virus during the first trimester of pregnancy may suffer from congenital defects (see Chapters 6 and 23) and sometimes exhibit a wide range of signs and symptoms known collectively as the "rubella syndrome." Virus can be isolated from virtually any organ of such babies, although only a minority of cells in any given organ is infected. Maternal antibody may inhibit the spread of virus but fails to eliminate the virus-producing cells, which continue to divide at a slower rate than normal. Usually the resulting clones of infected cells are lost a few months after birth, but occasionally in special sites they may persist longer. There is no immunological tolerance; large amounts of IgM are synthesized by the fetus and continue to be made after birth.

Chronic Infections with Late Immunopathological Disease in Animals

Finally, there are a number of chronic infections of animals in which severe disease of either immunopathological or neoplastic nature appears as a late development (Table 8-3). It is instructive to examine the pathogenesis of representative examples of these diseases since there may well be human counterparts yet to be discovered.

Lymphocytic Choriomeningitis in Mice (LCM). LCM virus is transmitted congenitally to every mouse in an infected colony. The mice are born normal and appear to be normal for most of their lives, although they have persistent viremia and viruria; almost every cell in the mouse is infected and remains so throughout the life of the animal. Circulating free antibody cannot be detected, but immunological tolerance is not complete, and the virus circulates in the bloodstream as virus-IgG-complement complexes, which are infectious. Late in life such mice may exhibit "late disease" due to the deposition of viral antigen—antibody complexes in the glomeruli. Such immunopathology (glomerulonephritis and sometimes arteritis) characterizes a number of other viral diseases of animals, including "lactic dehydrogenase-elevating virus" (LDV) infection of mice, Aleutian disease of mink, and infectious anemia of horses (see Table 8-3).

Chronic Infections with Late Neoplasia

Leukemias and lymphomas induced by the RNA tumor viruses fall into this category, which is unique in animal virology in that these viruses are maintained as DNA copies of their genomes, integrated with the cellular genome and inherited genetically in all individuals of certain species, including mice and chickens (see Chapter 9). In certain strains of rodents and chickens and under conditions that vary according to the host animal, the virus is activated and infectious noncy-

TABLE 8-3

Chronic Infections: Virus Always Demonstrable, Late Immunopathological or Neoplastic Disease, Nonneutralizing Antibodies

EXAMPLES	HOST	SITES OF INFECTION	NONNEUTRALIZING ANTIBODIES	DISEASE
Lymphocytic choriomeningitis	Mouse	Widespread, including lymphoid tissue	+	Glomerulonephritis
Lactic dehydrogenase virus	Mouse	Macrophages	+	Immune complexes in glomeruli, but no disease
Aleutian disease	Mink	Macrophages	++	Arteritis, glomerulonephritis, hyperglobulinemia
Equine infectious anemia	Horse	Macrophages	+	Anemia, vasculitis, glomerulonephritis
Murine leukemia	Mouse	Widespread	+	Immune complexes in glomeruli; leukemia occasionally
Avian leukosis	Chicken	Widespread	+	Leukemia occasionally; sarcoma rarely

tocidal virions are then produced. These may infect neighboring cells, or if injected into young animals of the same species may infect them also. Spontaneous horizontal spread may sometimes occur.

Murine Leukemia. The natural history of this infection depends upon the genetic background of the murine hosts. In a "high leukemia" strain of mice no free virus is found in early embryos, but it appears spontaneously in late embryonic life, and thereafter persists for life. Free circulating antibody cannot be demonstrated, but the deposition of immune complexes in the kidneys late in the life of some of the animals shows that some antibody is formed. Thus "high leukemia" strains of mice have a persistent infection with murine leukemia virus, acquired by activation of the integrated viral genome and resulting in late, mild "immune" disease. In addition, most of them get leukemia within 10 months of birth.

"Low leukemia" strains of mice carry and transmit the integrated DNA provirus, but it is usually not expressed at all during the life of the mouse, i.e., the "infection" (if it can be called that) is latent rather than chronic and is transmitted genetically to succeeding generations.

SLOW INFECTIONS

Three groups of diseases are listed in this category (Table 8-4), which includes the original "slow-virus infections" of Sigurdsson. *Group A* comprises the slowly progressive infections of sheep: visna, maedi, and progressive pneumonia, caused by serologically related but nononcogenic retroviruses. We shall not discuss these further. *Group B* consists of four obscure infections of the central nervous system (scrapie, mink encephalopathy, kuru, and Creutzfeldt–Jakob disease) that have been designated the subacute spongiform viral encephalopathies. *Group C* consists of two unrelated diseases of the human CNS, subacute sclerosing panencephalitis and progressive multifocal leukoencephalopathy.

Although in many respects these diseases resemble some of those described in the previous section as "chronic infections," they share a feature not found in the latter, namely that in all cases disease progresses slowly and inexorably to death.

Subacute Spongiform Viral Encephalopathies

It has been suggested that one of the many synonyms of Creutzfeldt–Jakob disease, namely subacute spongiform encephalopathy,

TABLE 8-4

Slow Infections: Long Incubation Period, Slowly Progressive Fatal Disease

GROUP	EXAMPLES	HOST	SITES OF INFECTION	ANTIBODIES	DISEASE
A. Nonneoplastic retrovirus	Visna Maedi Progressive pneumonia	Sheep	Brain Lung Lung	+	Slowly progressive, recurrent viremia, immunopathological component
B. Subacute spongiform encephalopathy	Scrapie Mink encephalopathy Kuru Creutzfeldt–Jakob disease	Sheep Mink Man Man	CNS and lymphoid tissue	−	Slowly progressive encepha-lopathy
C. Other human CNS diseases	Measles: subacute sclerosing panencephalitis	Man	CNS	++	Acute measles, recovery, then slowly progressive encephalitis years later
	Papovavirus: progressive multifocal leukoencephalopathy	Man	CNS	+	Progressive encephalopathy, following immunosuppression

TABLE 8-5

Natural and Experimental Host Range of Subacute Spongiform Viral Encephalopathies[a,b]

HOST	KURU	CREUTZFELDT–JAKOB	SCRAPIE	MINK ENCEPHALOPATHY
Man	+	+	NT	NT
Chimpanzee	+	+	−	−
Spider Monkey	+	+	NT	NT
Sheep	−	−	+	−
Goat	−	−	+	+
Mink	−	−	+	+
Mouse	−	−	+	+
Incubation period				
Natural disease	5–10 years	Unknown	3–5 years	8–12 months
Experimental disease	14–38 months	12–14 months	4–6 months	4–8 months

[a] Partial list from C. J. Gibbs and D. C. Gajdusek, Immunological disorders of the nervous system. *Res. Publ. Ass. Res. Nerv. Men. Dis.* **49**, 383 (1971).

[b] +, Clinical disease and confirmatory histopathological lesions; −, no disease or histopathological lesions; NT, not tested.

should be used as a generic name for four diseases that have strikingly similar clinicopathological features and causative agents, namely, scrapie of sheep and goats, mink encephalopathy, and kuru and Creutzfeldt–Jakob disease in man (see Table 8-5).

Scrapie. Scrapie is a natural infection of sheep in which transmission occurs with difficulty by contact, more commonly vertically from ewe to lamb. Infection was widely disseminated in Britain by the inoculation of sheep with louping-ill vaccine that was contaminated with the scrapie agent. The incubation period is very long, up to 3 years, and once symptoms have appeared the disease progresses slowly but inevitably to paralysis and death. The basic neurocytological lesion, as in all the subacute spongiform encephalopathies, is a progressive vacuolation in the dendritic and axonal processes of the neurons, and to a lesser extent, in astrocytes and oligodendrocytes, an extensive astroglial hypertrophy and proliferation, and finally a spongiform change in the gray matter.

Experimental studies in sheep and mice reveal that scrapie behaves as a typical infectious disease, and filtration shows that the causative agent is the size of a small virus. Unusual features are the absolute absence of any sign of an immune response and the lack of sensitivity to either interferon or measures that depress the immune response.

Tests on the inactivation of infectivity by a variety of physical and chemical treatments show that the agent has a high degree of resistance unlike that of conventional viruses. To the extent that they have been tested, these unusual biological and physicochemical properties are shared by the agents of the other three subacute spongiform encephalopathies. It has been suggested that scrapie virus may be a small molecule of free nucleic acid protected by close association with cellular membranes.

Kuru and Creutzfeldt–Jakob Disease. Kuru is a disease confined to a group of 50,000 highland New Guineans; it is thought to have been spread by ritualistic cannibalism and is now disappearing as that habit has disappeared. Its significance lies in the fact that it was the first human degenerative disease of the central nervous system to be shown to have a viral etiology, a finding whose importance has now been amplified by the demonstration that the rare but cosmopolitan presenile dementia of the Creutzfeldt–Jakob type can also be transmitted to chimpanzees. Insofar as they have been studied, the causative agents and the histopathology of these two human diseases closely resemble those of scrapie. They are discussed at greater length in Chapter 23.

Subacute Sclerosing Panencephalitis and Progressive Multifocal Leukoencephalopathy

We conclude with two other infections of the human brain, but they are caused by typical viruses quite unlike the bizarre agents just discussed. In many respects, both of these might more properly be designated chronic infections but they are tentatively placed alongside the slow infections of the CNS to emphasize common clinical and pathological features of this ever-growing assemblage of diseases of man.

Subacute Sclerosing Panencephalitis (SSPE). Measles was long thought to be an acute self-limited disease never associated with viral persistence or late complications. Most cases of this universal human ailment probably conform to this description, but recently a long-recognized but rare chronic disease of the CNS, subacute sclerosing panencephalitis (SSPE) was shown to be a late sequel to measles. SSPE is characterized by the slow development of neuronal degeneration, and is invariably fatal. The interval between recovery from measles and the onset of frank SSPE is always several years. Although brain cells from affected individuals show serological and electron microscopic evidence of measles virus infection, infectious measles virus cannot be recovered directly, but only by cocultivation of brain tissue with indicator cells, i.e., the virus in the brain is genetically complete but normal multiplication is suppressed. Measles virus has also been re-covered by cocultivation from lymph node biopsies of early cases of SSPE.

The levels of antibody to measles virus in the serum in cases of SSPE are extremely high, and antibody is also found regularly in the cere-brospinal fluid, the latter finding being pathognomonic of SSPE. An-tibody fails to arrest the progress of the disease. The critical question, yet to be answered, is why this slow complication of measles develops only in rare individuals (see Chapter 21).

Progressive Multifocal Leukoencephalopathy (PML). This rare sub-acute progressive demyelinating disease has only been recognized in individuals whose immunological responsiveness has been severely depressed by malignant disease or immunosuppressive therapy. Elec-tron microscopic study of brain biopsies reveals intranuclear ac-cumulations of viral particles morphologically indistinguishable from polyoma virus and these have now been cultured. Some isolates show strong serological cross-reactivity with SV40, whereas others show only slight cross-reactivity with this virus. A third serotype has been recovered from the genitourinary tract of patients with transplanted kidneys.

Serological surveys show antibody to the SV40-related virus in a substantial proportion of adults. PML, which is very rare, may repre-sent the activation of a persistent infection in immunosuppressed individuals.

FACTORS IN THE PATHOGENESIS OF PERSISTENT INFECTIONS

It is clear from the foregoing account that a wide variety of different conditions is included under the umbrella of persistent infections. However it is worthwhile enquiring whether there are any common mechanisms whereby the viruses that cause these infections bypass the host defenses that ensure the elimination of virus in the acute viral infections. Several mechanisms appear to be involved, but in the cur-rent state of knowledge we can only speculate about why some of these factors play a dominant role in certain persistent infections but not in others. They include factors related (a) to the virus and (b) to the host defense.

Unique Properties of the Virus

Nonimmunogenic "Viroids." The unknown agents that cause the subacute spongiform encephalopathies seem to be completely nonim-munogenic, they fail to induce interferon, and they are not demonstrably susceptible to interferon action. There appears to be no mechanism whereby the host can control the multiplication and pathological effects of these agents.

Integrated Genomes. RNA tumor viruses persist as DNA copies of their genomes integrated into cellular genomes. Until activated, the viral genome is only partially expressed, if at all, and it can be regarded as part of the cellular genetic material. When activated, the RNA tumor viruses multiply in lymphoid tissue and induce the production of nonneutralizing rather than neutralizing antibodies.

Among DNA viruses, the papovavirus of PML may well persist as an integrated genome until it is activated. Among the herpetoviruses, those associated with or thought to be associated with malignancies probably persist as integrated, perhaps defective, genomes. The state in which herpes simplex, varicella-zoster, cytomegalovirus, and EB virus persist intracellularly during prolonged latent infections is unknown, but integration (of complete genomes) is a likely possibility.

Temperature-sensitive (ts) Mutants. The strains of virus responsible for several of the persistent infections of animals or cultured cells with "orthodox" viruses have turned out to be ts mutants, although most ts mutants do not cause persistent infections. The significance of this discovery has yet to be determined.

Inadequate Host Defenses

Growth in Protected Sites. Herpes simplex virus and varicella virus avoid immune elimination by remaining within cells of the nervous system, in an occult form in the ganglion cells during the intervals between disease episodes, and within the Schwann cells of the nerve sheaths prior to acute recurrent episodes of disease. Likewise, other herpetoviruses such as human cytomegalovirus and EB virus appear to bypass immune elimination, but in this instance they persist in lymphocytes. Other viruses grow in cells on epithelial surfaces, e.g., kidney tubules, salivary gland, or mammary gland, and are persistently shed in the appropriate secretions and excretions. Most such viruses are not acutely cytopathogenic, and perhaps because they are released on the lumenal borders of cells they do not provoke an immunological inflammatory reaction, hence the cells are not destroyed by T lymphocytes or macrophages. Secretory IgA, which does have access to the infected cells, does not cause complement activation and therefore fails to induce complement-mediated cytolysis or an inflammatory response.

Growth in Macrophages. As Table 8-3 indicates, in many chronic infections the virus appears to grow mainly in lymphoid tissue, especially in macrophages. This may have two effects relevant to persistence: (a) modification of the antibody response, and (b) impairment of the phagocytic and cytotoxic potential of the reticuloendothelial system.

Nonneutralizing Antibodies. Viruses that cause persistent plasma-associated viremia usually multiply in lymphoid tissue and macrophages (see above) and they characteristically induce production of nonneutralizing antibodies. These antibodies combine with viral antigens and virions in the serum to form "immune complexes" which may (a) produce "immune complex disease" and (b) block immune cytolysis of virus-infected target cells by T lymphocytes or complement-fixing antibodies.

Tolerance. Many persistent infections are associated with a very weak antibody response, especially in congenitally infected animals. Immunological tolerance is rarely complete, but there is a severe degree of specific hyporeactivity in conditions like congenital LCM and retrovirus infections. Tolerance to a viral antigen may be genetically determined, and the immune response to several specific antigens has been shown to be under genetic control.

Another kind of "tolerance," about which very little information is available, is what might be called "interferon-tolerance." Little or no interferon is produced in mice congenitally infected with LCM, LDV, or murine leukemia virus, but in each case the causative virus is sensitive to interferon. Interferon "tolerance" appears to be specific to the virus involved, possibly involving a recognition mechanism at the mRNA level. Experiments with varicella cast doubt on the suggestion that low interferon production and low sensitivity are invariably characteristics of viruses that produce persistent infections, but this would hardly be expected in such a diverse group.

Defective Cell-Mediated Immunity. Persistent infections could be caused by partial suppression of the host's CMI response, as a result of any one or combination of several factors: immunodepression by the causative virus, immunological tolerance, the presence of "blocking" antibodies or virus-antibody complexes, failure of immune lymphocytes to reach target cells, or inadequate expression of viral antigens on the surface of the target cell. These factors are probably important in persistent infections such as visna, SSPE, PML and in those caused by herpetoviruses. Finally, we may again note that many persistent viruses multiply extensively in macrophages and lymphocytes, and thus affect several parameters of the immune response. Indeed, depressed CMI may diminish the rate of destruction of infected cells and thereby prolong the release of viral antigens. The resulting protracted immunogenic stimulus would explain the very high antibody levels found in SSPE and Aleutian disease of mink, for example, which may produce a cascade effect by blocking the already inadequate T cell defenses.

SUMMARY

Persistent viral infections, which are more common than previously recognized, may be classified as follows:

(a) *latent infections,* in which virus persists for long periods after an acute infection but is only demonstrable during intermittent attacks of recurrent endogenous disease. Examples in man include herpes simplex and herpes zoster.

(b) *chronic infections,* in which demonstrable virus persists for years after an acute (clinical or subclinical) infection. Examples in man include cytomegalovirus and EB virus infections, and hepatitis B. Examples in animals include two groups: chronic infections with late immunopathologic disease (like lymphocytic choriomeningitis in mice) or chronic infections with late neoplasia (e.g., the leukemias of rodents).

(c) *slow infections,* which produce slowly progressive diseases after a very long incubation period. Three subgroups are recognized: (i) slow retrovirus infections of sheep without neoplasia, (ii) the subacute spongiform viral encephalopathies, including scrapie in sheep and kuru and Creutzfeldt–Jakob disease in man, and (iii) subacute sclerosing panencephalitis (following measles) and progressive multifocal leukoencephalopathy (associated with a human papovavirus).

Factors affecting the pathogenesis of these diseases include certain unique properties of some of the viruses concerned, and inadequate host defenses associated with immunological and other responses.

Study of cultured cells has revealed three types of persistent infections.

(a) In *steady-state infections,* which are found most commonly with RNA viruses that mature by budding, cellular and viral multiplication may proceed concurrently for long periods; eventually some of these viruses produce malignancy.

(b) *Carrier cultures* are laboratory artifacts which may arise during serial passage of infected cultures; only a few cells in the culture yield virus, the rest being protected by antibody or interferon.

(c) Cells transformed by oncogenic DNA viruses carry the defective or complete viral genome integrated with the DNA of the cell. Mouse leukemia and avian leukosis viruses are RNA viruses whose DNA "provirus" is carried in the genome in an integrated form, which can under certain conditions be derepressed to yield infectious virus.

FURTHER READING

Brody, J. A., Henle, W., and Koprowski, H. (1967). Chronic infectious neuropathic agents (CHINA) and other slow virus infections. *Curr. Top. Microbiol. Immunol.* **40,** 1.

Choppin, P. W., Klenk, H.-D., Compans, R. W., and Caliguiri, L. A. (1971). The parainfluenza virus SV5 and its relationship to the cell membrane. *Perspect. Virol.* **7,** 127.

Cole, G. A., and Nathanson, N. (1974). Lymphocytic choriomeningitis: Pathogenesis. *Progr. Med. Virol.,* **18,** 94.

Fenner, F., McAuslan, B. R., Mims, C. A., Sambrook, J. F., and White, D. O. (1974). "The Biology of Animal Viruses," 2nd ed., pp. 452–478. Academic Press, New York.

Fuccillo, D. A., Kurent, J. E., and Sever, J. L. (1974). Slow virus diseases. *Annu. Rev. Microbiol.* **28,** 231.

Gajdusek, D. C., and Gibbs, C. J. (1973). Subacute and chronic diseases caused by atypical infections with unconventional viruses in aberrant hosts. *Perspect. Virol.* **8,** 279.

Gajdusek, D. C., Gibbs, C. J., Jr., and Alpers, M. (1965). "Slow Latent and Temperate Virus Infections," Nat. Inst. Neurol. Dis. Blindness, Monogr. No. 2. US Govt. Printing Office, Washington, D.C.

Hotchin, J., ed. (1974). "Slow Virus Diseases," *Progr. Med. Virol.* No. 18. Karger, Basel.

Kalter, S. S., and Heberling, R. L. (1971). Comparative virology of primates. *Bacteriol. Rev.* **35,** 310.

Oldstone, M. B. A. (1975). Virus neutralization and virus-induced immune complex disease. *Progr. Med. Virol.* **19,** 84.

Porter, D. D. (1971). A quantitative view of the slow virus landscape. *Progr. Med. Virol.* **13,** 339.

Rawls, W. E. (1974). Viral persistence in congenital rubella. *Progr. Med. Virol.* **18,** 273.

Sever, J. L., and Zeman, W., eds. (1968). Conference on measles virus and subacute sclerosing panencephalitis. *Neurology* **18,** Part 2.

Stevens, J. G., and Cook, M. L. (1973). Latent herpes simplex virus in sensory ganglia. *Perspect. Virol.* **8,** 171.

Ter Meulen, V., Katz, M., and Müller, D. (1972). Subacute sclerosing panencephalitis: A review. *Curr. Top. Microbiol. Immunol.* **57,** 1.

Walker, D. L. (1968). Persistent viral infection in cell cultures. In "Medical and Applied Virology" (M. Sanders and E. H. Lennette, eds.), p. 99. Green, St. Louis, Missouri.

Zeman, W., and Lennette, E. H., eds. (1974). "Slow Virus Diseases." Williams & Wilkins, Baltimore, Maryland.

Zu Rhein, G. M. (1969). Association of papova-virions with a human demyelinating disease (progressive multifocal leukoencephalopathy). *Progr. Med. Virol.* **11,** 185.

CHAPTER 9

Oncogenic Viruses

INTRODUCTION

Some herpetoviruses produce cancer under natural conditions in animals, while several other DNA viruses of the Papovaviridae and Adenoviridae families produce malignant tumors after injection into baby rodents (see Table 9-1). RNA viruses of the family Retroviridae (oncoviruses) cause leukemia or leukosis in many species of animal, and have been implicated in naturally occurring sarcomas of birds and rodents, and in mammary tumors of mice. However, in spite of these precedents there is as yet no definitive evidence for any human cancer virus, though herpetoviruses have been associated with certain types of malignancy, and oncoviruses have been recovered from breast cancer and leukemia. Positive, rather than suggestive, proof of a causal relationship may be very difficult to obtain. For obvious reasons, study of oncogenic viruses is currently a very active field of research, and the subject is highly complex. Our objective here is merely to give sufficient background information to provide an understanding of the processes involved.

ONCOGENIC DNA VIRUSES

DNA viruses of the Papovaviridae and Adenoviridae families have been shown to produce malignant tumors in baby rodents under laboratory conditions, while certain herpetoviruses cause malignancy in birds and animals under natural conditions and are under suspicion as the causative agents of lymphoid tumors and cancers in man.

Papovaviridae

Almost all members of this family can cause tumors. Most of those occurring naturally are benign, e.g., warts in man and in many other

162

162

animals; others, e.g., rabbit papilloma, are initially benign but may become malignant. Two papovaviruses, polyoma and SV40, which rarely if ever cause malignant tumors in nature, regularly produce them when inoculated into newborn rodents.

Polyoma Virus. The discovery of this virus in 1956 initiated the resurgence of interest in viral carcinogenesis, a hypothesis that had been in disrepute for many years since the enthusiasm generated by Rous' discovery of the virus of Rous sarcoma in 1911 had gradually waned. Under natural conditions polyoma virus causes no demonstrable disease, being spread as an inapparent infection in colonies of laboratory or wild house mice. But, when artificially inoculated into infant mice or other rodents, it produces a wide variety of histologically diverse tumors, in various parts of the body, hence "poly-oma."

Although some aspects of neoplastic change are impossible to study in any system less complex than the whole animal, it soon became apparent that advances in the understanding of the molecular basis of neoplasia, whether induced by viruses or by other agents, must be sought in simpler systems. It has been found that most viruses capable of inducing tumors in experimental animals can also "transform" cells *in vitro* (the properties of such transformed cells were described in Chapter 5; see Plate 5-2 and Table 5-1). Polyoma virus was the first oncogenic DNA virus for which a workable *in vitro* system of cellular transformation was developed, using primary mouse or hamster cells, or certain continuous lines (hamster, BHK21; mouse, 3T3). Transformation by this virus is a rare event (depending somewhat on the cell type involved), for even with inputs of 1000 pfu per cell only a minority of the cells become transformed. Moreover, by no means do all the cells transformed *in vitro* produce malignant tumors when inoculated into syngeneic animals. Though the primary virus-induced change can produce a cancerous cell, it more commonly produces a premalignant cell which, as a result of subsequent mutations in the proliferating clone, becomes progressively more malignant. The most common result of all is "abortive transformation," in which transiently transformed cells resume their normal growth characteristics after a few generations.

From the time of the earliest demonstrations of oncogenesis and transformation by polyoma virus an apparent anomaly was observed. Neither tumor cells induced by viruses *in vivo*, nor cells transformed in cell culture, produced infectious virus, nor could infectious viral DNA be extracted from them. Moreover, all attempts to induce the production of infectious virus by treatments effective in inducing lysogenic bacteria to yield bacteriophages were unsuccessful. Yet the malignant cells always contained two types of viral antigen, quite distinct from the structural proteins of the virion. These became known as the T (for tumor), and TSTA (tumor-specific transplantation) antigens, respectively. The *T antigen(s)* are identical with protein(s) produced in the early stages of cytocidal infections. Immunofluorescence studies show that they occur in the nucleus (and sometimes in the cytoplasm) of every transformed cell. *Transplantation antigens* occur in the plasma membrane of transformed cells (and indeed any malignant cell) and can be recognized by transplant rejection tests, and in some cases by membrane immunofluorescence. The continued synthesis of these proteins through an indefinite number of cell generations was strongly suggestive of the permanent retention of viral genetic material within the cell. This was unequivocally demonstrated by molecular hybridization tests. Two techniques were used. In the first, transformed cells were shown to contain virus-specific mRNA, by hybridization of radioactively labeled RNA from virus-transformed cells with viral DNA extracted from purified virions. In the second type of experiment, labeled mRNA transcribed *in vitro* from viral DNA was shown to hybridize with the DNA of transformed cells. Thus the malignant cell carries several molecules of viral DNA (not necessarily complete) integrated into the cell's chromosomes. This viral DNA is transcribed into mRNA specifying at least the viral T and transplantation antigens, but not all the proteins required for complete viral synthesis.

Simian Virus 40 (SV40). This papovavirus was discovered in apparently normal cultures of monkey kidney cells during the production of poliovirus vaccine. Tests on baby hamsters showed that it was oncogenic, and subsequent investigations have revealed a picture very similar to that already described for polyoma virus, which it resembles closely in its physical and chemical properties. As in the case of polyoma virus, the complete viral genome is not required for carcinogenesis. Indeed, the probability of transformation is increased by the use of UV-irradiated SV40, or defective hybrids of SV40 and adenovirus. Yet, the complete genome of SV40 is often present in virus-transformed cells and in virus-induced tumors, as demonstrated by the induction of infectious virus production, either spontaneously (on rare occasions), or as a result of cell fusion experiments. In the latter, monkey cells, which support cytocidal infection by SV40, are fused with SV40-transformed cells by treatment with UV-irradiated paramyxovirus; induction of infectious SV40 occurs in about 10% of the resulting heterokaryons. Molecular hybridization experiments reveal that transformed cell lines contain several "viral DNA equivalents" per cell. The T and transplantation antigens coded by SV40 are virus-specific; they show no cross-reactivity with the corresponding antigens of polyoma virus or adenovirus.

SV40 has virtually replaced polyoma virus as the model for molecular biological studies. Very sophisticated technology has been applied to the characterization of the small circular DNA genome of this virus. Temperature-sensitive and host-cell dependent mutants, as well as a comprehensive range of adenovirus-SV40 hybrids, have been employed in an attempt to pinpoint the gene(s) responsible for cancer. "Restriction endonucleases" from bacteria have been used to cleave the DNA at particular sites yielding "gene-sized" fragments whose function can be assessed *in vitro,* and whose situation in the genome can be determined by direct visualization in the electron microscope of artificial hybrids with SV40 DNA ("heteroduplex mapping"). A complete map of the SV40 genome is now available and the functions of certain genes have been identified, as well as the sites of initiation and termination of DNA replication and transcription. Nevertheless, the exact nature of the "early" genes that determine cell transformation remains elusive.

Adenoviridae

In 1962 Trentin showed that human adenovirus type 12 produced sarcomas when inoculated into newborn hamsters; subsequent work has shown that several other adenoviruses, of human and animal origin, are also oncogenic, although the capacity of different adenoviruses to produce tumors varies considerably. As judged by the latent period and the minimum effective dose, human types 12, 18, and 31 have the highest oncogenic potential. They are regularly oncogenic for a number of species of newborn, but not adult, rodents. In spite of intensive studies by immunological and molecular hybridization techniques, no evidence of oncogenicity for man has emerged.

No infectious adenovirus can be extracted from the tumors induced in hamsters by oncogenic adenoviruses, nor from cells transformed to the malignant state *in vitro,* even after subjection to various chemical and physical treatments (including cell fusion) that might be expected to induce an integrated viral genome to multiply. However, the cells always contain T and transplantation antigens. Although only a tiny percentage of DNA from the malignant cells will anneal specifically with DNA extracted from purified adenovirus, the viral DNA must be preferentially transcribed since up to 5% of the mRNA recovered from the polyribosomes of transformed cells is virus-specific. Competitive hybridization experiments show that this viral mRNA is

TABLE 9-1
Some Characteristics of Malignant Disease Induced by Viruses

VIRUS	NATURAL MALIGNANT DISEASE		ARTIFICIAL MALIGNANT DISEASE			TRANSFORMATION OF CULTURED CELLS	
	HOST	TUMOR	HOST	TUMOR	VIRUS PRODUCTION	SPECIES	VIRUS PRODUCTION
Adenovirus	—		Baby rodents	Sarcoma	—	Rodent	—
Papovavirus							
Polyoma	—		Baby rodents	Several	—[a,b]	Rodent	—
SV40	—		Baby rodents	Sarcoma	—	Rodent Man	—[a,b]
Rabbit papilloma	Rabbit	Papilloma → carcinoma	Rabbit	Papilloma → carcinoma	—	—	
Herpetovirus							
Marek's virus	Chicken	Lymphomatosis	Chicken	Lymphoma	—	—	
Lucké virus	Frog	Adenocarcinoma	Tadpole	Adenocarcinoma	—	—	
Herpesvirus saimiri	Monkey	Lymphoma	Primates Rabbits	Lymphoma	—[b]	—	
EB virus	? Man	Lymphoma Nasopharyngeal carcinoma	—			Man	—
Herpes simplex 2	? Man	Cervical carcinoma	—			Man	—

Oncovirus								
Avian	Chicken	Leukosis	+	Chicken	Leukosis	+	Chick[a]	+
		Sarcoma	+[c]	Chicken, Rodent, Monkey	Sarcoma	+[c]	Chick, Rodent, Man	+[c]
Murine	Mouse	Leukemia	+	Rodents	Leukemia	+	Mouse[d]	+
		Sarcoma	+[c]	Rodents	Sarcoma	+[c]	Rodents, Man	−[c]
Feline	Cat	Leukemia	+	Kitten	Leukemia	+	Cat[d]	+
		Sarcoma	+	Dog, Rabbit	Sarcoma	+	Man	
Mammary tumor virus	Mouse	Mammary carcinoma	+	Mouse	Carcinoma	+	—	−

[a] Occasional cells spontaneously yield virus.

[b] Virus production can be induced by cocultivation.

[c] But some sarcoma virus strains are defective and cells of foreign species are nonpermissive.

[d] The "transformed" cells appear normal.

transcribed *in vivo* from only a short sequence of "early" viral genes, and recent data indicates that the adenoviral DNA molecules integrated with the cell's chromosomes are incomplete.

Herpetoviridae

Since the first edition of this book was published, suspicions have grown that several different herpetoviruses may be associated with malignancies in man, as well as in experimental animals (Table 9-1). We shall say little about the situation in lower animals, except that (a) the evidence for herpetovirus oncogenesis is strong, and (b) oncogenesis seems to be associated with incomplete expression of the functions of the viral genome. *Herpesvirus saimiri* represents perhaps the closest analogy to a human oncogenic herpetovirus; it was isolated from a malignant lymphoma in a monkey and causes lymphomas or reticulum cell sarcomas on inoculation into various primates. Marek's disease of fowls is a particularly interesting model because, not only has a herpetovirus clearly been proven to be the etiological agent of what is equally clearly a typical malignant lymphoma, but the disease can be very effectively controlled in flocks of chickens by immunization with an attenuated live-virus vaccine.

Epstein-Barr Virus (EBV). One kind of cancer accounts for about one-half of all malignancy in children in East Africa, namely Burkitt's lymphoma, named after Dennis Burkitt, who first recognized the abnormally high incidence of the tumors and suggested that their peculiar geographical distribution might be the result of infection with an oncogenic virus transmitted by an insect. This tumor is also common in Papua-New Guinea, and cases of Burkitt's lymphoma have now been reported at low frequency in virtually every area of the world in which an adequate search has been made. This finding, together with the virological evidence discussed later, persuaded Burkitt to postulate that the lymphoma may be caused by a ubiquitous virus in combination with an environmental factor which is restricted geographically, and which may be vector-associated.

In 1964 Epstein and his colleagues recovered a herpetovirus (now called EBV) from cultured tumor cells. There are multiple copies of EBV DNA integrated into the chromosomes of all lines of EB tumor cells. Some of these yield active virus spontaneously; the others can be activated to yield viral particles by chemical treatment (e.g., 5-bromodeoxyuridine). A further surprise was the demonstration by Henle and Henle that EBV is in fact the etiological agent of infectious mononucleosis.

The central question remains whether EBV causes Burkitt's lymphoma. There is no doubt that (a) EBV causes infectious mononucleosis which, although a self-limiting disease, in other respects resembles quite closely the early stages of lymphoma; (b) EBV causes transformation of lymphocytes *in vitro;* (c) all cases of Burkitt's lymphoma show serological evidence of EBV infection; and (d) EBV antigens are detectable in all Burkitt tumors and virions can be recovered following cultivation of the cells *in vitro.* None of this circumstantial evidence is conclusive proof of an etiological (rather than a fortuitous) association, but we can hope that the question will be resolved when the results of a large prospective epidemiological study conducted by WHO become available in a few years' time.

Serological associations have also been found between EBV and nasopharyngeal cancer seen in the Chinese population in Southeast Asia and between EBV and Hodgkin's disease; furthermore, EBV DNA has been demonstrated regularly in biopsies of nasopharyngeal carcinoma. Whether EBV is merely a passenger in these tumors or whether there is a causal relationship is unknown.

Herpes Simplex Type 2. A strong association between the possession of antibodies to herpes simplex virus (HSV) type 2 and cervical carcinoma of women has now been widely reported. Cervical smears taken from women with carcinoma contain cells in which herpes simplex specific antigens can be demonstrated, and, more significantly, herpes simplex DNA is detectable in the chromosomes of the malignant cells. In addition, UV-inactivated herpes simplex virus type 2 will transform cultured cells. The circumstantial evidence indicates no more than that herpes simplex virus type 2 infection is often a covariable of cervical cancer; both diseases are correlated with a high frequency of sexual intercourse and especially with sexual promiscuity. Much more work is required before we can decide whether there is a direct link between the virus and the neoplasm.

In summary, it is clear that in a wide variety of animals, herpetoviruses enter into a permanent association with lymphocytes which results in little or no virus production but a great proliferation of the cells. In some cases (for example, *Herpesvirus saimiri* in marmoset monkeys, Marek's disease virus in chickens) we know the relationship is oncogenic; in others (EBV and HSV in humans) we cannot be sure. In all systems, the herpetovirus genome seems to enter into a stable relationship with that of the cell, and in every case the cells die when virus production is induced. The balance of the system is delicate and fascinating; it is now being intensively investigated by molecular biologists.

ONCOGENIC RNA VIRUSES

Only a small number of RNA viruses, all belonging to one family, have been unequivocally associated with neoplastic disease, but their importance is heightened by the fact that they are oncogenic under natural conditions. These are the oncoviruses, the largest subfamily of the family Retroviridae (see Chapter 1), of which three subgroups have been extensively investigated: the avian leukosis viruses, the murine leukemia (and sarcoma) viruses, and the mammary tumor virus of mice. Oncoviruses are associated with leukemia in several other animal species.

Our knowledge of the oncoviruses has only recently emerged from a chaotic state which resulted from the lack of any method of cloning them, and from the unavoidable need to maintain stocks of virus by serial passage in animals that were themselves latently infected with related viruses. A vast amount of detailed information is now available on their very complex relationships with vertebrate cells.

Avian Oncoviruses

Naturally occurring avian leukosis viruses are always transmitted genetically as an integrated DNA "provirus," and may also be transmitted as virions, congenitally, to produce tolerated latent infections. Rarely, some are transmitted postnatally from one bird to another. Usually they cause no disease, but they may induce a variety of pathologic responses rather late in the life of the affected birds. Most of these diseases are related to the hemopoietic system: visceral (and ocular) lymphomatosis, erythroblastosis, myeloblastosis, and osteopetrosis; solid tumors (sarcomas, carcinomas, and endotheliomas) are rarer manifestations. All these types of neoplastic disease can be produced by the same or closely related viruses, but disease occurs in only a minority of infected birds. The best studied variant of avian leukosis virus is the Rous sarcoma virus (RSV).

Most "wild" strains of avian leukosis virus multiply readily in chick embryo fibroblasts, producing noncytocidal steady-state infections with no demonstrable effect on the cells other than to render them resistant to superinfection with RSV of the same antigenic subgroup ("virus-attachment interference"; see Chapter 4). Rous sarcoma virus, on the other hand, produces proliferative foci in chick fibroblasts, and focus production provides a convenient method of assay of the virus. The morphology of the transformed cell differs with the strain of RSV, but all RSV-transformed cells are demonstrably malignant. Some

strains of RSV can also transform cultured mammalian cells and produce tumors in newborn mammals.

Much interest has been generated recently by the *in vitro* synthesis
of the DNA "provirus" of Rous sarcoma virus. Using the virion-associated reverse transcriptase, plus virion RNA as template, in association with the transfer-RNA which serves as the primer, the 6 million
dalton circular DNA which is eventually integrated with cellular DNA,
has been manufactured in a test-tube. It is not difficult to foresee a
new era of cancer research in which the virus and the virologist will be
largely irrelevant while the molecular biologists work on synthetic
molecules representing the provirus of the various oncoviruses.

Murine Oncoviruses

Most mouse leukemias and lymphomas are induced by viruses;
indeed several different viruses produce histologically distinguishable
diseases of the lymphoid or erythroid cells. The viruses that cause
murine leukemia are strikingly similar to avian oncoviruses in their
physical, chemical, and biological properties (Plate 9-1). Most strains
produce steady-state noncytocidal infections in mouse fibroblasts,
hence they are difficult to detect. The *murine sarcoma* viruses, on the
other hand, appear to be defective oncoviruses which, though sar-
comagenic on their own, are capable of multiplication only in the presence of a "helper" murine leukemia virus. Infection of cultured cells
by murine sarcoma virus alone causes transformation, analogous to
that seen with Rous sarcoma virus. However, dual infection with
murine sarcoma and leukemia virus is necessary if the defective sarcoma genome is to be "rescued," i.e., infectious sarcoma virions are to
be produced; the leukemia genome codes for the envelope protein of
both viruses.

Our understanding of the murine oncoviruses has been greatly
complicated by the discovery that a complete DNA copy of the RNA
oncovirus genome is present in the chromosomes of most or all
normal mouse cells. The oncoviral DNA must, in this context at
least, be regarded as a group of normal cellular genes which are transmitted vertically from cell to cell and from mother to offspring in perpetuity. In so-called "low leukemia strains" of mice, virions are rarely
if ever produced, whereas mice of "high leukemia strains" begin to
shed virus and develop leukemia later in life. Cultured cells taken from
virus-free embryos of high or low leukemia strains of mice can be induced to synthesize murine oncoviruses by treatment with certain
carcinogens or mutagens. A strictly analogous situation applies to the

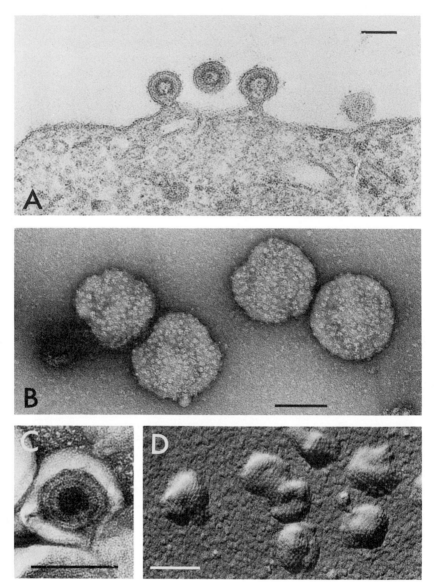

Plate 9-1. *Retroviridae, subfamily Oncovirinae, murine leukemia virus* (*bars = 100 nm*). *(A) Budding of virions from a cultured mouse embryo cell. (B) Virions negatively stained with uranyl acetate, showing peplomers on the surface. (C) Virion somewhat damaged and penetrated by uranyl acetate, so that the concentric arrangement of core, shell, and nucleoid becomes visible. (D) Cores isolated by ether treatment of virions, freeze dried and shadowed. The hexagonal arrangement of the subunits of the shell around the core is recognizable. (Courtesy Drs. H. Frank and W. Schäfer.)*

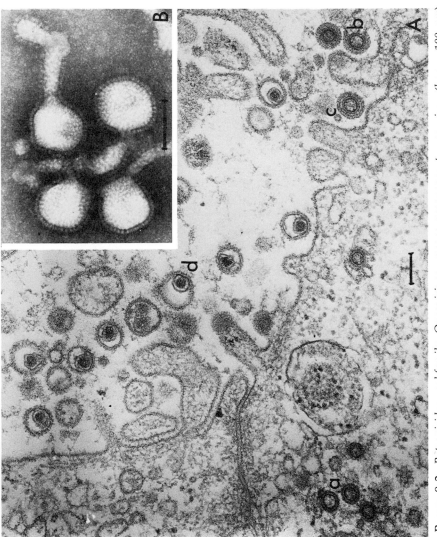

PLATE 9-2. *Retroviridae, subfamily Oncovirinae, mouse mammary tumor virus (bars = 100 nm). (A) Thin section of mammary adenocarcinoma of mouse; a, intracellular A particle; b, budding A particle; c, enveloped A particle; d, B particle. (B) Negatively stained preparation, showing peplomers and "tails." (Courtesy Dr. D. H. Moore.)*

avian oncoviruses; DNA provirus is present in every normal chick cell.

The mammary tumor virus of mice resembles the leukemia viruses in several properties, but the morphology (B-type) and morphogenesis of the virus in infected cells are different (Plate 9-2). It was demonstrated as long ago as 1936 that the mammary tumor virus could be milk-transmitted. Because of the lack of a satisfactory cell culture system, little work has been possible concerning the relation of this virus to cellular transformation. However, there is evidence to suggest that every "normal" mouse contains in its own genome at least one DNA copy of the RNA genome of some strain of mammary tumor virus, thus accounting for the genetic transmission of these viruses. Conceivably, carcinogens or hormones are responsible for derepression of the integrated provirus, leading to virus synthesis and cancer. Presumably when mammary tumor virus indigenous to one strain of mouse is introduced into other strains it is transferred only in the mother's milk.

Feline Oncoviruses

This oncovirus was first isolated in 1964 from cats with lympho-sarcomas and leukemia which are the most common form of neoplasms in these animals. It is likely that all cat leukemias are of viral origin. Feline leukemia virus (FeLV) induces leukemia after inoculation into kittens, and can be grown in cell culture, where, like the murine and avian oncoviruses, it is produced continuously by infected cells without noticeable cytopathic effect.

Oncovirus particles have also been shown to be causative agents of two sorts of solid tumors of cats, fibrosarcomas and liposarcomas, and the same preparations of virus also induce sarcomas in dogs, rabbits, marmosets, and monkeys, and can transform human embryo cells. Whether these virus stocks contain only sarcoma virus genomes or whether they are mixtures of FeLV with a feline sarcoma virus is not yet clear.

Although the feline leukemia virus(es) has been much less intensively investigated than either the avian or murine oncoviruses, it may be more relevant to the human cancer virus problem. All investigations of murine leukemia viruses, and most of those on avian oncoviruses, have been carried out with highly inbred lines of experimental animals. Cats, like man, are highly outbred, are subject to many of the same environmental stresses and circumstances as man, and develop leukemia at roughly the same rates as in man, considering the difference in lifespan. All evidence points to the horizontal transfer in nature of the particular viruses that cause leukemia in cats, and suggests that the endogenous oncovirus genome copy plays a trivial role, if any, in leukomogenesis. This view is supported by the demonstrated excretion of the virus into the environment by infected cats, and by the transmission by contact of the infection and of the disease in the natural as well as artificial closed environments.

Primate Oncoviruses

Oncoviruses of two distinct classes have recently been discovered in nonhuman primates. Baboons carry endogenous oncoviruses which appear to be vertically transmitted and nononcogenic. By contrast, horizontally transmissible tumorigenic oncoviruses recovered from sarcomas, lymphomas, and leukemias of gibbon apes or woolly monkeys have been shown to induce malignant neoplasms when artificially inoculated into primates.

The Oncogene Hypothesis

It is certain that: (a) at least some RNA tumor viruses cause at least some of the cancers that occur in animals, (b) production of oncovirus particles can be induced in cells that are destined never to be malignant, and (c) the genetic information for these viruses is vertically transmitted as part of the genome of the host cell, although horizontal transmission of virions also occurs. A number of hypotheses have been proposed to bring these seemingly disparate observations together, one of which, the so-called "oncogene hypothesis" has received a great deal of recent publicity.

The oncogene hypothesis presented by Huebner and Todaro in 1969, and subsequently revised in 1972, proposes that every cell of most or all vertebrate species contains a DNA copy of the genome of one or more oncoviruses and that this virus-specific information is vertically transmitted from parent to offspring. Depending on a complex interplay between the host genotype and environmental conditions, viral production could be elicited at some stage in the life of the individual animal. Oncoviruses are supposed to carry oncogenic information ("oncogene(s)") and tumor formation that sometimes ensues after induction of these viruses is supposed to be due to these oncogenes. Much of the more recent molecular biological data is consistent with the oncogene hypothesis, but the oncogene itself has yet to be defined and demonstrated. Furthermore, recent data indicates that most of the oncoviruses arising from normal cells are not tumorigenic; indeed many are "xenotropic," i.e., incapable of multiplying in cells of the mammalian species from which they were originally induced!

The implication of the oncogene hypothesis, that spontaneous neo-plasms are a consequence of events occurring within the genome of cells, fits well with epidemiological data which shows that the major-ity of naturally occurring tumors of vertebrates do not occur in time-space clusters and do not, therefore, behave as infectious diseases. Of course, the hypothesis does not bear on the fact that some tumors are transmissible by viruses in an orthodox way, nor on the artificial production of cancers by tumor viruses in the laboratory, nor on on-cogenesis by DNA viruses. Moreover, observations with feline leu-kemia virus, and naturally occurring avian leukosis in domestic flocks and jungle fowl, suggest that horizontally transmitted or congenitally virion-transmitted oncoviruses may be much more relevant in natural oncogenesis than the respective endogenous viral genomes.

POSSIBLE VIRAL CAUSATION OF HUMAN CANCER

We do not know with certainty whether any single type of cancer in man is induced by a virus, although warts and molluscum con-tagiosum, two trivial benign tumors of the skin, certainly are. Yet it would be surprising if viruses were not responsible for some hu-man cancers, since viruses from several different families induce so many different types of malignancy in several species of animal.

One extreme view is that all human cancer is due to viruses; it would be consistent with the oncogene hypothesis to suggest that all other carcinogens (chemical, radiological) merely serve to trigger an endogenous viral oncogene. By contrast, some oncologists believe that cancer arises by somatic mutations, the background incidence of which is known to be increased by carcinogens such as hydrocarbons and radiation. However, the "somatic mutation" and "viral" theories of cancer etiology are clearly not mutually exclusive. Cancer is not one disease; it may not have one cause. The overwhelming evidence link-ing smoking with lung cancer, or radiation with leukemia, in no way negates the possibility that viruses may be causally implicated in other malignancies of man.

It will not be easy to establish a causal relationship between viruses and cancer in man. The mere recovery of a virus from a human tumor in no way incriminates that virus as the causative agent. The intrac-table problem is that vertebrates carry so many viruses not causing recognizable disease (Chapter 8). Such viruses may localize and multi-ply preferentially in tumor cells. A case in point is the Burkitt lym-phoma; EB virus, which is invariably present, *may* be etiologically responsible for the tumor (though usually, like the oncoviruses, producing only a subclinical infection) or it may be an irrelevant passenger. Similarly, the association between genital herpes simplex and carcinoma of the cervix may merely reflect the high incidence of both diseases in promiscuous women.

In 1882 Robert Koch formulated a series of postulates that should be fulfilled before a particular bacterium could be accepted as the cause of a disease. He proposed that the microorganism must (a) be isolated from most cases of the disease, (b) be grown in pure culture, and (c) reproduce the same disease when reinoculated into healthy animals. Koch's postulates still make good sense today. However, the first two of these requirements are difficult to achieve in the present case and the third impossible in the human situation. We have seen from close examination of viral carcinogenesis that often the causative virus is simply not demonstrable in the tumor. Logically, therefore, we would not expect to recover the causative virus from most virus-induced human cancers.

Koch's second postulate poses less of a problem, but even this cannot, in fact, be fulfilled in a literal sense, because viruses can only be grown in cells (or animals) which often, perhaps always, carry their own viruses and cannot be sterilized as can bacteriological media. Because of the nature of viruses, it is possible to demonstrate their presence, but it is never formally possible to prove their absence. This particular difficulty applies equally to attempts to establish a causal correlation between a cytocidal virus and the more conventional type of infectious disease, but it is not an insuperable obstacle to fulfilling Koch's second postulate in a practically adequate form. Many viruses grow to high concentrations, and after purification by plaque isolation at limit dilution they can be shown to reproduce a particular disease in animals. This has already been done with polyoma virus, for example.

The third of Koch's postulates cannot be fulfilled with material derived from human cancers. It is morally unjustifiable to attempt to induce cancer in healthy human subjects. Some experiments have been carried out on patients already suffering from advanced incurable malignant disease. They have not yielded much information, and since the immunological responses of these patients are known to be diminished, it is questionable how applicable the results of such experiments are to normal subjects.

How, then, are we to know whether a virus or other agent isolated from a human cancer was responsible for the induction of the cancer? No direct proof is possible, but indirect evidence can be obtained in various ways which may, in sum, be sufficient to establish a *prima facie* case. For example, an agent recovered from a human tumor may transform human cells *in vitro* (but by no means all transformed clones are malignant). Two other approaches which are now being vigorously exploited depend upon discoveries we have described earlier in relation to virus-induced tumors in animals. Human tumor cells are being searched for evidence of virus-coded nonstructural antigens (T antigens), mRNA, and integrated DNA; attention has recently turned from adenoviruses to herpetoviruses in this regard. The fact that all RNA tumor viruses contain a viral RNA-dependent DNA polymerase ("reverse transcriptase") has stimulated the search for a similar enzyme, as well as for oncovirus particles, in human tumor cells; as we shall see below, reverse transcriptase has turned out to be a sensitive — perhaps too sensitive — probe.

Oncoviruses have recently been recovered from a number of human leukemias. The best studied comes from a long-term culture of myeloid cells from a patient with acute myelogenous leukemia. The virus, which grows well in several normal lines of human cells *in vitro*, closely resembles the oncovirus associated with woolly monkey sarcoma, as indicated by nucleic acid hybridization and by serological resemblance of their reverse transcriptases and of other antigens of the virions. The virus is widespread in normal humans, as evidenced by the fact that almost all adults have antibody against it, and virions can be induced by chemical treatment of various malignant human cell lines. It is important to note however that the virus has not yet been shown to transform human cells *in vivo* or *in vitro* to the malignant state. The relevance of an oncovirus found in human breast cancers is at least equally uncertain at the present time.

These experiments are certainly exciting and suggestive, and they show a close connection between viruses and certain sorts of human tumors. However they do not resolve whether the relationship is causal or coincidental. Nevertheless, discoveries during the last 5 years have greatly heightened interest in oncoviruses as well as herpetoviruses as possible causative agents in some human cancers; it seems likely that eventually both will be shown to be naturally oncogenic in man. Ironically, prevention of the disease by a specific viral vaccine may offer the only definitive means whereby a viral etiology of cancer will be established.

SUMMARY

Several oncoviruses and herpetoviruses cause cancer under natural conditions in animals and birds, while several papovaviruses and adenoviruses are oncogenic when injected into baby rodents.

There is as yet no unequivocal proof of a viral etiology for any human cancer.

Oncogenic viruses also transform cultured fibroblasts *in vitro*. Abortive transformation is reversible, but stable transformation results in permanent changes in the cell's phenotype which often progress to malignancy. Biochemical analysis of such transformants has taught us much about the nature of the cell-virus interaction that leads to cancer.

Whereas the oncoviruses of mice, chickens, cats, monkeys, and several other species establish noncytocidal steady-state infections with continuous production of virions, malignant transformation by the DNA tumor viruses is not accompanied by virus production. In the case of these viruses only a minority of virus-cell encounters lead to transformation (the remainder being cytocidal); the incidence can be increased by infection with defective mutants or UV-inactivated virus. Clearly then, the whole viral genome is not required for malignant transformation, but certain virus-coded functions are necessary both for the establishment and for the maintenance of the malignant state.

The DNA of herpetoviruses, papovaviruses, or adenoviruses is integrated into the cell's chromosomes, and is replicated in perpetuity as part of the genome of the malignant cell. However, only a limited number of "early" viral genes are transcribed, hence no virions are produced, though several nonstructural viral proteins known as T (tumor) antigens and TSTA (transplantation antigens) are continuously synthesized. The presence of TSTA in the plasma membrane alters the surface properties of the cell, including contact inhibition, but it is not known whether this is the fundamental lesion in virus-induced cancer. Viral replication can be induced in some, but not all, virus-free cancers of viral origin, by fusion of cancer cells to permissive cells or by exposure to chemical carcinogens or irradiation.

Oncogenesis by the oncoviruses is more complex. First, a DNA copy of the RNA genome is transcribed by reverse transcriptase and integrated with the cellular genome. However, most, or all, normal cells in normal chickens, mice, and presumably other animals have been found to carry in their chromosomes the complete genetic information for at least one oncovirus. In "low leukemia strains" of animal this information is never (or rarely) expressed, but it may be derepressed by irradiation or chemical carcinogens. The "oncogene hypothesis" postulates that cancer results from the triggering of such oncogenes which are present in all normal people and transmitted vertically as part of the normal genetic makeup of any individual to his offspring.

So far there is no conclusive evidence for human tumor viruses, but large sums of money are being spent in their pursuit. By analogy with

what we know of oncogenic viruses of other animals, the oncoviruses and the herpetoviruses are most suspect. Since oncoviruses have been recovered from numerous species representing
three orders of vertebrates (reptiles, birds, and mammals) and have
been shown to cause leukemias, lymphomas and sarcomas under natural conditions, it seems almost inconceivable that viruses of this subfamily are not responsible for some or all leukemias in man. Oncovirus
particles containing reverse transcriptase have been recovered from
human leukemias, and particles analogous to those of mammary carcinoma virus of mice have been found in human breast cancers. Moreover, the herpetovirus, EBV, is present in all cases of Burkitt lymphoma and most nasopharyngeal carcinomas of man, while herpes
simplex type 2 has been associated with carcinoma of the cervix. It
must be stressed however that a causal correlation has not been established in any of these instances, and, in view of the ethical
problems of testing Koch's postulates in man, it is quite conceivable
that definitive proof of a viral etiology for human cancer may only
come from the development of a successful vaccine.

FURTHER READING

Bader, J. P. (1975). Reproduction of the RNA tumor viruses. In "Comprehensive Vi-
    rology" (H. Fraenkel-Conrat and R. R. Wagner, eds.), Vol. 4, p. 253. Plenum, New
    York.
Bauer, H. (1974). Virion and tumor cell antigens of C-type RNA tumor viruses. Advan.
    Cancer Res. 20, 275.
Benjamin, T. (1972). Physiological and genetic studies of polyoma virus. Curr. Top.
    Microbiol. Immunol. 59, 107.
Biggs, P. M., de Thé, G., and Payne, L. N., eds. (1972). "Oncogenesis and Herpes-
    viruses." Int. Agency Res. Cancer, Lyon.
Cold Spring Harbor. (1974). Tumor viruses. Cold Spring Harbor Lab. Symp. Mol.
    Biol. 39.
Eckhart, W. (1974). Genetics of DNA tumor viruses. Annu. Rev. Genet. 8, 301.
Emmelot, P., and Bentvelzen, P., eds. (1972). "RNA Viruses and Host Genome in On-
    cogenesis." North-Holland Publ., Amsterdam.
Fenner, F., McAuslan, B. R., Mims, C. A., Sambrook, J. F., and White, D. O. (1974). "The
    Biology of Animal Viruses," 2nd ed., pp. 479–542. Academic Press, New York.
Green, M. (1970). Oncogenic viruses. Annu. Rev. Biochem. 39, 701.
Green, M. (1972). Molecular basis for the attack on cancer. Proc. Nat. Acad. Sci. U.S. 69,
    1036.
Gross, L. (1970). "Oncogenic Viruses," 2nd ed. Pergamon, Oxford.
Hirsch, M. S., and Black, P. H. (1974). Activation of mammalian leukemia viruses. Advan.
    Virus Res. 19, 265.
Klein, G. (1972). Herpesviruses and oncogenesis. Proc. Nat. Acad. Sci. U.S. 69, 1056.
Kurstak, E., and Maramorosch, K., eds. (1974). "Viruses, Evolution and Cancer." Aca-
    demic Press, New York.
McAllister, R. M. (1973). Viruses in human carcinogenesis. Progr. Med. Virol. 16, 48.

Merkow, L. P., and Slifkin, M., eds. (1973). "Oncogenic Adenoviruses," *Progr. Exp. Tumor Res.* No. 18. Karger, Basel.

Sambrook, J. (1972). Transformation by polyoma virus and simian virus 40. *Advan. Cancer Res.* **16**, 141.

Silvestri, L. G., ed. (1971). "The Biology of Oncogenic Viruses." North-Holland Publ., Amsterdam.

Temin, H. M. (1971). Mechanism of cell transformation by RNA tumor viruses. *Annu. Rev. Microbiol.* **25**, 609.

Temin, H. M. (1972). The RNA tumor viruses — background and foreground. *Proc. Nat. Acad. Sci. U.S.* **69**, 1016.

Temin, H. M., and Baltimore, D. (1972). RNA-directed DNA synthesis and RNA tumor viruses. *Advan. Virus Res.* **17**, 129.

Todaro, G. J. (1973). Detection and characterization of RNA tumor viruses in normal and transformed cells. *Perspect. Virol.* **8**, 81.

Todaro, G. J., and Huebner, R. J. (1972). The viral oncogene hypothesis: New evidence. *Proc. Nat. Acad. Sci. U.S.* **69**, 1009.

Tooze, J., ed. (1973). "The Molecular Biology of Tumor Viruses." Cold Spring Harbor Lab., Cold Spring Harbor, New York.

Weiss, R. A. (1975). Genetic transmission of RNA tumor viruses. *Perspect. Virol.* **9**, 165.

Wyke, J. A. (1974). The genetics of C-type RNA tumor viruses. *Int. Rev. Cytol.* **38**, 68.

zur Hausen, H. (1972). Epstein-Barr virus in human tumor cells. *Int. Rev. Exp. Pathol.* **11**, 233.

CHAPTER 10

Epidemiology of Viral Infections

So far we have considered viruses, as such (Chapter 1), in relation to the individual cells in which they multiply (Chapters 3, 4 and 5) and in relation to the integrated assembly of cells that constitutes the animal body (Chapters 6, 7, 8, and 9). Viruses can survive in nature only if they are able to pass from one animal to another, whether of the same or another species. The epidemiology of viral infections consists of the study of the transfer and persistence of viruses in populations of animals.

ROUTES OF ENTRY AND EXIT

The vertebrate body presents three large epithelial surfaces to the environment — the skin, the respiratory mucosa, and the mucosa of the alimentary tract — and two lesser surfaces — the eye and the genitourinary tract. To gain entry to the body viruses must infect a cell on one of these surfaces, or otherwise breach the surfaces (by trauma, including insect bite) or bypass them by congenital transmission. The same considerations apply to the escape of virus from the body.

The Skin

The tough protective covering of the skin, the stratum corneum, offers an effective barrier to infection by viruses. A number of viral infections are nevertheless initiated by infection of the skin, or more rarely of the oral mucosa, when the surface is breached by trauma, or by inoculation by arthropod vectors or a hypodermic needle (Table 10-1).

TABLE 10-1

Viruses That Initiate Infection in Man via the Skin or Mucosa

ROUTE	FAMILY	VIRUS
Minor trauma	Papovaviridae	Warts
	Herpetoviridae	Herpes simplex types 1 (oral) and 2 (genital) EB virus
	Poxviridae	Molluscum contagiosum, cowpox, orf, milkers' nodes
Arthropod bite,		
Mechanical	Poxviridae	Tanavirus
Propagative	Togaviridae	Alphaviruses; flaviviruses
(arboviruses)	Bunyaviridae	Bunyaviruses
	Reoviridae	Orbiviruses
Bite of vertebrate	Herpetoviridae	B virus
	Rhabdoviridae	Rabies virus
Injection	Herpetoviridae	Cytomegalovirus
	Unclassified	Hepatitis B

Trauma. Molluscum contagiosum, a specifically human poxvirus infection, orf, cowpox and milkers' nodules, which are sometimes acquired from sheep and cows, respectively, and human warts, caused by a papovavirus, are spread by direct contact through minute abrasions; the viruses escape from the lesions in like manner.

Two rare but highly lethal human infections, rabies and B virus encephalomyelitis, are transmitted by the bites of animals. The likelihood of rabies occurring after bites by animals whose saliva contains the virus is related to the severity of the trauma produced by the bite. Human infection with B virus (the simian equivalent of herpes simplex virus type 1) is usually due to the bite of a monkey with infected saliva. Rabies and B virus infections of man, though usually fatal, are dead-end infections, since they are not transmitted by arthropods and no virus is released from the infected individual into the environment.

Artificial Inoculation. Experiments in animals conducted by virologists over many years attest to the ease of transmitting many viral infections by inoculation, either subcutaneously or more effectively by some other route (intracerebral, intranasal, or intraperitoneal). Accidental inoculation can be important epidemiologically in both human and veterinary medicine. The classic human example is hepatitis B, transmitted in serum or blood transfusions or by contaminated syringes and needles. More recently, cases have been reported of the transmission of two herpetoviruses, cytomegalovirus, and EB virus, by blood transfusions.

Transmission by Arthropods. By far the most important means of producing viral infection by breach of the skin is by the bite of an arthropod. Transmission may be *propagative,* i.e., involving multiplication of the virus in its arthropod vector, or simply *mechanical.*

Mechanical transmission by arthropods is important in a number of diseases of animals in which virus is readily accessible in prominent skin lesions, such as myxomatosis of rabbits and fowlpox in chickens, but is not known to be important in viral diseases of man.

Propagative transmission by an arthropod vector is the principal mode of transfer of a large number of viruses, which are therefore called arbo-(*arthropod-bo*rne) viruses. The requirement that these viruses must multiply in their arthropod vector imposes an additional barrier of specificity, which many different viruses, from several taxonomic groups (see Table 10-1), have successfully overcome. Multiplication of the ingested virus and its spread through the insect take some time, so that an interval of several days necessarily elapses between the acquisition and transmission feeds—this interval is called the *extrinsic incubation period.* Arthropod transmission provides a very effective way for a virus to cross species barriers, since the same arthropod may bite birds, reptiles, and mammals that rarely or never come into direct contact in nature (Figure 10-1).

Furthermore, arthropods may provide a reservoir of the virus in nature that is independent of the infection of vertebrates, since some viruses, e.g., flaviviruses and bunyaviruses may be transovarially transmitted by arthropods (see Figure 10-1) in which they produce inapparent infections. The epidemiology of human arbovirus infections is discussed in more detail in Chapter 19.

Shedding. Although lesions are commonly produced in the skin in several localized and generalized diseases, relatively few viruses are shed from the skin lesions in a way that leads to transmission. Smallpox, herpes zoster, and herpes simplex are exceptions. Smallpox virus has been shown to be viable in dried crusts or scabs for more than a year at room temperature. The dust and bedlinen in smallpox hospitals are heavily contaminated with infective virus, some of which comes from scabs and some from the lesions inside the mouth. Herpes zoster constitutes an important reservoir for the maintenance of varicella in communities; virus shed from the vesicular skin lesions of zoster can produce varicella in exposed susceptible individuals.

The Respiratory Tract

Inhalation. Although the concentration of virus in air may be very low, the average adult human samples about 600 liters of air hourly.

Particles larger than 6 μm in diameter are retained in the nose, and even with mouth breathing they rarely penetrate further than the secondary bronchi. A large proportion of the particles less than 2 μm in diameter pass through the nose, and a small proportion reach the alveoli. The normal nasal mucociliary blanket rapidly removes particles that are deposited in the nose and transfers them to the pharynx. Most of the particles on the pharynx are swallowed, although a few can be found in the saliva. Viruses that are infectious by the respiratory route attach to cells rapidly, as shown by the fact that a few particles suffice to produce infection when given as a small intranasal drop, or as an aerosol.

Some viruses, commonly grouped as the respiratory viruses, do not spread beyond the respiratory tract; others produce generalized disease (Table 10-2).

Shedding. Many different viruses, causing localized disease of the respiratory tract or generalized infections, are expelled from the respiratory tract. High-speed photography shows that large numbers of airborne particles are generated from the saliva around the lips and teeth during speech, and much larger numbers are emitted after a sneeze, again mostly from the mouth. Coughing produces fewer particles than sneezing, but there is a greater chance that they have been formed from respiratory secretions dislodged from the mucous membranes of the lower respiratory tract rather than from saliva. Large droplets fall rapidly to the ground. Smaller droplets sediment slowly, and because of their relatively large surface area rapid evaporation reduces them to

TABLE 10-2

Human Viruses That Initiate Infection of the Respiratory Tract

1. With the production of local respiratory symptoms
     Orthomyxoviridae (influenza A and B)
     Paramyxoviridae (parainfluenza, respiratory syncytial virus)
     Coronaviridae (many serotypes)
     Adenoviridae (many serotypes)
     Picornaviridae
         Rhinovirus (many serotypes)
         Enterovirus (a few serotypes)

2. Producing generalized disease, usually without initial respiratory symptoms

Herpetoviridae	Varicella
Poxviridae	Smallpox
Togaviridae	Rubella
Paramyxoviridae	Mumps, measles
Arenaviridae	Lymphocytic choriomeningitis, Lassa virus

small dry *droplet nuclei* which can remain airborne indefinitely. Most viruses are inactivated by this rapid desiccation. Hence the effective spread of respiratory viruses is dependent upon close contact, and they cause infections of gregarious rather than solitary animals. A small number of efficient "shedders" may be of considerable importance in their dissemination.

Although large quantitities of all respiratory viruses are swallowed by infected individuals, viruses of the orthomyxovirus, paramyxovirus, coronavirus, or rhinovirus groups are not excreted in the feces in a viable state, doubtless because they are inactivated by the acid or bile. Adenoviruses and some enteroviruses which cause primary infection of the respiratory tract are readily recovered in the feces, since they are acid- and bile-resistant.

Epidemiological evidence shows that the viruses responsible for the acute exanthemata (measles, rubella, varicella) are not excreted from the respiratory tract until late in the incubation period, usually just before or as symptoms occur, despite the fact that this is the site of their primary implantation. Hence, these diseases are highly infectious by the respiratory route at the end of the incubation period and during the first few days of illness.

The Alimentary Tract

Ingestion. The third major surface of the body exposed to material from the environment is the alimentary tract, which constitutes a very large surface area of cells: stratified epithelium in the mouth and esophagus and columnar epithelium in the intestinal mucosa. In general, infection of the intestinal tract is initiated by viruses from feces, which are resistant to the acids, bile salts, and enzymes that occur in the gut. Enteroviruses and hepatitis A virus are usually spread by the alimentary route; adenoviruses and reoviruses are sometimes transmitted in this way (Table 10-3).

Excretion. Although viruses transmitted by the alimentary route may be inactivated in feces dried in sunlight, they are in general more

TABLE 10-3

Human Viruses That Initiate Infection of the Alimentary Tract

Picornaviridae	Many enteroviruses, including polioviruses
Reoviridae	Including rotaviruses
Adenoviridae	Several serotypes
Unclassified	Hepatitis viruses, especially A

resistant to inactivation by environmental conditions than the enveloped respiratory viruses, especially when suspended in water, e.g., in water supplies contaminated with sewage. Thus, unlike the respiratory viruses which must spread directly from infected to susceptible individuals, viruses transmitted by the alimentary route may persist for some time outside the body, and can therefore cause waterborne or foodborne epidemics.

The Eye and the Genitourinary Tract

Neither of these mucous surfaces is of great importance as a route of entry of viruses into the body. Viruses of several families can occasionally produce localized infections of the conjunctiva or the cornea. Indeed, herpes simplex virus is one of the most common infectious causes of blindness, after trachoma. On the other hand, when conjunctivitis is seen as part of a systemic illness, the virus has reached the eye via the bloodstream.

Venereal transfer of viruses is uncommon. In man it is restricted to type 2 herpes simplex virus (now the second most common venereal agent in many western countries) and condyloma accuminatum (genital warts). The urinary tract is not known to be a portal of entry of virus, but excretion in the urine may be an important mode of contamination of the environment. Viruria occurs in many generalized infections, such as mumps and measles, and is often prolonged in congenital rubella and cytomegalovirus infections (Table 10-4).

TABLE 10-4

Viruses That Initiate Infection of the Eye or Genitourinary Tract of Man or are Excreted in Urine

INFECTION	FAMILY	VIRUS
1. Ocular	Adenoviridae	Human type 8 and several others
	Herpetoviridae	Herpes simplex type 1
	Poxviridae	Accidental vaccinia
	Picornaviridae	Enterovirus type 70
2. Venereal	Herpetoviridae	Herpes simplex type 2
	Papovaviridae	Warts (condyloma accuminatum)
3. Excretion in urine	Herpetoviridae	Cytomegalovirus
(viruria)	Togaviridae	Rubella
	Paramyxoviridae	Measles, mumps
	Unclassified	Hepatitis B

VERTICAL TRANSMISSION

The routes of transmission described so far apply to postnatal individuals of the same or different animal species. This is sometimes called *horizontal transmission*, to differentiate it from *vertical transmission*, which refers to the transfer of virus from an individual of one generation to its offspring, thence to the offspring's progeny, and so on. The most important routes of vertical transmission are via the ovum, across the placenta to the fetus, or to the newborn infant via the mother's milk. Several viruses cross the human placenta and multiply in the fetus; the consequences range from fetal death and abortion (smallpox), through teratogenic effects (rubella), to severe neonatal disease (rubella and cytomegalovirus) (see Chapter 6). Presumably congenital infections may also be asymptomatic.

Herpes simplex virus can be transmitted "vertically," in a sense, by a totally different mechanism; virus can spread from a nursing mother to her susceptible infant by salivary contamination. Then, because of its long latency and periodic recurrence, the same virus may again be transferred to the next human generation. In small isolated human populations, zoster-chickenpox may constitute a similar cycle of "vertical" transmission.

The classical examples of vertical transmission in animals involve the oncoviruses: avian leukosis, murine leukemia, and the mammary tumor virus of mice. All three groups may be transmitted with the germ plasm, as an integrated DNA copy of the viral RNA genome, and in birds, as infectious virions via the egg. Important investigations on the pathogenesis of congenital transmission of virions have been carried out with lymphocytic choriomeningitis virus of mice.

THE VIRAL ZOONOSES

The term zoonosis is used to describe an infection naturally transmissible from animals to man. Table 10-5 lists the viral zoonoses. By far the largest group are the arboviruses (see Chapter 19). A wide variety of animal reservoir hosts and arthropod vectors play a role in the maintenance of arboviruses in nature and most human arbovirus infections originate from these reservoirs. Figure 10-1 illustrates the natural cycles of some togaviruses. Other zoonoses listed in Table 10-5 are primarily viral infections of domestic or wild mammals transmissible only under exceptional conditions to humans engaged in particular occupations involving close contact with animals.

TABLE 10-5

Multiple-Host Viruses, Including Those Responsible for Viral Zoonoses

FAMILY	SPECIES	"RESERVOIR" HOST	"SENTINEL" HOST	MODE OF TRANSMISSION
Poxviridae	Cowpox, milkers' nodes	Cattle	Man	Contact, through skin abrasions
	Orf	Sheep, goats	Man	Contact, through skin abrasions
Togaviridae				
Alphavirus	Several species	Birds and mammals	Man, domestic animals	Mosquitoes
Flavivirus	Several species	Birds and mammals	Man, domestic animals	Mosquitoes, ticks, rarely milk
Bunyaviridae	Several species	Birds and mammals	Man	Mosquitoes, phlebotomus, culicoides
Reoviridae				
Orbivirus	Several species	Mammals	Man, domestic animals	Mosquitoes, culicoides, ticks
Rhabdoviridae	Rabies	Canines, felines, bats	Man, cattle	Animal bite
Orthomyxoviridae	Influenza A	Horse, swine, birds	Man	Respiratory
Paramyxoviridae	Newcastle disease	Birds	Man	Contact through conjunctiva
Arenaviridae	LCM, Machupo, etc.	Rodents	Man	Respiratory

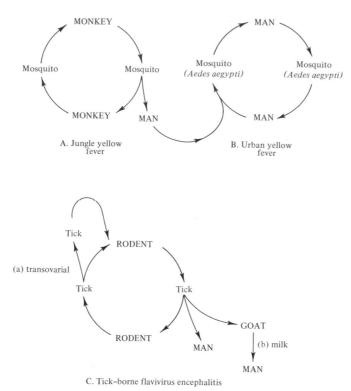

FIG. 10-1. *Cycles of arbovirus infection. (A) Jungle yellow fever, which may lead to (B) urban yellow fever. (C) Tick-transmitted flavivirus infection, with (a) transovarial transmission in ticks and (b) transmission to man in goat's milk.*

Overwintering of Arboviruses

The survival of arboviruses at times other than when they are actively spreading, constitutes a major problem in the ecology of viral diseases. Transovarial infection of ticks provides a natural mode of indefinite survival of the tickborne flaviviruses irrespective of infection of vertebrates (see Fig. 10-1). Similarly, transovarial transmission of bunyaviruses has recently been demonstrated in mosquitoes, which may explain how these viruses "overwinter" in temperate climates where mosquitoes do not breed in winter. Other possibilities include the recrudescence of latent infections in birds or unusually long survival of infected mosquitoes. In addition, attention has recently been directed to the potential importance of hibernating animals. In cool temperature climates several mammals, including bats and small rodents, hibernate during the winter months, as do poikilothermic vertebrates like snakes and frogs. During hibernation their body temperature falls to a level not much higher than that of their environment. Bats can be infected with certain arboviruses, maintain an inapparent infection throughout a long period of hibernation, and become viremic when transferred to a warm environment. Bats may also play an important role as a natural reservoir of rabies virus, and the migratory habits of certain species may be important in spreading the virus to resident wild animals in other geographical locations. Snakes also appear to be potentially important as overwintering hosts for arboviruses. Both naturally and experimentally infected snakes exhibit cyclic viremia over periods of many months, and many of the offspring of naturally infected female snakes are found to carry virus. Such congenital transfer offers an alternative mode of survival of the virus in nature.

THE EPIDEMIOLOGICAL IMPORTANCE OF IMMUNITY

Although less importance is now attached to antibodies in the process of recovery from viral infections (see Chapter 7), there is no doubt about their importance in the prevention of reinfection. Humoral immunity therefore plays a major role in the epidemiology of viral diseases. As well as considering the immunity of individuals to infection or reinfection, epidemiologists are concerned with *herd immunity* as a property of a population. Because of the barriers to serial transmission of viruses posed by immune individuals, a high level of community protection may be achieved even when a significant minority of the individuals in the "herd" are not immune.

Active Immunity in Generalized Infections

Monotypic Viruses. The viruses responsible for several important generalized diseases are *monotypic,* i.e., only a single antigenic type is known. The evidence that a single infection from such a virus produces lifelong immunity has been presented in Chapter 7. On the other hand, those generalized diseases that may be produced by any of several antigenically distinct viruses result only in *homotypic* immunity; other antigenic types of virus may subsequently produce the same disease syndrome.

Measles, mumps, rubella, chickenpox, and smallpox are all monotypic generalized diseases. Two factors influence the relation of immunity in these diseases to the maintenance of the virus in nature: (a) the rate of entry of new susceptibles into the population, and (b) the recurrence of viral activity (and excretion) in subjects who have re-covered from the original infection. Data derived from the study of measles show that the disease can be maintained without periodical reintroduction only in communities of at least 500,000 persons, i.e., large enough populations to provide an annual addition of some 20,000 new susceptibles by birth alone. Mumps, rubella, and smallpox follow a very similar pattern. On the other hand, chickenpox has a very much smaller "critical community size"; although the virus is monotypic, and lifelong immunity to reinfection follows a single at-tack, the critical community size for varicella is less than 1000 persons. This is explained by the fact that varicella virus often persists in the body following the initial attack and may remain latent for many years. It may then be reactivated to cause herpes zoster; virus from this source can produce chickenpox in susceptible children (see Chapter 8).

Few populous and gregarious animals are as long-lived as man; the much more rapid turnover of population in other animals diminishes the epidemiological importance of actively acquired immunity. Most wild rabbits, for example, live less than a year; hence in Australia the summer epidemics of myxomatosis (a generalized monotypic poxvirus infection) can infect almost the whole population of rabbits, since most have been born since the previous summer outbreak. In man, by contrast, the annual crop of new susceptibles comprises only a small proportion of the total population, except in unusual situations in-volving mass migration, e.g., wars.

Viruses with Several Serotypes. Many clinical syndromes, including some generalized infections, can be caused by several antigenically different serotypes of the same species of virus, e.g., poliomyelitis and dengue in man. If there is no cross-protection between the serotypes in question, the epidemiological situation with each one, analyzed in-dependently, is comparable with that found in monotypic infections with prolonged immunity.

Active Immunity in Superficial Infections

In contrast to their importance in generalized viral infections, circu-lating antibodies are largely irrelevant in infections that are strictly localized to superficial membranes, like the mucosae of the respiratory tract and the gut. As pointed out in Chapter 7, local antibody, mainly IgA, is of major importance in preventing infection in these situations.

Superficial infections of the mucous membranes of the gut and the upper respiratory tract may be caused by a very large number of serotypes, mainly enteroviruses and rhinoviruses respectively. With few exceptions the different serotypes show neither antigenic cross-reactivity nor cross-protection. The seemingly endless succession of common colds suffered by urban man reflects a series of minor epidemics due to different serotypes of rhinoviruses, coronaviruses, or other respiratory viruses. Homotypic immunity persists for a few years but then falls off, in the absence of boosting by homologous reinfection, but in heavily exposed individuals and populations there may be some short-term heterotypic resistance due to nonspecific factors, possibly interferon induced by repeated clinical or subclinical infections.

The effects of repeated exposure to the common cold viruses in maintaining herd immunity is well illustrated by experience in Arctic regions. Explorers, for example, are notably free from respiratory illnesses during their sojourn in the Arctic, despite the freezing weather, but invariably contract severe colds when they reestablish contact with their fellow men. Even in larger communities it is clear that herd immunity to respiratory viruses falls rapidly when the community is isolated from the outside world. As Fig. 10-2 illustrates, there were al-

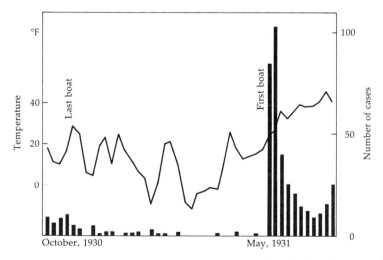

Fig. 10-2. *Acute respiratory disease in Longyear City in Spitzbergen, during 1930–1931. The tracing shows the mean weekly minimum environmental temperature; the histogram shows the number of cases of respiratory disease, which virtually disappeared during the long Arctic winter, when the harbor was icebound, but reappeared in epidemic incidence shortly after the arrival of the first boat in spring.* [From J. H. Paul and H. L. Freese, Amer. J. Hyg. **17**, 517 (1933).]

most no colds in Longyear City throughout the long Arctic winter of 1930–1931, when the harbor was closed by ice, but an epidemic began as soon as the first boat arrived at the island in the spring.

Serological Surveys

The proportion of immune individuals in a population, which provides an index of the *herd immunity,* can be assessed by serological surveys. These may give results of considerable value, especially for viruses with one or a few well-recognized serotypes. Nevertheless, prudent interpretation of the results of serological surveys must take into account a number of complicating factors:

(a) A particular syndrome may be caused by any one of several serologically different strains of a particular virus, or indeed by viruses belonging to completely different families.

(b) Only neutralizing antibody provides a satisfactory index of prior infection over a period of several years; titers of complement-fixing antibody drop to a low level within a few months of infection.

(c) The serum neutralizing antibody level detectable by the techniques used may fall more rapidly than the level of protection.

(d) Serum antibody may be irrelevant to protection, especially in localized respiratory virus infections.

(e) Because of cross-reactivity, the presence of "neutralizing antibody" does not necessarily indicate protection, even with viruses that have a viremic stage. For example, a group of human volunteers had no protection against dengue type 1 virus, even though they had neutralizing antibody against this virus as a result of prior infection with Japanese encephalitis or yellow fever vaccine.

(f) The immunological response to infection with a given viral serotype may be directed mainly against the antigens of a cross-reacting serotype encountered earlier in life, hence an anamnestic response to the original serotype may mark the response to a more recent infection. This phenomenon, called "original antigenic sin" (see Chapter 7), is particularly important in influenza, and also increases the difficulty of achieving successful vaccination. A similar phenomenon may complicate the interpretation of serological surveys for togavirus infections.

THE EPIDEMIOLOGICAL IMPORTANCE OF INAPPARENT INFECTIONS

We have repeatedly emphasized the fact that infection is not synonymous with disease. Many viral infections are inapparent. This poses insuperable difficulties, in many diseases, in tracing case-to-case in fection without recourse to laboratory aids. Although clinical cases may be somewhat more productive sources of environmental contamination than inapparent infections, the latter are more numerous (see Table 10-6), and because they do not incapacitate the individual they do not restrict his movements, and thus provide a major source of viral dissemination. Other viral infections may be latent for long periods and are either never associated with disease, or produce disease only after a long period of latency, or are associated with recurrent but infrequent attacks of disease. Such intermittent excretors play an important role in the perpetuation of the virus.

THE EPIDEMIOLOGICAL IMPORTANCE OF THE WEATHER

A feature of some types of viral infection is the association of cases of disease with meterological events. Arbovirus infections, especially in temperate climates, are almost restricted to the summer months, when their vectors are active. Enteroviruses are more prevalent in the summer, respiratory viruses in the winter. If the infectious agents are present in a community, the incidence of colds increases when the weather gets colder; a change in weather conditions is probably more important than any particular level of temperature, rainfall, or humidity. Such a change in weather may convert some inapparent infections into overt disease, and the increased excretion of virus that follows may amplify this effect by producing further infections. The crowding into restricted areas that occurs in winter may also promote the exchange of respiratory viruses. However, the experiences in the Arctic (Fig. 10-2) show that herd immunity to common colds is maintained by repeated exposure, which must often cause unrecognized infections.

The annual winter outbreaks of respiratory syncytial virus infections in infants are a feature of western communities in both northern and southern hemispheres (Fig. 10-3). The major impact of each epidemic lasts for only a couple of months each winter. Likewise, in spite of seeding of virus throughout the community during the summer months, epidemic influenza rarely occurs except during the winter. In contrast to the regularity of epidemics of respiratory syncytial virus, epidemics of influenza vary greatly in extent from year to year, large epidemics usually occurring after a major antigenic shift (see Chapter 11).

Respiratory disease in military recruits due to adenoviruses likewise shows a winter incidence. Surveys have shown that adenoviruses spread equally well among recruits throughout the year, but the

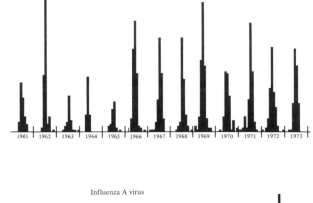

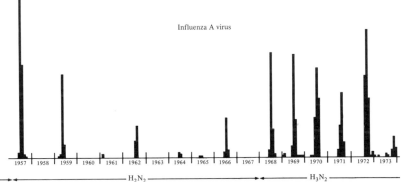

FIG. 10-3. *Epidemic occurrence of influenza A2 and respiratory syncytial viruses. The histograms show the monthly isolations of these two viruses from patients admitted to the Fairfield Hospital for Infectious Diseases, Melbourne, over the periods indicated. Compare the regular annual epidemics of respiratory syncytial virus (causing significant disease mainly in infants) with the irregular occurrence of winter epidemics of influenza. There were major peaks in 1957 (first experience of A2 virus) and again in 1968 and 1969 (Hong Kong variant of A2 virus). (The Australian winter runs from June to August.) (Courtesy Drs. A. A. Ferris, F. Lewis, and I. D. Gust.)*

number developing feverish illness as a result is twice as great in winter as in summer. Furthermore, the stress effects of vaccination against smallpox and typhoid fever were sufficient, in one study, to cause a significant increase in the incidence of respiratory disease due to adenovirus infection.

EPIDEMIOLOGICAL RESEARCH

Epidemiology, the study of diseases in populations, stands at the top of the pyramid of virological research, in the sense that it draws upon research from every other level. This is obvious in relation to etiology and pathogenesis, and the example of influenza virus described in Chapter 11 illustrates how molecular biology may directly illuminate epidemiological problems. In addition, there are some kinds of research that are unique to epidemiology itself.

Epidemiological Studies in Human Communities

Several different kinds of epidemiological study are carried out in human communities. In relation to attempts to determine etiology we can distinguish *retrospective* (case history) studies, of which *cohort* studies are an example, from *prospective* studies.

Retrospective and Prospective Studies. The causation of congenital defects by rubella virus provides examples of each kind of approach. Gregg was struck by the large number of cases of congenital cataract he saw in 1940–1941, and by the fact that many of the children also had cardiac defects. By interviewing the mothers he found that the great majority of them had experienced rubella early in the related pregnancy. The hypothesis that there was a causative relation between maternal rubella and congenital defects quickly received support from other retrospective studies, and prospective studies were then organized. To do this, groups of pregnant women were sought who had experienced an acute exanthematous disease during pregnancy, and the occurrence of congenital defects in their children was compared with that in women who had not experienced such infections. Gregg's predictions were confirmed and the parameters defined more precisely.

Retrospective studies are valuable for providing clues but by their nature they are not conclusive, in that they lack suitable control groups; prospective studies are much more difficult to organize but can establish etiological relationships in a more convincing way.

Human Volunteer Studies. Epidemiological facets of several specifically human diseases that have not been reproduced in other animals have been studied in human volunteers, e.g., hepatitis and common colds. Many major discoveries that have led to the control of viral diseases were possible only with the use of human volunteers, but serious ethical problems now arise because of an increased appreciation of the ubiquity of "occult" viruses and our ignorance of possible long-term effects of some of the agents being tested.

Studies in Closed Communities. A special kind of epidemiological study in human beings is the long-term observation of naturally occurring infections in selected institutional communities, such as "Junior Village" in Washington, which has yielded much useful information on human respiratory and enteric viruses.

Long-Term Family Studies. Another kind of investigation that provides the opportunity for epidemiological studies is the long-term family study, like the Tecumseh Study and the Virus Watch program. Because of the present advanced state of diagnostic virology, such studies now yield a much greater array of valuable data than was possible a few years ago, but they are very expensive and require long-term dedication of both personnel and money.

Vaccine Trials. Having begun with monkeys if possible, then individual human volunteers, and then small-scale human trials, viral vaccines must be tested on a very large scale before their worth and safety can be properly evaluated. Perhaps the best known vaccine trial was the famous "Francis Field Trial" of inactivated poliovirus vaccines, carried out in 1954; similar trials were necessary for live poliovirus vaccines, yellow fever vaccines before this, and measles and rubella vaccines since. There is no alternative way to evaluate a new vaccine or indeed a new drug, and the design of field and clinical trials has now been developed so that they yield maximum information with minimum risk and cost.

SUMMARY

The three great epithelial surfaces of the body—skin, respiratory tract, and gastrointestinal tract— are the important portals of entry and routes of exit for viruses.

The skin has a cornified protective layer which must be breached if infection is to be established; this can occur by trauma or by arthropod bite. Arthropods may transfer virus mechanically, via contaminated mouthparts, or by propagative transmission when virus inbibed with blood multiplies in the vector and is transmitted at a subsequent feed.

The most important viruses, as far as the health of man and his domestic animals is concerned, are the respiratory viruses. These belong to several taxonomic groups: rhinoviruses, orthomyxoviruses, paramyxoviruses, coronaviruses, and adenoviruses. They are transmitted in droplets of secretion which are commonly expelled by sneezing or coughing, and are highly infectious. Some viruses that cause generalized infections also enter the body via the respiratory tract, and are shed by this route very late in the incubation period and in the first few days after the appearance of symptoms.

The acid secretions of the stomach and the bile salts excreted into the duodenum inactivate enveloped viruses but not enteroviruses, adenoviruses, reoviruses, and hepatitis viruses. These viruses commonly cause inapparent infections of the small intestine, and may be excreted in the feces for weeks. Being relatively stable, they may survive for long periods in water contaminated by human sewage.

Animal reservoirs are crucial to the survival of some of the viruses affecting man (the viral zoonoses), many of which are transmitted by arthropods.

Humoral immunity plays a major role in the epidemiology of viral diseases. Immunity to viruses that cause generalized infections is usually lifelong, although the occurrence of multiple serotypes may obscure this at the clinical level. The continued survival of such viruses depends upon their circulation in a large urban community,

TABLE 10-6

Epidemiological Features of Some Common Human Viral Diseases

DISEASE	MODE OF TRANSMISSION	INCUBATION PERIOD[a] (DAYS)	PERIOD OF COMMUNICA- BILITY[b]	INCIDENCE OF SUBCLINICAL INFECTIONS[c]
Influenza	Respiratory	1–2	Short	Moderate
Common cold	Respiratory	1–3	Short	Moderate
Bronchiolitis, croup	Respiratory	3–5	Short	Moderate
A.R.D. (adenovirus)	Respiratory	5–7	Short	Moderate
Dengue	Mosquito bite	5–8	Short	Moderate
Herpes simplex	Salivary	5–8	Long	Moderate
Enteroviruses	Alimentary	6–12	Long	High
Poliomyelitis	Alimentary	5–20	Long	High
Measles	Respiratory	9–12	Moderate	Low
Smallpox	Respiratory	12–14	Moderate	Low
Chickenpox	Respiratory	13–17	Moderate	Moderate
Mumps	Respiratory	16–20	Moderate	Moderate
Rubella	Respiratory	17–20	Moderate	Moderate
Mononucleosis	Salivary	30–50	?Long	High
Hepatitis A	Alimentary	15–40	Long	High
Hepatitis B	Inoculation	50–150	Very long	High
Rabies	Animal bite	30–100	Nil	Nil
Warts	Contact	50–150	Long	Low

[a] Until first appearance of prodromal symptoms. Diagnostic signs, e.g., rash or paralysis may not appear until 2–4 days later.

[b] Most viral diseases are highly transmissible for a few days before symptoms appear. Long = >10 days; short = <4 days.

[c] High = >90%; low = <10%.

or their persistence in an animal reservoir, or their capacity to set up a latent infection. On the other hand, the multiplicity of viral serotypes causing superficial infections of the alimentary and respiratory tracts are ensured of survival, because of the short-lived immunity they provoke.

The principal epidemiological features of the common human viral infections are listed in Table 10-6. The major factors of epidemiological importance in viral infections can be summarized as follows:

Factors influencing transmissibility (communicability)
   Number of virions shed into the environment
   Duration of shedding
   Stability of virus
   Infectivity of virus
   Probability of contact
   Herd immunity

Period of communicability
   Depends on pathogenesis:
      Arboviruses—while viremic
      Respiratory viruses—shed from 2–3 days before until a few days after symptoms develop
      Enteric viruses—excreted in feces for up to a few weeks

Degree of communicability ("infectiousness")
   Generally high in infections spread by respiratory route, e.g., "secondary attack rate" in siblings exceeds 50% with measles and varicella and approaches 50% with common colds

Seasonal factors
   Respiratory viruses spread more in winter
   In temperate climates, enteric and mosquito-borne viruses spread more in summer

FURTHER READING

Andrewes, C. H. (1965). "The Common Cold." Weidenfeld & Nicolson, London.
Andrewes, C. H. (1967). "The Natural History of Viruses." Weidenfeld & Nicolson, London.
Bell, J. A., Huebner, R. J., Rosen, L., Rowe, W. P., Cole, R. M., Mastrota, F. M., Floyd, T. M., Chanock, R. M., and Shvedoff, R. A. (1961). Illness and microbial experiences of nursery children at Junior Village. *Amer. J. Hyg.* **74,** 267.
Benenson, A. S., ed. (1970). "Control of Communicable Diseases in Man," 11th ed. Amer. Pub. Health Ass., New York.
Berg, G., ed. (1966). "Transmission of Viruses by the Water Route." Wiley (Interscience), New York.

Beveridge, W. I. B. (1967). Epidemiology of virus diseases. *In* "Viral and Rickettsial Infections of Animals" (A. O. Betts and C J. York, eds.), Vol. 1, pp. 335–364. Academic Press, New York.

Burrows, R. (1972). Early stages of virus infection: Studies *in vivo* and *in vitro*. *Symp. Soc. Gen. Microbiol.* **22**, 303.

Chanock, R. M., Mufson, M. A., and Johnson, K. M., (1965). Comparative biology and ecology of human virus and mycoplasma respiratory pathogens. *Progr. Med. Virol.* **7**, 208.

Christie, A. B. (1974). "Infectious Diseases: Epidemiology and Clinical Practice." 2nd ed. Livingstone, Edinburgh.

Dingle, J. H., Badger, G. F., and Jordan, W. S. (1964). "Illness in the Home. A Study of 25,000 Illnesses in a Group of Cleveland Families." Western Reserve Univ. Press, Cleveland, Ohio.

Fenner, F., McAuslan, B. R., Mims, C. A., Sambrook, J. F., and White, D. O. (1974). "The Biology of Animal Viruses," 2nd ed., pp. 587–617. Academic Press, New York.

Fox, J. P., Elveback, L. R., Spigland, I., Frothingham, T. E., Stevens, D. A., and Huger, M. (1966). The Virus Watch program: A continuing surveillance of viral infections in metropolitan New York families. I. Overall plan, methods of collecting and handling information and a summary report of specimens collected and illnesses observed. *Amer. J. Epidemiol.* **83**, 389.

Fox, J. P., Hall, C. E., and Elveback, L. R. (1970). "Epidemiology: Man and Disease." Macmillan, New York.

Hope-Simpson, R. E., and Higgins, P. G. (1969). A respiratory virus study in Great Britain: Review and evaluation. *Progr. Med. Virol.* **11**, 354.

Monto, A. S., Napier, J. A., and Metzner, H. L. (1971). The Tecumseh study of respiratory illness. I. Plan of study and observations on syndromes of acute respiratory disease. *Amer. J. Epidemiol.* **94**, 269.

Paul, J. R., and White, C., eds. (1973). "Serological Epidemiology." Academic Press, New York.

Schwabe, C. W. (1969). "Veterinary Medicine and Human Health," 2nd ed. Williams & Wilkins, Baltimore, Maryland.

Tyrrell, D. A. J. (1965). "Common Colds and Related Diseases." Arnold, London.

Tyrrell, D. A. J. (1967). The spread of viruses of the respiratory tract by the airborne route. *Symp. Soc. Gen. Microbiol.* **17**, 286.

World Health Organization. (1967a). Arboviruses and human disease. *World Health Organ., Tech. Rep. Ser.* **369**.

World Health Organization. (1967b). Joint FAO/WHO Expert Committee on Zoonoses, Third Report. *World Health Organ., Tech. Rep. Ser.* **378.**

World Health Organization. (1969). Report of Scientific Group on Respiratory Viruses. *World Health Organ. Tech. Rev. Ser.* **408.**

CHAPTER 11

Evolutionary Aspects of Viral Diseases

INTRODUCTION

Viruses are ubiquitous and have probably played an important role in the evolution of bacteria, plants, and animals. Not only may the diseases they cause be powerful selective forces, but the DNA viruses that can integrate with the cellular genome may have played a role in the increase in cellular DNA that has characterized evolutionary progress.

The origin and evolution of viruses is a fascinating subject that we can only explore superficially here. We shall look briefly at the origin of viruses, and in greater detail at conditions that may lead to evolutionary changes in present-day viruses, which, like their hosts, are also subject to natural selection. Natural infections are usually initiated by very small doses of virus, perhaps by a single virus particle, but during the course of infection thousands of millions of progeny particles, which must include thousands of mutants, are produced. The vast majority of these progeny particles are not transferred to another host, but are destroyed. Natural selection operates very effectively at the stage of the transmission of a virus from one animal to another, for small differences in the properties of different mutants, as well as chance, play a major role in determining which particles are transmitted and therefore multiply again.

THE ORIGIN OF VIRUSES

All viruses depend for their replication on the energy sources, the ribosomes and some of the enzymes of their host cells. They must therefore have evolved after cells and be derived from them, either by "parasitic degeneration" or by the sequestration of fragments of cellular nucleic acid which acquire the capacity to infect other cells.

Parasitic degeneration, i.e., the progressive loss of genetic information as an organism becomes progressively more parasitic upon a
higher organism, probably explains the origin of Rickettsiae and Chlamydiae, and conceivably the Poxviridae, but not the majority of
viruses. The most plausible theory is that most viruses originated from
cellular nucleic acids. What we now define as viral genomes may once
have been cellular genes, present either in the chromosomes or in
cellular organelles (mitochondria and plasmids). The mechanism by
which this occurs is obvious with the DNA viruses. With the RNA
viruses two situations can be distinguished. First, let us consider the
RNA of retroviruses. Rather than thinking of it as viral RNA that is
capable of transcription into a DNA copy which becomes integrated
into cellular DNA, we could consider it to be an RNA transcript of a
segment of cellular DNA. The other RNA viruses may have originated
from the ribonucleoprotein complexes which comprise the form in
which mRNA is transported from nucleus to cytoplasm. Subsequent
mutation and selection may have led to the evolution of double-
stranded or negative-stranded or segmented RNA viral genomes.

The Distribution of Viral Families among Vertebrates

Viruses resembling those that infect vertebrates are found in plants
(e.g., reoviruses, rhabdoviruses, and picornaviruses), and many
viruses of vertebrates, belonging to several different families, also
multiply in arthropods. This raises the interesting possibility that arboviruses arose primarily in invertebrates. Other families apparently
occur only among vertebrate animals. If the viruses are host-specific
and the allocation of viral species to a particular family is reliable,
some information on the possible antiquity of the viral family can be
derived from consideration of the distribution of viral species among
different orders of vertebrates.

Herpetoviruses. Nahmias has applied this approach to herpetoviruses (Fig. 11-1). Assuming that the agents found in lower vertebrates are indeed host-specific herpetoviruses, the wide distribution
of the family suggests an ancient origin. The very wide G + C range
found in the nucleic acids of herpetoviruses is compatible with this interpretation. Tests for genetic relatedness of viral nucleic acids might
also be used to give a measure of the genetic distances. Electron
microscopy of heteroduplexes, the molecular weight of the nucleic
acid, its base composition, and polynucleotide maps provide additional useful data.

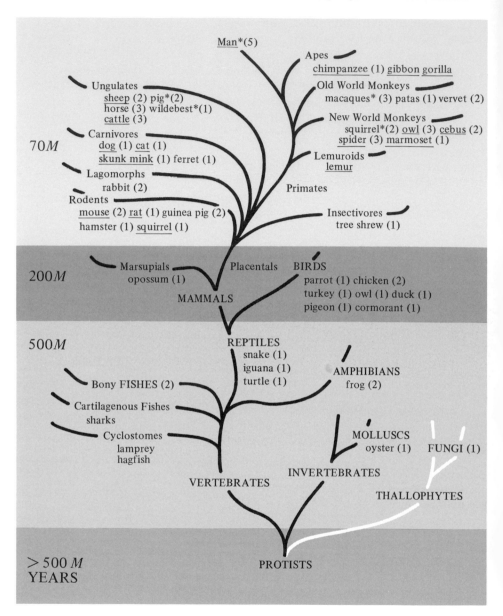

FIG. 11-1. *Evolution of organisms from which herpetoviruses have been recovered. Number in brackets is the number of herpetoviruses identified in each species. Species underlined are those susceptible to herpetoviruses from other species; * denotes species with herpetoviruses that have been found to infect other species under natural conditions.* [From A. J. Nahmias, in "Viruses, Evolution and Cancer" (K. Maramorosch and E. Kurstak, eds.), p. 605. Academic Press, New York, 1974.]

Oncoviruses. The oncoviruses may also be an ancient group that has evolved with the vertebrates, as part of their genome. Table 11-1 sets out the serological relationships between some of the internal antigens of different viral species. By analogy with the orthomyxoviruses, these internal antigens, unlike the peplomers, would not be subject to selection for antigenic change. Except for the species-specific antigen *gs* the crossreactivity between the internal antigens of oncoviruses derived from a variety of vertebrate species reflects the closeness of the relationship of the host species. The mammary tumor virus is seen to be antigenically as well as morphologically quite different from the leukemia–sarcoma viruses. Thus, as far as the investigations have been carried they are compatible with the hypothesis that the genetic material of all these viruses is maintained as part of the cellular genome. Clearly, more data is needed on the distribution of oncoviruses among other vertebrates and their antigenic properties. Groups that would be particularly significant for such studies are the lower vertebrates, and the Australian marsupials, for Australia has been geographically isolated for almost a hundred million years. Characterization of the amino acid composition and sequence of the purified proteins might shed light on the genetic relatedness of the viruses and their hosts.

CHANGES OF VIRUS AND HOST IN MYXOMATOSIS

Observations made on myxomatosis in wild European rabbits in Australia and Britain illustrate particularly well the occurrence of evolutionary changes in an animal virus and its mammalian host. They will be described at some length because they provide the only available example of a process which may have occurred many times in evolutionary history.

Since 1896 myxomatosis has been recognized as a highly lethal "spontaneous" disease of domesticated European rabbits (*Oryctolagus cuniculus*) in various parts of South America, and it still occurs there sporadically in colonies of these animals. The natural reservoir of myxoma virus is however the tropical forest rabbit (*Sylvilagus brasiliensis*) of South America, in which it produces a benign localized fibroma of the skin. Infection of *Oryctolagus* is due to mechanical transfer of the virus from the skin tumors in *S. brasiliensis* by mosquitoes.

The European rabbit was introduced into Australia in 1859 and rapidly spread over the southern part of the continent, where it became the major animal pest of the agricultural and pastoral industries. Myxoma virus was successfully introduced into Australia in 1950 to help control the rabbit pest. It spread rapidly over the continent,

TABLE 11-1
Cross-Reactions between Internal Antigens of the Oncoviruses

ANTIBODY	ONCOVIRUS								
	MOUSE	RAT	HAMSTER	CAT	WOOLLY MONKEY	GIBBON APE	CHICKEN	VIPER	MOUSE MAMMARY TUMOR VIRUS
Anti-transcriptase									
Mouse	++	+	+	+	−	−	−	−	−
Cat	+	+	+	++	−	−	−	−	−
Chicken	−	−	−	−	−	−	++	−	−
Anti-gs₃ (interspecific)									
Cat	+	+	+	++	+	+	−	−	−
Chicken	−	−	−	−	−	−	+	−	−
Anti-gs₁ (specific)									
Mouse	+	−	−	−	−	−	−	−	−
Cat	−	−	−	+	−	−	−	−	−
Chicken	−	−	−	−	−	−	+	−	−

causing enormous mortality among rabbits. The virus was spread mechanically by insect vectors, principally mosquitoes. The occurrence of this very lethal new disease offered an opportunity for the study of evolutionary changes in the virus, the host, and the disease.

Changes in Myxoma Virus

The virus originally liberated in Australia was very lethal for European rabbits; mortality rates exceeded 99%. This highly virulent virus spread readily from one animal to another during the summer, when the mosquito population was abundant; farmers ran "inoculation campaigns" to introduce virulent myxoma virus into the wild rabbit population every spring and summer. It might have been predicted that the disease would disappear after each summer, due to the absence of susceptible rabbits and the greatly lowered opportunity for transmission during the winter because of the scarcity of vectors. This must often have occurred in localized areas, but the outcome proved to be very different throughout the continent as a whole. The capacity to survive the winter conferred a great selective advantage on viral mutants able to cause a less lethal disease that enabled an infected rabbit to survive in an infectious condition for weeks instead of days. In the event, mutants that caused such a modified disease appeared within the first year of introduction of the virus, and within 3 or 4 years they became the dominant strains throughout the country. From about 1955 to 1965 well-timed inoculation campaigns produced localized highly lethal outbreaks, but in general, the viruses that spread through the rabbit populations each year were the somewhat attenuated strains, which offer more prolonged opportunities for mosquito transmission. The better transmission of such strains even during summer epizootics was attested by field experiments; their advantages for overwintering are obvious. Thus the single highly lethal introduced virus has now been replaced by a heterogeneous collection of strains, all of lower virulence.

Changes in the Genetic Resistance of Rabbits

Rabbits that recover from myxomatosis are immune to reinfection for the rest of their lives, and immune mothers transfer passive immunity of short duration to their young. However, since the outbreaks of disease are seasonal and in the wild state most rabbits have a life-span of less than a year, herd immunity, due to the presence of a substantial number of actively immunized rabbits in the population, is not a very important feature in the epidemiology of the disease. This contrasts sharply with the importance of herd immunity in a long-lived animal like man.

However, the fact that even in the initial outbreaks some rabbits recovered from the infection suggested that selection for genetically more resistant animals might operate rapidly. Sometimes recovery was due to environmental factors, e.g., the high ambient temperatures that often occur in the interior of the Australian continent greatly reduce the severity of myxomatosis. In general, however, it was found that in areas of continued annual exposure to epizootics of myxomatosis the genetic resistance of the surviving rabbits steadily increased. The early appearance of the somewhat attenuated virus strains which allowed 10% of genetically unselected rabbits to recover was an important factor in allowing such genetically resistant rabbit populations to build up. In areas of annual exposure to myxomatosis, the genetic resistance

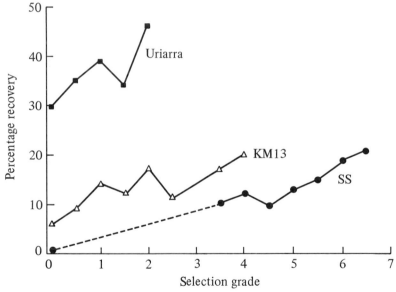

FIG. 11-2. *Changes in the genetic resistance of rabbits to myxoma virus that occurred during breeding from animals that had recovered from infection with various strains of virus. "Selection grade" indicates the parentage of rabbits under test: 0.5, one parent had recovered from myxomatosis; 1, both parents had recovered; 2, both parents and both grandparents had recovered, etc. For the initial selection two attenuated strains were used: Uriarra, from selection grade 0 to grade 2, ■ — ■; KM13, from grade 0 to grade 4, △ — △. At grade 3.5 a selection line was begun with virulent myxoma virus, SS, ● — ●, a strain that killed all unselected rabbits [data from W. R. Sobey, J. Hyg. **67,** 743 (1965).]*

of wild rabbits changed so that the mortality rate after infection with a particular strain of virus under laboratory conditions fell from 90% to 25% within 7 years. Experiments with laboratory rabbits confirmed the observations made on wild rabbits, and also showed that the increased resistance developed against somewhat attenuated field strains (Uriarra and KM13) was effective against a highly virulent strain (SS) as well (Fig. 11-2).

The Continuing Evolution of Myxomatosis

In myxomatosis of the European rabbit in Australia natural selection has operated to produce a type of disease which ensures that there is an adequate susceptible population of host animals, and that the virus can survive through periods of low vector density, i.e., that it can "overwinter." These requirements are met, in genetically unselected rabbits, by a virus less rapidly lethal than that originally introduced. Such attenuated mutants will allow infectious rabbits to survive for prolonged periods, and enough rabbits will recover and breed to ensure that there is an adequate population of new susceptible animals for the next summer epizootic.

Evolutionary changes in the virus and the host, and therefore in the disease, are still in progress, and two types of climax situation can be envisaged. It is possible that myxomatosis in *Oryctolagus* will eventually become very like the disease we know in *Sylvilagus* in the Americas: a benign disease characterized by localized skin tumors with overwintering of the virus in the persistent tumors produced in juvenile animals. Such a situation would arise primarily as a result of progressive increase in the resistance of the host, and it is probable that the virus obtained from such lesions would still prove highly virulent for genetically unselected laboratory rabbits. An alternative climax situation may be comparable to variola major in man: a moderately severe disease with an appreciable mortality. In the latter situation, as with the "benign" fibroma, it is possible that, as the genetic resistance of the rabbits increases, the virulence of the virus (for genetically unselected laboratory rabbits) will increase again. Too mild a generalized disease, like too severe a disease, is not conducive to successful overwintering of the virus, as is shown by experiments with a "laboratory" mutant of myxoma virus ("neuromyxoma") which produces a very mild generalized skin disease in normal rabbits. This virus multiplies to such a low titer in the skin, and the lesions heal so rapidly, that mosquito transmission occurs rarely and only over a brief period of time.

ANTIGENIC DRIFT AND EPIDEMIC DISEASE

The usual pattern of established infectious diseases is endemicity with periodic fluctuations in intensity, which may be seasonal or may occur at longer intervals. Widespread epidemics result from some unusual ecological situation, such as the introduction of large numbers of susceptible individuals into a population where a disease is endemic, e.g., invasion or occupation during wartime, or the introduction of a "new" virus into a population with no immunity, e.g., myxomatosis in the Australian rabbit, or measles into human populations in Greenland. Influenza provides an important exception. Among the many viruses that cause respiratory disease in man only influenza and respiratory syncytial viruses regularly cause epidemics. The epidemics of influenza every year or so are associated with minor changes in the envelope antigens of the virus, a phenomenon described by Burnet as *antigenic drift* or *immunological drift*, by analogy with the concept of genetic drift (a process by which new characters tend to become normal for a species as a result of phases of isolation and reduction in numbers). Larger epidemics at longer intervals are associated with major changes in one or both of the envelope antigens, a process that has been called *antigenic shift*.

Influenza

Human influenza virus was first isolated in 1933. Since that time influenza viruses have been recovered from all parts of the world, especially since the establishment of the Influenza Surveillance Program of the World Health Organization, and their antigenic constitution has been exhaustively studied. This has provided the opportunity for observing continuing evolutionary changes in the influenza viruses. The mortality from influenza is too low, and the generation time of man too long, for any significant observable change to have occurred in the genetic resistance of man.

The two common types of influenza virus (A and B) are distinguished on the basis of their internal ribonucleoprotein and "membrane protein" (see Figure 20-1), which are different in viruses of different "*types*," but constant for all *strains* within any one type. New strains of influenza type A and influenza type B are isolated from year to year; these differ somewhat from earlier strains of the same serotype in either their hemagglutinin or neuraminidase antigens (see Figure 20-1), or both.

Influenza A, unlike influenza B, characteristically causes occasional pandemics (worldwide epidemics), those of 1889–1890, 1918–1919,

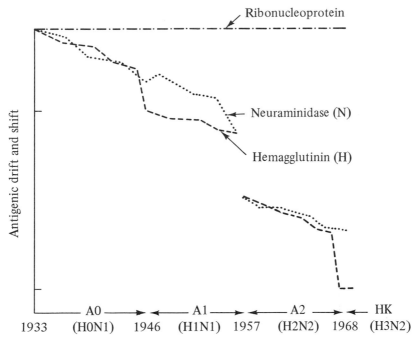

Fig. 11-3. *Diagram illustrating antigenic drift, antigenic shift, and the appearance of new subtypes of human influenza A virus. "Antigenic drift and shift" (ordinate) represents the serological relatedness of the antigens of influenza A virus recovered from man between 1933 and 1975. The internal (ribonucleoprotein) antigen (-·-·-) has not changed over the whole period. Both the hemagglutinin (--) and the neuraminidase (····) have shown antigenic drift, i.e., annual but independent changes in their antigenicity, and antigenic shift, i.e., larger changes such as the change in antigenicity of the hemagglutinin antigen of influenza A in 1946 (H0 to H1). This led to the 1947 and subsequent strains being called a new subtype, A1. There was no parallel change in the neuraminidase in 1946; it remained N1 from 1933 to 1957. The appearance of Asian influenza virus (subtype A2) in 1957 was related to major changes in both the hemagglutinin and the neuraminidase antigens (antigenic shift H1N1 → H2N2). Since 1957 these antigens of influenza A2 underwent antigenic drift until 1968, when the hemagglutinin underwent antigenic shift from H2 to H3, whereas the neuraminidase continued to show antigenic drift as N2. The new strain is designated the Hong Kong strain (HK: H3N2).*

1957–1958 and 1968–1969 being the most dramatic. Between pandemics scattered smaller epidemics occur every 2 or 3 years, usually in winter, and sporadic cases every winter (see Figure 10-3). Antigenic analyses of the strains causing each successive epidemic have shown that there are usually minor, independent changes in the antigenicity

TABLE 11-2
Antigenic Characteristics of Pandemic Strains of Influenza Virus

TIME OF PANDEMIC	PREVIOUS DESIGNATION OF STRAIN	CURRENT DESIGNATION OF STRAIN	NEW PANDEMIC ASSOCIATED WITH ANTIGENIC SHIFT IN	
			H	N
1918–1919	A0	A/London/1/35 (H0N1)		
1946	A1	A/FM/1/47 (H1N1)	+	−
1957	A2	A/Singapore/1/57 (H2N2)	+	+
1968	Hong Kong	A/Hong Kong/1/68 (H3N2)	+	−

(antigenic drift) of the hemagglutinin, the neuraminidase, or both (Figure 11-3).

In addition to this continuing antigenic drift, influenza A underwent much greater changes in the hemagglutinin antigen in 1946, 1957, and 1968. Analysis of these major "antigenic shifts" shows that two different processes have occurred (Table 11-2). In 1946, and again in 1968, there was a major change in the hemaglutinin antigen (H) but no significant change in the antigenicity of the neuraminidase (N). On the other hand, the new A2 subtype (H2N2) appearing in 1957 differed greatly from the preexisting A1 virus (H1N1) in respect of the antigenicity of both hemagglutinin and neuraminidase, as well as in several other properties. From an evolutionary point of view the secular changes in the influenza viruses recovered from man can be interpreted as follows. Recovery from an attack of influenza is accompanied by antibody production against all antigens of the virus concerned. The important antibodies from the point of view of protection against reinfection are those directed against the hemagglutinin and to a lesser extent, the neuraminidase. Antibody can protect, however, only if present on the mucous surfaces exposed to infection. Only a small amount of locally produced IgA antibody, and perhaps some transported across the membrane from the serum, is found on the respiratory mucosa. This local antibody usually prevents reinfection with the strain of virus that produced the original infection, but since its concentration is low, it does not protect against viruses whose antigenic pattern is slightly altered. This strong selection for antigenic novelty produces the observed annual antigenic drift in the hemagglutinin and neuraminidase but not in the internal ribonucleoprotein antigen.

The large changes in the hemagglutinin antigen in 1946 and in 1968 were examples of antigenic shift. It has been postulated that these were due to genetic recombination (reassortment) between human and animal influenza A viruses. Even more striking was the appearance of subtype A2 in 1957, for the A2 virus differed from all previous viruses in many different characters, including the antigenicity of both the envelope antigens. Such a virus could have originated from some extra-human source, for we know that there are many "animal" strains of influenza type A (in birds, swine, and horses), or it could have resulted from recombination. Presumably some such virus, possibly from an animal closely associated with man, produced infection in man. Since the new virus encountered a population with no immunity, the disease spread widely — a pandemic occurred, affecting half the population of the world. The mortality was not particularly high, in striking contrast to the 1918–1919 pandemic which may also have been derived in a similar way. The human mortality rate is irrelevant as far as viral survival and spread are concerned.

An interesting and unexplained phenomenon is the rapid and universal replacement of the strain of influenza virus current at a particular time by the strain that succeeds it. For example, when subtype A1 virus spread around the world in 1946–1947, viruses belonging to subtype A0 virtually disappeared; only one subsequent epidemic (in Alaska in 1949) yielded a strain of the A0 subtype. In like manner A1 strains disappeared almost completely after 1958, only one A1 strain being subsequently recovered (in England in 1960). The apparent disappearance of the earlier strains may be partly due to chance factors in sampling, for the antigenically novel viruses are more likely to cause small epidemics attracting the attention of public health workers, and isolated cases of disease would rarely be tested for virus. Nevertheless, it is rather surprising that a situation has not arisen like that commonly found with pathogenic enterobacteria and streptococci, or with the rhinoviruses, i.e., the coexistence of many different serological types, rather than the regular replacement of one by another.

Other Viruses

Antigenic drift is probably an important evolutionary mechanism in all those viral infections of long-lived vertebrates in which infections are limited to a mucous surface, such as the respiratory epithelium or the gut. Such situations provide an environment with small amounts of antibody on the surface of the susceptible cells, so that there is a strong selective pressure for antigenic novelty. Relatively minor changes in antigens may enable a virus to bypass antibody and establish a transmissible infection. The reason for the lack of evidence with other viruses of clearcut antigenic drift like that found with the influenza viruses is not apparent. Perhaps new rhinoviruses or en-

teroviruses must undergo major antigenic changes to become established in the body, thus accounting for the multiplicity of different
serotypes of these viruses.

CULTURAL CHANGES AND VIRAL DISEASES

Man is distinguished from other creatures by the speed with which
major changes have occurred in the form of his social organization.
Within a few thousand years isolated bands of a few hundred hunters
and food gatherers have given way to the vast and complex conurbations of modern man (see Table 11-3). By comparison, the social organizations of nondomesticated animals are relatively static. The pattern
of viral infection sustained by any type of animal is greatly dependent
upon its social contacts with its fellows, and by its life-span.

Viral Diseases in Primitive Man

By "primitive man" we mean the nomadic hunters and food
gatherers found today only in a few isolated parts of the world, but
characteristic of man as a species before the development of agriculture or the domestication of other animals. Primitive man lived in
closely knit groups, and we can assume that the viral "flora" characteristic of man as an animal was shared among all members of his group.
Under these circumstances most of the specifically human viruses of
modern man could not have existed. We have already discussed
(Chapter 10) the basic community size needed to maintain measles
virus; exactly the same requirement holds for all generalized human
viral infections, except those in which latency and recurrent infection
occur. Acquired immunity is not as durable in many of the diseases of
mucous surfaces (intestinal and respiratory tracts) as it is in the generalized infections, but even the rhinoviruses and enteroviruses could
not survive in a long-lived animal like man in small societies of a few
hundred individuals. We have ample evidence of the disappearance of
respiratory viruses and the diseases associated with them, influenza
and the common cold, in isolated communities of modern man (see
Figure 10-1); such isolation must have been universal in primitive
man.

Indeed the only "specifically human" viral diseases that we could
expect to have survived in primitive man are those marked by latency
or congenital infection. Herpes simplex and varicella–zoster for example, could have survived in primitive man, even in isolated family
units. Most of the viral diseases of primitive man were probably not

TABLE 11-3

The Time Scale of Cultural Changes in Man in Relation to the Number of
Generations, World Population, and the Size of Human Communities

NUMBER OF YEARS BEFORE 1975	WORLD POPULATION (MILLIONS)	NUMBER OF GENERATIONS	CULTURAL STATE	SIZE OF HUMAN COMMUNITIES
500,000	0.1	25,000	Hunter and food-gatherer	Scattered nomadic bands of <100 persons
10,000	5	500	Development of agriculture	Relatively settled villages of <300 persons
6,000	50	300	Development of irrigated agriculture	Few cities of ~100,000; mostly villages of <300 persons
250	600	10	Introduction of steam power	Some cities of ~500,000; many cities of ~100,000; many villages of ~1000 persons
130	1,000	5	Introduction of sanitary reforms	
0	3,500	—	Modern urbanized man	Some cities of ~5,000,000; many cities of ~500,000; fewer villages of ~1000

specifically human diseases; they were zoonoses, caused by viruses of some other animal which accidentally infected man. The principal viral zoonoses are the arbovirus infections, and their incidence in primitive man, as in modern man, would obviously have differed greatly in different parts of the world, because of climatic effects on the activity of vectors.

There still exist today, in certain remote and isolated areas of the world, a very small number of tribes of "Stone Age" people, living much as their ancestors did many thousands of years ago. With few exceptions, even these peoples have now been "contaminated" by contact with Western culture and Western parasites, but some interesting conclusions can be drawn from an analysis of their viral experience by seroepidemiology. For instance, the Tiriyo Indians of Northern Brazil revealed evidence of a high (though not universal) incidence of past infection with various arboviruses and herpetoviruses which can be assumed to have been endemic for millenia. By contrast, the viruses of influenza, measles, and rubella are unable to maintain themselves in such a small isolated community; in a recent survey no member of the tribe had antibodies to influenza or measles, but all had been infected by rubella during a recent epidemic which followed a rare contact with the outside world.

Viral Diseases in Societies Based on Agriculture

The development of agriculture led to a great change in human habits. No longer were men nomads, and no longer was the population size of a group limited by the availability of food gathered or killed. Villages developed, and when irrigation made large-scale agriculture possible man became urbanized. The pattern of viral infections was greatly influenced by the development of these large societies, for now community size had exceeded the minimum level needed to maintain diseases like smallpox, measles, and rubella, and the close-knit society permitted the ready spread of enteric and respiratory viruses. The origins of the "prototype" rhinoviruses, orthomyxoviruses, and paramyxoviruses remain a matter for speculation; almost certainly they were acquired from some animal source. In contrast to man many animals do not need to live in very large groups to sustain such viruses because their turnover rate (leading to the accession of new susceptibles) is so much more rapid than in man. Smallpox virus may well have been derived from a related virus in an animal host, possibly monkeypox; measles virus may have been derived from a virus related to distemper or rinderpest viruses. Human enteroviruses, rhinoviruses, reoviruses, adenoviruses, orthomyxoviruses, and paramyxoviruses are all similar to corresponding serotypes found in wild and domesticated animals.

Once the "prototype" rhinoviruses and enteroviruses had been successfully established in human communities they were subjected to natural selection for survival, which operated most powerfully at the time of transmission. Other things being equal, antigenically novel viruses had a better opportunity of becoming established and multiplying to a sufficiently higher titer to be transmitted. This may have been accomplished initially by antigenic drift, but with the picornaviruses a large number of the serotypes that evolved appear to have persisted. Sampling has not yet been extensive enough, nor on a sufficiently large scale, to know whether some serotypes of these viruses actually become extinct and are replaced by new serotypes.

Urbanization: Infectious Diseases in the Cities

Urbanization is a recent phenomenon, dating back only a few hundred years and caused initially, in the Western world, by the recruitment of country folk to the cities, still a powerful force in urbanization in developing countries. It was, and is, accompanied by intense squalor and poverty, and a greatly increased incidence of the associated infectious diseases. Ultimately, in more affluent countries, the dwellings improved and crowding decreased; sanitation in the form of adequte disposal of excreta and a safe water supply, and, finally, the development of effective antibacterial drugs diminished the importance of bacterial diseases as causes of death and severe morbidity in the great conurbations. For reasons that are not clear, viral infections of the gastrointestinal tract have not disappeared as rapidly as the enteric bacterial pathogens; enteroviruses, reoviruses, and hepatitis A virus still circulate widely. Perhaps this is a relative matter. We take no notice of intestinal bacteria unless they cause disease. Many enteric viruses appear to be nonpathogenic in the vast majority of cases, but because of the sort of laboratory technique required for viral isolation, all positive results are recorded. We know that there has been a great diminution in the circulation of enteric viruses in modern cities of western man compared with places like Karachi or Calcutta.

The number and variety of respiratory viral infections of man is probably in a stage of explosive expansion at the present time. As outlined earlier, man the hunter-gatherer lived in aggregations too small to maintain respiratory viruses that produced acute infection without prolonged or recurrent viral excretion. The great increase in the number of human beings, their crowding into ever larger cities, and the increasing communication between these cities all over the world, are tending to make the "human" world into a single ecological unit. The respiratory viruses of the Northern and Southern Hemispheres, and of East and West, are mingled by air travelers, so that the population of almost any large city in the world today is potentially exposed to human respiratory viruses from all over the world. With this increased exposure there is, of course, increased opportunity for evolutionary radiation, especially in the antigenicity of the viral protein coats. This may not only increase the likelihood of antigenically novel viruses with an "orthodox" pathogenic potential emerging, but may result in novel pathogenic effects. For example, some strains of fowl plague virus (an influenza type A virus) cause severe generalized disease with high mortality in some avian species. If a human influenza virus evolved which combined such a pathogenic potential with novel coat antigens that allowed it to spread as readily as did influenza A2 (H2N2) in 1957 or influenza A2 (H3N2) in 1968, a pandemic of unequalled severity could result.

It is not easy to see how to control the human respiratory viral infections. Vaccines are practicable and sensible for only a few relatively serious diseases; the vast majority of viral infections do not justify either the cost or the risk of vaccination. Yet in the aggregate these "minor" infections are important in producing general ill health and potentiating the adverse effects of inspired pollutants. "Air sanitation" is effective only under very special and local conditions; we can never expect to produce "sanitized" air as we do expect to produce safe drinking water, although considerable improvement should be possible in the quality of air provided to crowded public places. The most logical approach to the prevention of respiratory viral infections would be chemoprophylaxis. If effective specific antiviral drugs are produced, however, their use will carry all the risks associated with existing antibacterial chemotherapeutic agents, including toxic side effects and the emergence of drug-resistant viral mutants.

Human Colonization and Viral Disease

Until the development of ocean-going ships in Europe during the fourteenth century, urban and rural man were largely confined to the continents in which their progenitors had lived as primitive agriculturists and nomads. The explosive period of European colonization of all the other continents had profound effects on the disease patterns, as well as social and demographic patterns, in both the migrants and the indigenes. European man took his endemic diseases to new virgin populations; there were explosive outbreaks of measles and smallpox, for example, in the native inhabitants of America and Oceania. On the other hand, European man intruded on situations in which the indigenous inhabitants had acquired, by natural selection, considerable genetic resistance to diseases lethal to the European, e.g., falciparum malaria and yellow fever in Africa. The slave ships of the sixteenth century took the vector of urban yellow fever, *Aedes aegypti,* as well as the yellow fever virus itself to South America, where both rapidly established themselves and the virus became enzootic in its jungle reservoir hosts.

The European colonists also took their domesticated animals with them. Cattle, sheep, and horses, in the large herds used for pastoral purposes, encountered almost disease-free environments in Australia, where the native animal population was novel and sparse; but in Africa they were exposed to a multitude of viruses and other parasites which had evolved with the great herds of ungulates. Many of the viruses were arthropod-borne, and in the new host animals they caused devasting epizootics of diseases like bluetongue, Rift Valley fever, Nairobi sheep disease, African horse sickness, African swine fever, and so on. Having become established in the sheep and cattle of the colonists in Africa, some of these diseases were then transported to other continents, just as yellow fever had been; in quite recent times bluetongue virus became established in Europe and the United States and African horse sickness spread to India. The livestock industries of Australia, the last continent colonized and one free of serious indigenous viral diseases, have been maintained free of the great scourges of sheep and cattle only by strict quarantine.

The alteration of the environment by man may also lead to other major alterations in the incidence of viral diseases. A good example of this is furnished by the history of arbovirus infections in southern California. The floor of the San Joaquin Valley was largely a desert until man transformed it by irrigation into a rich agricultural area. Under natural conditions the arboviruses were either absent or persisted only in limited areas, but the irrigation channels and vegetation that grew around them provided an ideal environment for mosquitoes, birds, and two arboviruses, St. Louis and Western equine encephalitis virus. Both these viruses cause disease in man and in one of his domestic animals, the horse.

Air Travel and Viral Disease

As long as travel from the endemic and enzootic centers of disease to disease-free areas was relatively slow, quarantine restrictions operated reasonably effectively in keeping out diseases such as smallpox from the United States and the countries of western Europe. Air travel poses much greater problems, since movement from any part of the world to another takes less than 24 hours, and the volume of air transport is increasing at a rapid rate. In 1957 pandemic influenza spread around the globe at the speed of ship and train travel; in 1968 air travel played an important role in speedily distributing the Hongkong A2 strain throughout both hemispheres.

The threats posed by the rapid transport of viruses and their vectors from one part of the world to another have led to a series of countermeasures which have been approved by most nations. These include compulsory smallpox vaccination for all international travelers from endemic areas, the designation of "yellow fever free" airports, and the spraying of aircraft to kill insects. These measures and quarantine are reasonably effective in preventing the spread of the few serious viral diseases of man, and of most viral diseases of veterinary importance. They are completely ineffective in controlling the respiratory viruses and most of the enteric viruses.

Iatrogenic Viral Diseases

Medical innovations since World War II have increased the incidence of some viral diseases. Perhaps the most disturbing trend is the greatly increased spread of hepatitis viruses. Hepatitis B virus was spread by the use of contaminated syringes in prewar venereal disease and diabetic clinics, but it became a much more common disease with its unwitting inclusion in some vaccines (e.g., early batches of yellow fever vaccine) and with the greatly expanded use of human blood and blood products for therapeutic purposes. More recently, serum hepatitis has produced serious difficulties in the operation of hemodialysis units, affecting both patients and staff. Finally, it has increased dramatically in association with the great increase in drug abuse involving injected drugs.

Advanced surgical procedures, notably organ transplantation, have also produced a crop of novel viral problems, due, in part, to massive blood transfusions with fresh blood (EB virus and cytomegaloviruses, as well as hepatitis, see Chapter 8). The prolonged immunosuppressive therapy needed to control transplant rejection renders such patients subject to severe disease after normally trivial infections, and may precipitate recurrent endogenous disease in some persistent viral infections.

Hospital-acquired ("nosocomial") infections are not confined to staphylococci, pseudomonas, and other bacteria; viruses can also spread rapidly through susceptible populations, especially young children. For example, respiratory syncytial virus has often caused epidemics in nurseries for premature infants, as have parainfluenza 3 and several echoviruses. Even the familiar chickenpox and measles viruses can be devastating if they get into a ward of leukemic children on therapy that is immunosuppressive.

The discovery in the 1950's and 1960's of the widespread occurrence of previously unrecognized viruses in "normal" cell cultures used for vaccine virus production caused considerable consternation, since millions of people had by that time been inoculated with living SV40 virus in poliovaccine, or with avian leukosis virus in yellow fever vaccine. Fortunately no iatrogenic disease seems to have resulted from this potentially disastrous situation and today rigorous precautions are taken to ensure that all cells used for live virus vaccine production, including the strains of human embryonic fibroblasts now favored for this purpose, are demonstrably virus-free.

However, recent advances in the technology of viral genetics and oncology have raised the specter of new hazards to human health that may arise if molecular virologists fail to exercise great caution in the experiments they choose to undertake. The availability of restriction endonucleases and ligases now makes it possible to insert particular viral genes (e.g., from tumor viruses) into the genomes of other viruses, or bacteria such as *Escherichia coli,* in which the transplanted genes may replicate indefinitely. Responsible scientists and scientific organizations are moving promptly to impose a moratorium on such extremely hazardous work.

NEW VIRUSES AND NEW VIRAL DISEASES

The whole pattern of public health administration, in relation to viruses, is based upon the assumption that every virus is derived from a preexisting virus, although it may have undergone mutation during the process, i.e., that viruses are no longer evolving from nonviral material. This view is reinforced by the observation that virtually all the known animal viruses belong to a small number of families (see Chapter 1) which must have originated a very long time ago. As discussed earlier in this Chapter, at some time viruses must have evolved from cellular components, but such events must be excessively rare on a human time-scale. New human viral diseases, on the other hand, have arisen relatively commonly over the last few centuries. Most have been due to human intervention in some natural situation, or to changes in the social habits of man of the kinds described in the previous section. On the other hand certain clinical entities that were clearly described in the older literature, such as the "English Sweats" prevalent in Tudor times, and encephalitis lethargica, seems to have disappeared.

Although paralytic poliomyelitis was probably present in ancient

Egypt, judging by illustrations uncovered in the pyramids, it appears that extensive epidemics of the disease first appeared in Sweden less than a century ago and became increasingly prevalent in Western countries as standards of sanitation improved. This seeming paradox is thought to be due to the fact that increasing numbers of children escaped the mild, immunizing, subclinical infections characteristic of infancy, hence remained susceptible to the paralyzing effects of first infection acquired in adolescence or adult life. A similar explanation probably applies to the dramatic increase in incidence of hepatitis A since World War II.

A somewhat different situation obtains with the "new" disease, "dengue shock syndrome" (or dengue hemorrhagic fever), which first appeared in the Philippines, Thailand, and Malaysia in the mid 1950's. Epidemiological and serological observations suggest that this "new" disease is usually associated with a second infection with dengue virus of a serotype different from one previously experienced by the same individual. The shock syndrome is therefore thought to represent some sort of hypersensitivity reaction. Just why it has emerged as a clinical entity so recently may be a consequence of the increased opportunity for infection of children with unfamiliar serotypes of dengue as a result of increasing movement between villages and towns due to better transport facilities.

Contact with exotic animals carries unknown risks of serious human disease. Two recent episodes gave cause for considerable alarm. A substantial number of laboratory personnel died in Germany and Yugoslavia in 1967 when they contracted a previously unknown virus, now known as the Marburg agent, from a consignment of African monkeys purchased by pharmaceutical companies to provide kidney tissue for poliovaccine production. In 1969, nurses working in a mission in Nigeria contracted, from African rodents, a lethal infection now known as Lassa fever, as did a laboratory worker attempting to identify the new agent at Yale University. The episodes simply underline the fact that successful evolution of a satisfactory host-parasite relationship requires thousands of years; an unscheduled encounter between man and a virus that the human species has not met before may have lethal consequences, although in the great majority of cases the virus doubtlessly fails to establish infection in man at all.

ERADICATION OF VIRAL DISEASES

Monotypic, single-host viruses, that do not persist to produce recurrent disease in the infected individual and do not have an alternative animal host, can survive only in communities with a relatively high annual input of new susceptibles. Measles in man provides the best example, but smallpox and rubella are exactly comparable; infections with rhinoviruses and enteroviruses are also probably in the same category, although the critical population size may be smaller because homotypic immunity is less durable. Each of the many serotypes causing identical symptoms constitutes a separate problem in herd immunity, so that elimination of a particular disease would require the eradication of many viruses.

Since this dependence of the virus on a critical community size is due to herd immunity, it should be possible to render even very large communities immune by actively immunizing most, but not necessarily all, susceptible individuals. This has encouraged some public health workers to think in terms of the total eradication of disease. The concept is feasible for some viral diseases, on a national, regional, or even worldwide scale. The first and only example so far, achieved late in 1975 following years of dedicated effort, was the global eradication of smallpox by means of a WHO-sponsored vaccination campaign throughout all countries in which the virus was known to be endemic. Smallpox was everywhere a severe and feared disease, there was no animal host (detailed studies of monkeypox appear to have exonerated it), it was monotypic, immunity was lifelong, and vaccination was easy and effective, i.e., it had all the prerequisites for a successful eradication campaign. Yet the WHO campaign met with repeated setbacks, because of social rather than technical difficulties, often of a novel and unexpected kind. Major social upheavals like the wars and famines in Bangladesh and Ethiopia set back the program for several years, but the final triumph now seems to have been achieved.

Global eradication of diseases that produce recurrent activity after long periods of quiescence, or that have an animal reservoir, is impossible and constitutes an unreal goal. Yellow fever provides an excellent example of the latter situation because it was the first disease for which eradication was proposed as a deliberate program. At first success seemed assured, but then the discovery of jungle yellow fever, with the monkey as the primary host (see Figure 10-1) made it obvious that eradication of the virus was impossible. If hemorrhagic dengue fever continues to be an important disease in tropical areas, an *Aedes aegypti* eradication campaign based on the destruction of breeding places might prove worthwhile, but in many tropical areas it might founder because monkeys may constitute a nonhuman reservoir of dengue viruses, or it might show that mosquitoes we now believe to be minor vectors are capable of maintaining the disease in human communities.

Local eradication of poliomyelitis also has been achieved in many developed countries, and could probably be accomplished with measles, and eventually with rubella and mumps by nation-wide vaccination, although the difficulties of achieving nation-wide vaccination for relatively mild or infrequent diseases should not be underestimated. Experience in the U.S.A. and some other advanced industrialized nations with supposedly "universal" immunization against poliomyelitis and measles has shown that even a herd immunity in excess of 90% nation-wide does not protect unimmunized pockets of ethnic minority groups in inner city slums and other poverty areas.

Even if a virus is eradicated from a particular country, a major public health problem is raised concerning the continued protection of that population. Unless eradication is global the virus will sooner or later be reintroduced by a traveller. Protection could probably be maintained by continued large-scale immunization, but the psychological difficulties of achieving this in the absence of disease have been made apparent by experience with compulsory vaccination against smallpox. Some measure of protection of a virus-free nonvaccinated community can be achieved by quarantine and by the demand for evidence of recent vaccination of travelers re-entering the country. The latter measure could well be extended to include immunization against the viruses of poliomyelitis and measles.

SUMMARY

The topics discussed in this chapter do not call for a summary; instead an attempt will be made to set out some factors that are important in the evolution of viral diseases.

Survival of virus in populations is facilitated by
   High transmissibility, low mortality
   Large urban populations
   Latent infections
   Vertical transmission
   Antigenic drift
   Animal reservoirs, with or without arthropod transmission

Natural balance of virus and host may be disturbed by
   Introduction of a new virus, leading to virgin soil epidemics, e.g.,
      (i) Smallpox and yellow fever to the Americas
      (ii) Measles to the Pacific Islands
      (iii) Myxomatosis to European rabbits in Australia and Europe
   Introduction of a new host animal, e.g.,
      (i) White man to Panama or West Africa—yellow fever

    (ii) Domestic livestock to Africa—bluetongue, Rift Valley fever, etc.

Control of viral diseases may be achieved by
  Quarantine
  Immunization
  Destruction of vectors and nonhuman hosts
  Hygiene and sanitation

Eradication of viral diseases may be achieved by
  Natural circumstances, e.g., "English Sweats," encephalitis lethargica
  Isolation, e.g., measles and poliomyelitis in eskimos
  Immunization, e.g., smallpox, poliomyelitis, measles

FURTHER READING

Andrewes, C. H. (1967). "The Natural History of Viruses." Weidenfeld & Nicolson, London.

Black, F. L. (1971). Penetrance and persistence of virus in isolated communities. In "Viruses Affecting Man and Animals" (M. Sanders and M. Schaeffer, eds.), pp. 284–294. Green, St. Louis, Missouri.

Black, F. L., Woodall, J. P., Evans, A. S., Liebhaber, H., and Henle, G. (1970). Prevalence of antibody against viruses in the Tiriyo, an isolated Amazon tribe. *Amer. J. Epidemiol.* **91**, 430.

Burnet, F. M., and White, D. O. (1972). "Natural History of Infectious Disease," 4th ed. Cambridge Univ. Press, London and New York.

Cockburn, T. A. (1961). Eradication of infectious diseases. *Science* **133**, 1050.

Fazekas de St. Groth, S. (1970). Evolution and hierarchy of influenza viruses. *Arch. Environ. Health* **21**, 293.

Fenner, F., and Ratcliffe, F. N. (1965). "Myxomatosis." Cambridge Univ. Press, London and New York.

Fenner, F., McAuslan, B. R., Mims, C. A., Sambrook, J. F., and White, D. O. (1974). "The Biology of Animal Viruses," 2nd ed., pp. 618–641. Academic Press, New York.

Joklik, W. K. (1974). Evolution in viruses. In "Evolution in the Microbial World," 24th Symp. Soc. Gen. Microbiol. Cambridge Univ. Press, London and New York.

Nahmias, A. J. (1974). The evolution (evovirology) of herpesviruses. In "Viruses, Evolution and Cancer" (K. Maramorosch and E. Kurstak, eds.), pp. 605–624. Academic Press, New York.

Reanney, D. C. (1974). Viruses and evolution. *Int. Rev. Cytol.* **37**, 21.

Skehel, J. J. (1974). The origin of pandemic influenza viruses. In "Evolution in the Microbial World," 24th Symp. Soc. Gen. Microbiol. Cambridge Univ. Press, London and New York.

Temin, H. (1974). Origin of RNA tumor viruses. *Annu. Rev. Genet.* **8**, 155.

Todaro, G. J., and Huebner, R. J. (1972). The viral oncogene hypothesis: New evidence. *Proc. Nat. Acad. Sci. U.S.* **69**, 1009.

Webster, R. G. (1972). On the origin of pandemic influenza viruses. *Curr. Top. Microbiol. Immunol.* **59**, 75.

World Health Organization. (1972). WHO Expert Committee on Smallpox Eradication, Second Report. *World Health Organ., Tech. Rep. Ser.* **493**.

CHAPTER 12

Immunization against Viral Diseases

INTRODUCTION

There are three main approaches to the prophylaxis and therapy of viral infections: (a) chemotherapeutic agents that inhibit viral multiplication directly; (b) agents that induce the body to synthesize interferon or to mount other types of nonimmunological defence; and (c) vaccines that elicit an immune response. All three can be used to prevent infection (chemoprophylaxis or immunoprophylaxis) or to treat established or impending disease (chemotherapy or immunotherapy), but for obvious reasons all methods are more effective the earlier they are applied and some, notably active immunization, are usually ineffective after infection has begun. A number of viral vaccines have been outstandingly successful, especially those directed against smallpox, yellow fever, poliomyelitis, measles, and rubella, all of which are generalized diseases (Table 12-1). Vaccines against the many diseases caused by viruses localized to the respiratory tract have been much less effective; the containment of these relatively trivial but annoying infections is more likely to come via the eventual development of antiviral chemotherapeutic agents.

Vaccines are prepared by rendering viruses harmless without destroying their immunogenicity; this can be achieved either by inactivating the infectivity of the virion, or by selecting an avirulent mutant. A satisfactory vaccine must fulfill the dual requirements of safety and efficacy. The problems encountered with live attenuated vaccines are in the main concerned with safety, while those with inactivated vaccines are almost exclusively those of effectiveness, i.e., of adequate immunogenicity.

TABLE 12-1

Viral Vaccines Recommended for Use in Man[a]

DISEASE	VACCINE STRAIN	CELL SUBSTRATE	ATTENUATION	INACTIVATION	ROUTE
Smallpox	Vaccinia	Calf or sheep skin	+	—	Intradermal
Yellow fever	17D	Chick embryo	+	—	Subcutaneous
Poliomyelitis	Sabin 1, 2, 3	WI-38	+	—	Oral
Measles	Schwarz	Chick fibroblasts	+	—	Subcutaneous
Rubella	RA 27/3	WI-38	+	—	Subcutaneous
	Cendehill	Rabbit kidney	+	—	Subcutaneous
Mumps	Jeryl Lynn	Chick fibroblasts	+	—	Subcutaneous
Rabies	Pitman-Moore	WI-38	—	β-Propiolactone	Intramuscular
Influenza	A$_2$(H3N2)	Chick embryo	—	Formaldehyde and deoxycholate	Subcutaneous
Adenovirus	4, 7	WI-38	—	Live, in enteric-coated capsules	Oral

[a] A great variety of different viral strains and cellular substrates are used in different countries; the selection listed is not comprehensive.

EFFICACY OF VIRAL VACCINES

The long-term protection that follows systemic infection with such viruses as smallpox or measles is primarily a function of humoral (antibody-based) immunity which depends principally upon immunoglobulin G (IgG) (see Chapter 7). On the other hand, immunity after superficial infection of the respiratory mucous membrane is relatively short-lived, depending upon immunoglobulin A (IgA) synthesized by lymphoid cells in the epithelium and secreted into the mucus. The poor immunity that follows natural infection with most of the respiratory viruses does not augur well for the future of vaccines directed against them.

Attention has already been drawn in Chapter 7 to the fact that most of the viruses that cause generalized viral infections are monotypic, whereas great antigenic diversity is found among the viruses that cause localized infections of superficial mucous membranes. Such antigenic diversity amongst the rhinoviruses, for example, virtually rules out any possibility of an effective common cold vaccine. However, the severe lower respiratory tract infections of young children tend to be attributable to a relatively narrow range of viral serotypes: one respiratory syncytial virus, four parainfluenza viruses, two influenza viruses, and three adenoviruses. Effective vaccines against these agents might make an appreciable impact on the total spectrum of serious respiratory disease in children.

Antigenic drift constitutes an additional difficulty with vaccines against influenza virus; its basis is discussed in Chapters 7 and 11.

SAFETY OF VIRAL VACCINES

Throughout the history of vaccine development there has been a succession of accidents, some of which were disastrous, others potentially so (Table 12-2). Most of the problems were encountered with live vaccines and are discussed in detail under that heading; they include (a) insufficient attenuation of the vaccine strain, leading to disease and sometimes the spread of disease to contacts and/or back-mutation to high virulence and (b) the presence of adventitious viruses in the cells used to grow the vaccine virus. The risks have now been considerably diminished by the establishment of elaborate laboratories and government authorities to test vaccines for safety and efficacy and to evaluate reports of clinical trials in man.

There is one immunological problem that has arisen only with inactivated vaccines. It was found that children immunized with inac-

TABLE 12-2
*Some Problems Encountered during the Development
and Use of Viral Vaccines*

VACCINE	PROBLEM
Live vaccines	
Poliovirus (Sabin)	Contaminating viruses (e.g., SV40)
	Back-mutation to virulence (type 3)
	Interference by endemic enteroviruses
Smallpox	Inadequate attenuation leading to complications
	Overattenuation leading to lack of protection
Measles (Edmonston)	Inadequate attenuation leading to fever and rash
Rubella (HPV-77)	Inadequate attenuation leading to arthritis in adult females
Yellow fever	Contaminating hepatitis B in "stabilizing" medium
Inactivated vaccines	
Poliovirus (Salk)	Residual live virulent virus
	Contaminating virus (SV40) resisting formaldehyde
Influenza	Pyrogenicity
Rabies (Semple)	Allergic encephalomyelitis
Measles ⎱	Hypersensitivity reactions on subsequent natural infection
Respiratory syncytial ⎰	or live vaccine booster

tivated measles or respiratory syncytial vaccine, when subsequently exposed to live vaccine (in the first case) or to natural infection (in the second case), developed more serious illness than unimmunized controls. The immunological explanation of these unfortunate episodes is still a matter of debate but clearly some sort of hypersensitivity reaction was involved. The lesson to be learned is that inactivated vaccines must be carefully evaluated before being used at all in the future, and that living attenuated vaccines intended for delivery by nasal spray might conceivably produce bronchial spasm and exudation following repeated application in an atopic individual.

LIVE VACCINES

Almost all of the successful viral vaccines now used in man are living avirulent viruses (see Table 12-1). Their advantages over inactivated vaccines are summarized in Table 12-3. Multiplication in the host, leading to a prolonged immunogenic stimulus of similar kind and magnitude to that occurring in natural subclinical infection, gives rise to a substantial immunity. In some cases, the virus can be inoculated by the natural route, i.e., orally or nasally, where it can induce the continuing synthesis of immunoglobulin A by lymphoid cells

TABLE 12-3

Advantages and Disadvantages of Live and Inactivated Vaccines

ADVANTAGES	DISADVANTAGES
Live vaccines	
Single dose, given ideally by natural route, invokes full range of immunological responses, including local IgA as well as systemic IgG production	Reversion to virulence[a]
	Natural spread to contacts
	Contaminating viruses[a]
Possibility of local eradication of wild-type viruses.	"Human cancer viruses"[b]
	Viral interference
	Inactivation by heat in tropics
Inactivated vaccines	
Stability	Multiple doses and boosters needed, given by injection, therefore local IgA fails to develop
	High concentration of antigen needed: production difficulties

[a] With care these difficulties have been largely overcome.

[b] Theoretical objection that has been raised against use of human diploid cell strains as substrate for vaccine production, but vaccines prepared in such cells are now licensed in several countries including Great Britain.

present in tonsils, Peyer's patches, and throughout the respiratory epithelium, and the resulting local immunity may suffice to prevent reimplantation of wild strains.

Derivation of Vaccine Strains

A few vaccines still used in veterinary practice consist of fully virulent infectious virus injected in a site where its multiplication is of little consequence. This procedure has been considered too dangerous for use in man ever since immunization with live smallpox virus ("variolation") was abandoned more than a century ago. The ingenious technique of delivering respiratory pathogens, such as adenoviruses, to the lower reaches of the alimentary tract in enteric coated capsules is an interesting exception.

Most vaccine strains are mutants which are sufficiently attenuated to produce no symptoms (or trivial ones), but they are not so avirulent that they fail to multiply extensively in the body and elicit a lasting immunity. It is not always easy to strike this balance. The process of attenuation of most current vaccines has been empirical, based on the knowledge that prolonged passage of a virus in a foreign host tends to select mutants better suited to growth in the new host than in the old one (see Chapter 4). The emergence of rare mutants is facilitated by frequent passage of large viral inocula (by contrast with the practice of infrequent passage at limit dilution used for the maintenance of laboratory stocks of wild-type viruses in their original state). To demonstrate loss of virulence the vaccine virus is extensively tested in animals, but the ultimate test of avirulence for man can only come from experimental inoculation of volunteers. The acceptable level of virulence sometimes requires a degree of compromise about the immunogenicity of a vaccine. For example, the original Edmonston strain of live measles vaccine produced quite a high fever in a substantial proportion of children, and even occasional rashes. For a time it was recommended that human immunoglobulin be simultaneously inoculated into the opposite arm to minimize these side effects. Nowadays, however, it is considered wiser to use the more attenuated Schwarz vaccine which is slightly less immunogenic, but produces negligible side effects.

Today it is no longer necessary to rely on the purely empirical process of rapid serial passage in a foreign host to produce an avirulent mutant, for our knowledge of animal virus genetics is sufficiently advanced for us to be able to apply rational procedures to the selection of candidate strains for attenuated viral vaccines (see Chapter 4). Chanock and his colleagues have made good use of genetic principles to derive *ts* mutants of respiratory syncytial virus and influenza virus that grow poorly at 37°, the temperature of the lower respiratory tract of man, but well at 33°, the temperature of the nose. Recently, workers in two laboratories have gone further and used recombination of a *ts* mutant with a wild strain of influenza virus to produce a genetic recombinant (see Chapter 4) with the hemagglutinin of the virulent strain but the *ts* lesion of the attenuated mutant. These genetically reassorted viruses grow satisfactorily only in the upper respiratory tract, and one of them has been successfully used as a living intranasal vaccine to protect human volunteers against challenge with virulent influenza virus.

Potential Problems

In spite of their obvious advantages, live vaccines are subject to a number of problems that do not occur with inactivated vaccines (Table 12-2). These are: (a) genetic instability, (b) contamination by adventitious viruses, (c) interference by wild viruses, and (d) heat lability Now that the problems have been recognized, ways have been found to circumvent each of them, and most vaccines commercially available for human use today are in no way hazardous.

Genetic Instability. Some vaccine strains are not as avirulent as one would wish, but tests have shown that further attenuation leads to an unacceptable loss of immunogenicity or even to loss of the capacity to grow in man when administered by the desired route. Indeed, each of the human live vaccines listed in Table 12-1, with the exception of poliovirus vaccine, produces mild symptoms in a minority of recipients, and vaccinia very occasionally causes death. Of more concern is the theoretical possibility that live vaccine strains may back-mutate to, or toward, the virulent wild type. Only for poliovirus type 3 have we statistical evidence that this may ever have occurred in practice, albeit very rarely.

A related question is whether vaccine strains, unchanged or virulent, are capable of spreading to the vaccinee's contacts. This would be particularly dangerous if rubella vaccine, which is excreted in the throat, were teratogenic and could spread to pregnant women, but this has not occurred. Vaccinia virus is occasionally transmitted, by direct contact, to other people, or more commonly to other parts of the vaccinee, and can be dangerous if it enters the eye.

Reports that *ts* mutants of some viruses may be capable of setting up persistent infections in the brain of animals must cause us to be more than usually cautious before adopting *ts* mutants for use as human vaccines.

Contaminating Viruses. Many cells used for culture are contaminated with viruses (see Chapter 8). Because of methods of shipping and holding monkeys, monkey kidney cells are particularly liable to be contaminated, with any or several of over fifty known simian viruses. The risks are less with closed, specific-pathogen-free colonies of animals, such as the rabbits used for the production of Cendehill strain rubella vaccine, but even in this situation viruses transferred congenitally cannot be eliminated. For example, most of the older human viral vaccines were grown in chick embryos or cultured chick embryo fibroblasts which were contaminated with avian leukosis viruses. Fortunately no ill seems to have come of it. Contaminating viruses are clearly more likely to be dangerous in live vaccines, but inactivated vaccines are not free from this risk, for the adventitious virus may be more resistant to the inactivating agent than the vaccine strain. For example, many early batches of formaldehyde-inactivated poliovaccine were contaminated by live SV40.

The discovery of adenovirus-SV40 hybrids (see Chapter 4) has added a new dimension to the problem of contaminating viruses. Such recombinants, some of which are highly oncogenic, are indistinguishable serologically from the parent adenovirus since they have identical capsids. Up to the present time, this interfamilial hybridization remains a unique laboratory artifact, but in relation to live virus vaccines it represents a risk that will have to be borne in mind.

The problem of contaminating viruses has led to the abandonment of monkey kidney cells for the preparation of vaccines. Human diploid cells like WI-38 constitute a logical replacement. Their use in some countries has been prohibited until recently because of the hypothetical risk that they might carry a "human cancer virus." However, since several million people in many countries have now received WI-38-grown vaccines against a number of viral diseases and no untoward complications have been reported, it may confidently be forecast that most human viral vaccines will be prepared in diploid cell strains from now on.

Interference. If a live vaccine such as the oral poliovirus vaccine is given by the natural route of infection, preexisting enteroviral infection of the gut of the vaccinated individual may prevent the establishment of the vaccine strain, by interference. This presents a problem in some underdeveloped countries where enteroviruses are present in the intestines of the majority of children at any particular time. It is obviated by the practice of giving the vaccine on three occasions, 6–8 weeks apart. In spite of earlier doubts, interference between the three serotypes comprising the poliovirus vaccine itself does not occur to any appreciable extent under this regimen, but the vaccine strains may themselves interfere with the circulation of virulent strains of poliovirus, if given at the time of an epidemic.

Interference between live viruses in a polyvalent vaccine does not usually occur following parenteral administration, probably because the viruses are rapidly dispersed throughout the body. For instance, a live measles-mumps-rubella vaccine appears to give satisfactory antibody responses against all three viruses.

Heat Lability. Since a reasonable titer of active virus must be maintained, shipping and handling of live vaccines calls for good organization of public health services, including refrigeration and prompt use of the vaccine after reconstitution from the lyophilized state. The problem is greatest in tropical countries where the ambient temperature is high and health services are often inadequate.

INACTIVATED VACCINES

Inactivated vaccines are "dead" in the sense that their infectivity has been destroyed, usually by treatment with formaldehyde. The only inactivated viral vaccine now in widespread use in man is that directed against influenza. An inactivated rabies vaccine is employed in particular emergencies. Others, like Salk poliomyelitis vaccine and inactivated measles vaccine have been displaced by attenuated live-virus vaccines.

Most inactivated vaccines used today have been treated with 1:4000 formaldehyde for long periods at 37° until there is no residual infectivity but antigenicity and immunogenicity are little affected. However, the whole process is empirical and we know very little about the chemistry of the inactivation of viruses by formalin or any other agent.

Ideally an agent is needed that specifically inactivates the viral nucleic acid (which carries the infectivity) without affecting the capsid or envelope protein (which carries the immunogenicity). Formalin does not, of course, come into this category, since it denatures protein by cross-linking, as well as reacting with the amino groups of nucleotides. Likely candidates are β-propiolactone or oxidized spermine. One of the main difficulties with agents whose sole target is nucleic acid, e.g., hydroxylamine, is that mutants may result unless every nucleic acid molecule is subjected to several hits by the agent being used, or unless the action of the agent is such that mutants cannot be formed. Multiplicity reactivation, which is known to occur with several viruses (Chapter 4), is a strong contraindication to ultraviolet irradiation.

The major difficulty with inactivated vaccines lies in producing enough virus to provide the necessary quantity of the relevant antigens and in administering the vaccine in such a way as to promote local IgA synthesis where this is relevant, as well as systemic antibody production. With influenza vaccine the large dose of virus required often causes febrile reactions, especially in young children. This difficulty can be overcome by disrupting the virions with ether or deoxycholate, and such "split vaccines" are now in use.

At an experimental level efforts are being made to produce effective "polyvalent" vaccines to control the major viruses that cause respiratory disease: respiratory syncytial virus, parainfluenza types 1, 2, and 3, influenza type A2, and adenovirus types 3, 4, and 7. Inactivated vaccines may be better suited to this end than living vaccines because of potential difficulties with mutual interference if live viruses were to be administered together intranasally. In order to ensure that immunogenic amounts of each inactivated virus can be contained in a volume suitable for injection, modern techniques for concentrating and purifying viruses will need to be employed in the development of such "polyvalent" vaccines.

Subunit Vaccines

If inactivated vaccines are to be used, there is a good case to be made for vaccines consisting solely of the relevant immunogenic material, i.e., the capsid proteins of nonenveloped icosahedral viruses or the peplomers of enveloped viruses. Techniques are now available that make this a possibliity. For instance, influenza hemagglutinin has been purified to the point of crystallization but the cost of producing the material in immunogenic amounts may turn out to be prohibitive.

Should one of the more common human cancers be found to be "caused" by virus(es), there will be considerable pressure to produce a vaccine. The difficulty of safety-testing in man rules out the use of attenuated strains, however effective they may be in animals (as in vaccination of chickens against the herpetovirus of Marek's disease). Inactivated vaccines, or preferably subunit vaccines free of viral nucleic acid, are likely to be the only acceptable form of vaccine.

Adjuvants

The immunological response to antigens can be greatly enhanced if they are administered in an emulsion with adjuvants. The mechanism is complex, involving, among other things, stimulation of phagocytosis and other activities of the reticuloendothelial system and a delayed release and degradation of the antigen. Adjuvants may be particularly valuable in enabling a number of highly concentrated and purified viruses to be incorporated into a single polyvalent inactivated vaccine in a volume suitable for injection. However, conventional adjuvants (e.g., Freund's) based on mineral oils are not accepted for use in man because they are nonmetabolizable and are potentially carcinogenic. Suspicions about the carcinogenicity of Arlacel A (the emulsifying agent) have so far prevented United States licensing of the metabolizable, isomannide monooleate (peanut oil) adjuvant known as Adjuvant 65, although it is now licensed in Great Britain.

PASSIVE IMMUNIZATION

Instead of actively immunizing with viral vaccines it is possible to confer short-term protection by the administration of preformed antibody, as immune serum or concentrated immunoglobulin prepared from it. Human immunoglobulin is usually preferred in human medicine, because heterologous protein provokes an immune response to the foreign antigenic components. Pooled normal human immunoglobulin can be relied on to contain high titers of antibody against all the common viruses that cause systemic disease in man.

Although passive immunization should be regarded only as an emergency procedure for the immediate protection of unimmunized individuals exposed to special risk, it is an important prophylactic measure in several human viral infections. Human immunoglobulin has proved to be most effective in the short-term prophylaxis of hepatitis A; it can be administered with advantage to contacts of cases in schools and institutions or to travelers or military personnel about to visit parts of the world where hepatitis is prevalent. Experience with prevention of hepatitis B using human normal immunoglobulin has on the whole not been so favorable. Incidentally, the immunoglobulin does not itself contain hepatitis B virus or antigen because the latter is lost during the cold ethanol precipitation step in the purification procedure.

The importance of passive immunization is diminishing now that satisfactory vaccines for such diseases as poliomyelitis, measles, mumps, and rubella are widely used. Its major role today is in short term protection against hepatitis and measles, and in the treatment of complications of smallpox vaccination (see relevant chapters in Part II for details). There is also a special place for hyperimmune γ-globulin in protection of congenitally immunodeficient children, or leukemic children on immunosuppressive therapy, or immunosuppressed cancer or kidney transplant patients, following exposure to viruses such as varicella or measles.

Much interest has been aroused recently by claims that adoptive transfer of "transfer factor" from convalescent patients confers some protection against specific viral diseases, particularly in immunosuppressed individuals. At the time of writing the therapy is being extensively tested on patients gravely ill with overwhelming generalized herpetovirus infections.

THE PSYCHOLOGY OF IMMUNIZATION

Fear is probably the major factor motivating people to seek or accept immunization for themselves and their children, so that when a feared disease like poliomyelitis has almost disappeared and is no longer seen as a threat to the individual, it is difficult to maintain community-wide vaccination. This has led to a limited resurgence of measles in the United States and of poliomyelitis in some other countries. Disappearance of pathogenic viruses from a community as a result of widespread immunization removes the beneficial influence of repeated subclinical infections to boost immunity and leaves both the immunized, and particularly the unimmunized, unusually vulnerable should the agent ever reappear. Continuation of routine immunization after the threat appears to have vanished but the virus has not been totally eradicated calls for highly organized public health services.

An important but rather neglected aspect of this problem stems from the unnecessarily complicated immunization schedules to which mothers are required to adhere. Polyvalent vaccines, dead or alive, would be a major practical advantage. Greater safety would be assured by the use of presterilized plastic syringes already loaded with a single dose of vaccine. Jet guns which quickly and painlessly deliver a jet of vaccine through the skin under high pressure should do much to lessen patients' objections and also to speed up crash immunization programs. Nevertheless, as far as members of the public are concerned, there is little doubt that oral vaccines will continue to remain the most popular.

The acceptability of a vaccine against any given disease depends upon the public need for the vaccine and the degree of protection it affords. Where the need is clear, as in the case of lethal diseases such as smallpox or rabies, or even influenza, we have been inclined to accept lower standards of safety (in the first two cases) or efficacy (in the latter) than we would today for new vaccines directed against more trivial illnesses. This is relevant to current work on respiratory vaccines; they will only be acceptable if they are completely safe and reasonably efficacious, and if the disease is sufficiently debilitating to be worth preventing. As severe and lethal bacterial diseases progressively disappear, levels of public expectation concerning freedom from viral diseases may be expected to rise.

HUMAN VIRAL VACCINES

Many of the principles introduced for the first time in this chapter are illustrated at greater length in the detailed discussions of the vaccines currently in use against smallpox, rabies, yellow fever, poliomyelitis, measles, mumps, rubella, and influenza, to be found in the relevant sections of Part II of this book.

There remains an outstanding need for a vaccine against hepatitis, and if the poliomyelitis analogy is valid, there is good reason to be optimistic about the prospects, once satisfactory cell culture systems have been developed. Other possible future developments of somewhat lower priority, and/or lower likelihood of success, are vaccines for viral enteritis, various herpetoviruses, arboviruses, and looking even further ahead, perhaps viral degenerative diseases of the brain and virus-induced cancer.

SUMMARY

The immunological procedures of active and passive immunization constitute the only widely applicable specific methods available for the prevention of viral infection. Active immunization can be achieved either with live attenuated viruses, which produce a mild infection, or with inactivated viral vaccines. Table 12-4 lists the vaccines currently available for use in man, together with typical schedules for their administration.

Live vaccines are more widely used than inactivated vaccines, but care must be taken by the manufacturers to achieve an adequate degree of attenuation and to ensure the absence of contaminating viruses. When given by the natural route of infection, live vaccines stimulate local production of IgA which is probably important in preventing infection.

Inactivated vaccines are free from most of the potential hazards associated with live vaccines, but the amount of viral antigen must be large and several injections must be given. The efficacy of purified viral antigens and the safety of adjuvants have still to be determined.

TABLE 12-4

Schedules for Immunization against Human Viral Diseases[a]

VACCINE	FIRST DOSE	SUBSEQUENT DOSES	BOOSTER DOSES
Live vaccines			
Smallpox	Before travel to endemic area	—	Valid for 3 years
Yellow fever	Before travel to endemic area	—	Valid for 10 years
Poliomyelitis	6 months	8 months and 10 months	Every 5 years
Measles	1 year	—	Every 5–10 years ?
Rubella	1 year or 12 years	—	Prepuberty in females
Mumps	1 year or later	—	?
Inactivated vaccines			
Influenza	Autumn before expected epidemic	1 month later	Annually in elderly
Rabies	Immediately after bite	—	—

[a] See also chapters in Part II relating to particular diseases.

Passive immunization with normal human immunoglobulin is useful for the prophylaxis of viral hepatitis and in a limited number of other situations including the treatment of complications of smallpox vaccination.

FURTHER READING

Bachrach, H. L., and Breese, S. S. (1968). Cell cultures and pure animal virus in quantity. *In* "Methods in Virology" (K. Maramorosch and H. Koprowski, eds.), Vol. 4, p. 351. Academic Press, New York.

Chanock, R. M. (1970). Control of acute mycoplasmal and viral respiratory tract disease. *Science* **169**, 248.

Chanock, R. M. (1971). Local antibody and resistance to acute viral respiratory tract disease. *In* "The Secretory Immunologic System" (D. H. Dayton *et al.*, eds.), p. 83. Nat. Inst. Child Health Hum. Develop., Nat. Inst. Health, Bethesda, Maryland.

Chanock, R. M., Kapikian, A. Z., Perkins, J. C., and Parrott, R. H. (1971). Vaccines for non-bacterial respiratory diseases other than influenza. *In* "International Conference on the Application of Vaccines against Viral, Rickettsial and Bacterial Diseases of Man," Sci. Publ. No. 226, p. 101. Pan Amer. Health Organ., Washington, D.C.

Fenner, F., McAuslan, B. R., Mims, C. A., Sambrook, J. F., and White, D. O. (1974). "The Biology of Animal Viruses," 2nd ed., pp. 543–567. Academic Press, New York.

Gajdusek, D. C., Gibbs, C. J., and Lim, K. A. (1971). Prospects for the control of chronic degenerative diseases with vaccines. *In* International Conference on the Application of Vaccines against Viral, Rickettsial, and Bacterial Diseases of Man," Sci. Publ. No. 226, p. 566. Pan Amer. Health Organ., Washington, D.C.

Hayflick, L. (1965). The limited *in vitro* lifetime of human diploid cell strains. *Exp. Cell Res.* **37**, 614.

Hilleman, M. R. (1966). Critical appraisal of emulsified oil adjuvants applied to viral vaccines. *Progr. Med. Virol.* **8**, 131.

Hilleman, M. R. (1969). Toward control of viral infections of man. *Science* **164**, 506.

Hilleman, M. R. (1973). Viral vaccines and the control of cancer. *Perspect. Virol.* **8**, 119.

Hilleman, M. R., Weibel, R. E., Villarejos, V. M., Buynak, E. B., Stokes, J., Argeudeas, G. J. A., and Vargas, A. G. (1971). Combined live virus vaccines. *In* "Proceedings of the International Conference on the Application of Vaccines against Viral, Rickettsial and Bacterial Diseases of Man," Sci. Publ. No. 226, p. 397. Pan Amer. Health Organ., Washington, D.C.

Horstman, D. M. (1973). Need for monitoring vaccinated populations for immunity levels. *Progr. Med. Virol.* **16**, 215.

Isacson, P., and Stone, A. (1971). Allergic reactions associated with viral vaccines. *Progr. Med. Virol.* **13**, 239.

Kohn, A., and Klingberg, M. A., eds. (1972). "Immunity in Viral and Rickettsial Diseases," Advan. Exp. Med. Biol. No. 31. Plenum, New York.

National Cancer Institute. (1968). Cell cultures for virus vaccine production. *Nat. Cancer Inst., Monogr.* **29**.

Neurath, A. R., and Rubin, B. A. (1971). Viral structural components as immunogens of prophylactic value. *Monogr. Virol.* **4**, 1.

Pan American Health Organization. (1971). "International Conference on the Application of Vaccines against Viral, Rickettsial and Bacterial Diseases of Man," Sci. Publ. No. 226. Pan Amer. Health Organ., Washington D.C.

Pollock, T. M. (1969). Human immunoglobulin in prophylaxis. *Brit. Med. Bull.* **25**, 202.

Potash, L. (1968). Methods in human virus vaccine preparation. *In* "Methods in Virology" (K. Maramorosch and H. Koprowski, eds.), Vol. 4, p. 372. Academic Press, New York.

Rossen, R. D., Kasel, J. A., and Couch, R. B. (1971). The secretory immune system: Its relation to respiratory viral infection. *Progr. Med. Virol.* **13,** 194.

World Health Organization. (1966a). Report to Scientific Group on Human Viral and Rickettsial Vaccines. *World Health Organ., Tech. Rep. Ser.* **325.**

World Health Organization. (1966b). The use of human immunoglobulin. *World Health Organ., Tech. Rep. Ser.* **327.**

World Health Organization. (1967). "International Conference on Vaccines against Viral and Rickettsial Diseases of Man," Sci. Publ. No. 147. World Health Organ., Washington, D.C.

World Health Organization. (1972). WHO Expert Committee on Biological Standardization, 24th Report. *World Health Organ., Tech. Rep. Ser.* **486.**

CHAPTER 13

Chemotherapy of Viral Diseases

Until comparatively recently it was thought to be self-evident that there was such a close relationship between the cellular metabolic processes required for viral multiplication and those needed for the survival of vertebrate cells that it was fruitless to search for antiviral chemotherapeutic or chemoprophylactic agents. Now, as a result of recent advances in molecular virology, we know that there are in fact several biochemical processes that are essential to viral replication but of no consequence to the cell. This knowledge should have provided us, one would have imagined, with a new rationale on which to base a logical search for antiviral agents, yet there has not so far been a corresponding harvest of new drugs. Only three groups of agents, all of which have been available since the early 1960's, have emerged for practical use: the pyrimidine nucleosides, methisazone, and amantadine, and they have had a negligible effect on human morbidity or mortality from viral diseases.

Reports of new antiviral agents appear at regular intervals, but the great majority of such "potentially useful" substances have not even been shown to be nontoxic to cultured cells at the minimum virus-inhibitory concentration. Because of the great expense now involved in drug testing, even fewer have been successfully tested in animals, and very few ever get to the stage of properly controlled human trials. Most "antiviral agents" picked up by empirical screening have no more selective toxicity than substances like actinomycin D, fluoro-deoxyuridine, or cycloheximide, which inhibit the synthesis of viral macromolecules at concentrations no lower than those that block synthesis of the corresponding cellular macromolecules. Although some of these substances have proved valuable tools for analyzing the viral multiplication cycle, they have no potential in human medicine. The task of the pharmaceutical companies is to mount a logical search for agents thought likely to inhibit a known viral enzyme or process, select the most promising from *in vitro* assays, and demonstrate lack of toxicity for cultured cells and then for animals, at the minimum virostatic dose. The *chemotherapeutic index* of a compound can be defined as the ratio between the lowest effective antiviral concentration and the highest nontoxic concentration. Substances displaying an index no greater than unity in cultured cells are not worth pursuing further because, even though they may be nonlethal to animals at virostatic concentrations, it can usually be assumed that they are killing some cells, often the crucial dividing cells of the bone marrow, intestine, and liver, and therefore doing an unacceptable amount of damage. On the other hand, substances that fail to cause cytopathic effects or to slow cell division in cultured cells at virostatic concentrations often turn out to be disappointingly ineffective *in vivo*; guanidine and HBB are good examples.

THE PROSPECTS FOR ANTIVIRAL CHEMOTHERAPY

The ideal chemotherapeutic agent would have a high chemotherapeutic index, i.e., a high margin of safety in man, as well as certain other desirable pharmacological properties such as a reasonably long half-life *in vivo,* solubility, and the capacity to penetrate the target cells. Hopefully it would also be broad spectrum so that it could be used against a wide range of viruses without the delay involved in laborious laboratory diagnosis. Until rapid diagnostic techniques such as immunofluorescence and immune-electron microscopy become much more highly developed and generally available than they are today, the delays (and costs) involved in establishing a definitive virological diagnosis would invalidate chemotherapy oriented toward specific viruses, except in two situations: (a) the few viral diseases caused by monotypic viruses such as herpes simplex, varicella-zoster, smallpox, mumps, and measles where the causative agent may be diagnosed on clinical findings alone, and (b) large-scale epidemic outbreaks of a particular disease like influenza.

Another theoretical difficulty with antiviral chemotherapy is the fact that viral multiplication may be almost over by the time the patient presents with an established illness. In the case of most respiratory infections, this may mean that antiviral agents will have to be used prophylactically rather than therapeutically, a procedure that would be practicable only for the short-term protection of family and professional contacts of the sentinel case, or perhaps more widespread protection of whole communities in the face of impending epidemics of serious viral diseases like influenza. However, the success attending the use of tetracyclines in several generalized rickettsial infections whose pathogenesis parallels that of the generalized viral infections suggests that if satisfactory drugs existed they would be useful even in established generalized diseases. Further, there are many viral diseases characterized by prodromal symptoms that may last up to a couple of days and provide time for therapy to be initiated; the childhood exanthemata come into this category. Drugs could also be valuable in preventing the development of complications, such as orchitis or meningitis in mumps. Chronic diseases, such as warts or hepatitis, slow diseases such as kuru, or recurrent endogenous diseases, such as herpes simplex and zoster, might be very suitable candidates for chemotherapy. Embryopathy caused by rubella or cytomegaloviruses might also be preventable and antiviral chemotherapy might have a major role in human medicine if viruses were found to play a continuing determinative role in human cancer.

Further, it is apparent that even if effective antiviral chemotherapy does become a reality, we will have to face the same problems of drug resistance that have created such difficulties for antibacterial chemotherapy in recent years. In laboratory tests, mutants resistant to almost all the present antiviral agents emerge with great rapidity, and in the case of HBB and guanidine, drug-dependent mutants have also been obtained.

Finally, it is appropriate to comment on the use of antibacterial agents in the treatment of viral diseases. For a variety of reasons, physicians the world over tend to overprescribe antibiotics for the treatment of infections of the respiratory and alimentary tracts. Over 90% of respiratory infections and many gastrointestinal infections are of viral etiology and therefore totally refractory to treatment with existing antibacterial antibiotics. The only valid circumstances for the use of antibiotics in the treatment of viral infections are the following: (a) to prevent bacterial superinfection, e.g., otitis media or pneumonia in a bronchiectatic child with measles, or bronchopneumonia in an elderly cardiac or pulmonary invalid who contracts influenza, or perhaps in any infant with respiratory syncytial virus infection of the lower respiratory tract, and (b) to "play safe" in potentially serious illnesses where there is real diagnostic doubt about the possibility of a bacterial etiology until results come back from the laboratory, e.g., in meningitis or pneumonia.

A RATIONAL APPROACH TO THE SEARCH FOR ANTIVIRAL AGENTS

Although empirical screening led to the discovery of numerous antibacterial antibiotics, the same approach has produced a very disappointing yield of antiviral drugs. A more rational approach should be based on a logical search for agents capable of inhibiting those particular biochemical reactions now known to be unique to the multiplication of vertebrate viruses. It is instructive to consider these points of potential antiviral chemotherapeutic attack.

Attachment and Penetration

Attachment of virus to the cell's plasma membrane often necessitates the apposition of specific complementary receptors on the surface of virus and cell respectively (see Chapter 3). Agents that interfere with either could theoretically block infection. For example, neuraminidase destroys the glycoprotein receptors in the lungs of mice and renders them refractory to influenza infection until new receptors appear some hours later. Neuraminidase is not seriously contemplated as a potential chemotherapeutic agent for use in man, but illustrates the principle well. One class of drugs (amantadine and derivatives) that does not inhibit attachment appears to have some effect on the process of "penetration."

Transcription

The transcriptase carried by RNA viruses of most families (see Table 3-3) as well as the poxviruses, is essential for the transcription of "early" mRNA from the partially uncoated viral core. Since all such transcriptases are virus-specific and all except that of the poxviruses are RNA-dependent, there may very well be chemicals that will inhibit these enzymes without affecting the DNA-dependent cellular RNA polymerases. Evidence is accumulating that interferon may block transcription (see Chapter 5). The reverse transcriptase of retroviruses may render them particularly vulnerable to attack at this point; some of the newer rifamycin derivatives show promise, and the search is bound to be widened now that a novel rationale for anti-cancer therapy, at least in experimental animals, has emerged.

Posttranscriptional cleavage of the large molecules of polycistronic mRNA transcribed from most DNA viruses and perhaps some RNA viruses, and/or the subsequent addition to poly(A) may conceivably be susceptible to selective chemotherapy, though the latter process at least seems to be a regular occurrence in normal mammalian cells.

Translation

Translation of proteins from viral mRNA is the process affected by at least two of the very few successful antiviral agents known so far. Methisazone, a thiosemicarbazone, appears to interact with some early product of infection to block the translation of "late" vaccinia mRNA, while interferon (or another cellular protein derepressed by interferon) may block the translation of all viral mRNA's. Since the cell must continue to translate its own messenger RNA's it may seem paradoxical that chemicals could discriminate specifically against viral mRNA. However, the latter may carry a different initiation codon, or have a ribosomal attachment site that can be blocked either directly or by substances that bind to ribosomes. We know that the opposite phenomenon is a regular feature of the multiplication of most cytocidal viruses, namely a virus-coded protein is produced which specifically blocks the translation of cellular mRNA's without adversely affecting viral protein synthesis.

Replication of Viral Nucleic Acid

This process occurs very rapidly, and often in resting cells. Inhibitors of DNA synthesis, such as 5-iodo-2'-deoxyuridine, adenine arabinoside, and cytosine arabinoside, although toxic when administered systemically, have turned out to be quite useful agents against DNA viruses multiplying in the localized environment of the cornea, where relatively few cells are dividing. The RNA viruses should, in theory, be far more vulnerable at this particular point in their multiplication cycle, for they are dependent upon novel virus-specific, RNA-dependent RNA polymerases (replicases) that are not present at all in the normal mammalian cell. Certain benzimidazole derivatives, such as HBB, inhibit RNA replication by picornaviruses in cultured cells, though the effect is probably not directed specifically at the replicase.

Posttranslational Cleavage of Proteins

Although this process is not unique to viruses, all of the picornavirus (and probably togavirus) proteins arise in this way, and cleavage seems to constitute a vital step in the late stages of assembly of several other viruses. A number of toxic chemicals have been used experimentally to inhibit cleavage proteases, and rifampicin blocks the cleavage of a precursor of a vaccinia viral core protein; the evidence suggests that this may be secondary to inhibition of an early step in the maturation of the virion.

Assembly

Guanidine, which disrupts hydrogen bonds, may exert its very varied effects on the multiplication of picornaviruses by so distorting the tertiary configuration of the capsid protein precursor that mature virions are not assembled.

Regulation of Gene Expression

Regulation of viral RNA transcription, translation, and replication may be under the control of viral proteins. Quite subtle changes exerted by a simple drug like guanidine might act in this way to throw the whole cycle "out of gear."

In the following pages we make no attempt to review the multitude of substances for which some degree of antiviral activity *in vitro* has been claimed. Rather, we discuss in some detail the few antiviral agents that have been clearly shown to be of potential value in human medicine.

INTERFERON

Superficially, human interferon suggests itself as the ideal antiviral chemotherapeutic agent for use in man. It is a natural by-product of human viral infections, completely nontoxic and nonallergenic in man, and active against a broad spectrum of viruses (Chapter 5). Yet interferon has now been known for nearly 20 years during which period research workers in universities and pharmaceutical companies throughout the world have endeavoured without success to convert Isaacs' discovery into a practicable proposition for human use. The reasons for this failure are worth exploring.

Interferon has been shown to protect animals, usually mice or rabbits, against several systemic and localized viral diseases. Some important generalizations should be made about the results obtained. (a) Prophylaxis is invariably more successful than therapy; best results, and often the only positive results, are obtained by administering interferon some hours before viral challenge and continuing therapy for some days thereafter. Treatment of an established infection is rarely successful, and never if started late in the illness. (b) There is a clear dose-response relationship; higher doses of interferon are needed to combat higher doses of challenge virus. (c) Topical application of interferon to the eye, skin, or respiratory tract followed by challenge via the same route tends to be more successful than is systemic administration in the prevention of a generalized infection.

These promising results in animals have not been followed by comparable success in man. The Scientific Committee on Interferon set up by the Medical Research Council of Great Britain has published a number of reports on its own carefully controlled human trials. The latest of these is decidedly discouraging. Almost a million "research standard units" of human interferon delivered by spray gun over a 24-hour period failed to give any protection against intranasal challenge with influenza B virus. Only when the duration of the treatment was extended to 1 day before until 3 days after challenge with rhinovirus type 4, and the dose of interferon increased to impracticable levels (the total dosage per patient being the yield from the virus-infected leukocytes cultured from 15–25 liters of human blood) was a detectable reduction in severity of symptoms and virus-shedding obtained.

Even if future studies in man were to give cause for greater encouragement than we have had so far, there are forbidding problems in the commercial mass-production of the very large quantities of interferon that appear to be required. There is much evidence that man (and monkey) are not nearly as readily protected by homologous interferon as are the mouse and the rabbit. The use of leukocytes, available to blood banks as a by-product of the preparation of human plasma, or human amnions obtained from maternity hospitals, cannot be seriously contemplated in the long term as sources of human cells for interferon production. Diploid strains of human embryonic fibroblasts such as WI-38 could provide an adequate supply of cells, but precautions would still need to be taken to ensure that the virus used as interferon inducer (often Sendai or NDV), or indeed any adventitious viruses, e.g., hepatitis, were inactivated by the various procedures that have now become standard steps in the purification of interferon. Human skin fibroblasts can be repeatedly stimulated with poly(I)·poly(C) to produce a 6-hour burst of interferon about once every 24 hours, whereas most other cells become refractory to secondary stimulation by virus or any other inducer.

The problem of the therapeutic use of interferon is primarily quantitative, for the substance is harmless and the body can tolerate unlimited amounts. Nevertheless, present indications are that if interferon cannot be demonstrated to protect man against superficial infections of the respiratory tract under laboratory conditions designed to favor the drug, there does not appear to be much prospect for its use as a chemotherapeutic agent for treating established viral infection, systemic or superficial.

SYNTHETIC POLYNUCLEOTIDES AND OTHER
INTERFERON INDUCERS

Attention over the last few years has moved to the possibility of harnessing the body's capacity to manufacture its own interferon. Although the levels of interferon achieved in human serum following natural viral infection or artificial stimulation do not approach those found in mice ($<10^2$ compared with $>10^4$ units/ml), there is strong circumstantial evidence that these low levels are protective. For example, 9–10 days after inoculation of live measles vaccine, when interferon levels reach their peak, smallpox vaccine "fails to take." However, the mainstream of the current search for interferon inducers is directed not toward the use of live attenuated viruses but to synthetic chemicals such as the polynucleotide, poly(I)·poly(C), otherwise known as poly(rI)·poly(rC) or poly(rI:rC). Such double-stranded RNA's are powerful inducers of interferon (see Chapter 5).

Poly(I)·poly(C) has been convincingly shown to protect animals against a wide variety of viral infections. As with interferon itself, prophylaxis is much more successful than therapy, but there are some reports of effective treatment of established disease. In man, poly(I)·poly(C) administered intranasally in frequent divided dosage from 1 day before until 4 days after challenge with rhinovirus type 13, inoculated by the same route, has been shown experimentally to produce a marginal reduction in the number and severity of common colds. Topical application of synthetic polynucleotides to the respiratory tract may conceivably prove to have some value in human medicine since high concentrations of the drug can be attained by this route with no appreciable toxic effects. However, systemic administration of poly(I)·poly(C) seems to be contraindicated for any but grave viral illnesses because of unacceptable side effects on hemopoiesis and liver function.

Another major drawback of poly(I)·poly(C) as a potential antiviral agent is that, following a single dose, animals become temporarily refractory to interferon-induction by a second dose; this "refractory state," or "tolerance", or "hyporeactivity" is seen with all varieties of interferon inducer (Chapter 5). Assuming that hyporeactivity also occurs in man, the refractory state imposes a major, perhaps insuperable, flaw in the rationale of interferon induction, for unless much higher interferon titers than currently attainable can be achieved it is improbable that a single dose would provide enough interferon to prevent a viral infection, let alone cure one.

Current research is concentrated on attempts to synthesize new polynucleotides or other agents with higher antiviral potency and lower toxicity for man. For example, a substituted propanediamine, namely N, N-dioctadecyl-N',N'-bis (2-hydroxyethyl) propanediamine, administered as an oil-in-water emulsion intranasally, has recently been shown to induce high titers of interferon.

NUCLEOSIDE DERIVATIVES

Halogenated Nucleosides

As if to deny the logic of the principles upon which the rational search for antiviral agents is based, the most successful group of drugs available so far are highly toxic to mammalian cells. The halogenated pyrimidines are potent inhibitors of cellular DNA synthesis, yet one of them, 5-iodo-2'-deoxyuridine (idoxuridine, IUdR, IDU, see Fig. 13-1), has proved itself in practice to be a very useful chemotherapeutic agent when applied topically to the human eye in early cases of kera-toconjunctivitis or dendritic ulcer due to herpes simplex virus. This paradoxical situation is attributable to the localized nature of the disease; herpetic ulcers are initially quite superficial lesions, readily accessible to high concentrations of iododeoxyuridine applied topically at frequent intervals. The cornea is relatively avascular, hence the drug remains localized. Moreover, viral DNA synthesis which is proceeding rapidly, is more vulnerable than that of the slowly proliferating corneal cells.

Claims have been advanced for the efficacy of iododeoxyuridine in the therapy of other herpetic infections. Herpes labialis (recurrent fever blisters) in man has been treated with a 5% solution of iododeoxyuridine in dimethyl sulfoxide (DMSO), which greatly improves the penetration of the drug into the cells of the deeper layers of the skin. Presumably herpes genitalis should respond as well. Evidence has also been presented that continuous topical application of 40% iododexoyuridine in DMSO reduces the severity of herpes zoster in the elderly. Iododeoxyuridine has even been used systemically as "heroic therapy" in cases of herpes simplex encephalitis, a disease with a 50% mortality rate, but the rarity of the disease makes it impossible to conduct a large-scale controlled trial to assess whether the drug has any real value. Clearly, iododeoxyuridine should never be administered parenterally in anything less than such desperate circumstances, because it is highly toxic (and mutagenic).

The related thymidine analog, trifluorothymidine, has recently been reported to be a significantly superior alternative to iododeoxyuridine for the therapy of human ocular infections with herpes simplex virus.

TABLE 13-1

Mechanisms of Action of Antiviral Chemotherapeutic Agents

DRUG	PROBABLE POINT OF ACTION	PRINCIPAL VIRUSES INHIBITED
Amantadine	Penetration or uncoating	Influenza A2
Iododeoxyuridine ⎫ Ara-A, Ara-C ⎭	DNA replication	Herpetoviruses
Rifamycins	?Reverse transcriptase	Retroviruses
	Assembly	Poxviruses
Thiosemicarbazones	Translation of late viral mRNA	Poxviruses
Interferon	Transcription or translation	All viruses
Poly(I)·poly(C) ⎫ Statolon ⎪ Pyran copolymer ⎬ COAM ⎪ Tilorone ⎪ Propanediamine ⎭	Induction of interferon ?Stimulation of reticuloendothelial system	

Arabinofuranosyl Nucleosides

Though their full clinical potential has yet to be determined, there are indications that this group of compounds may be more valuable in human medicine than the halogenated nucleosides just discussed. The arabinofuranosyl nucleosides also interfere with both viral and cellular DNA synthesis but their chemotherapeutic index is higher.

Cytosine arabinoside (1β-D-arabinofuranosylcytosine), abbreviated to ara-C, has a similar, but not identical, antiviral spectrum to iododeoxyuridine and has been shown to be at least equally effective in treating herpetic and vaccinial infections, particularly keratitis, in man and animals. There are several controversial reports, difficult to evaluate in the absence of adequate numbers of controls, that the drug is also of benefit when administered by continuous intravenous infusion to gravely ill patients with overwhelming generalized herpes zoster or simplex infections.

The more recently synthesized adenine arabinoside, 9β-D-arabinofuranosyladenine (ara-A), which is not deaminated *in vivo* as rapidly as ara-C, has a higher chemotherapeutic index than ara-C or IDU against various herpetoviruses and vaccinia virus in cultured cells and animals, and promises to be a valuable agent in the treatment of infections by these viruses in man.

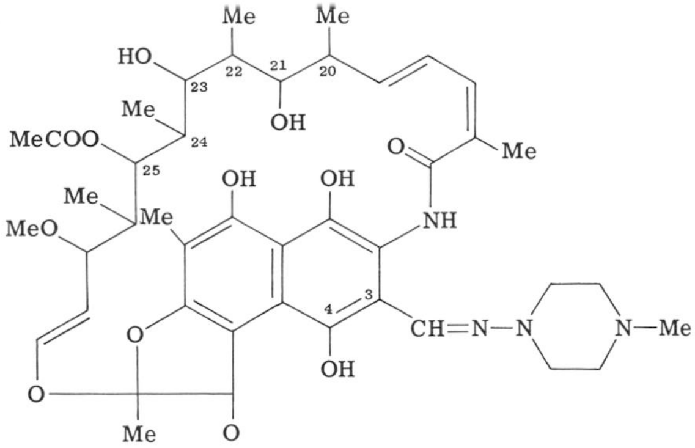

Rifampicin

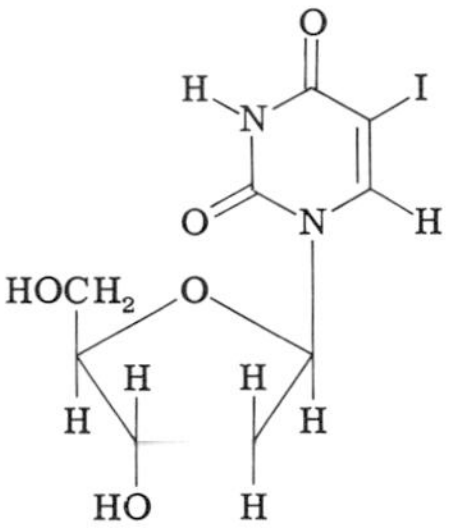

5-Iodo-2'-deoxyuridine

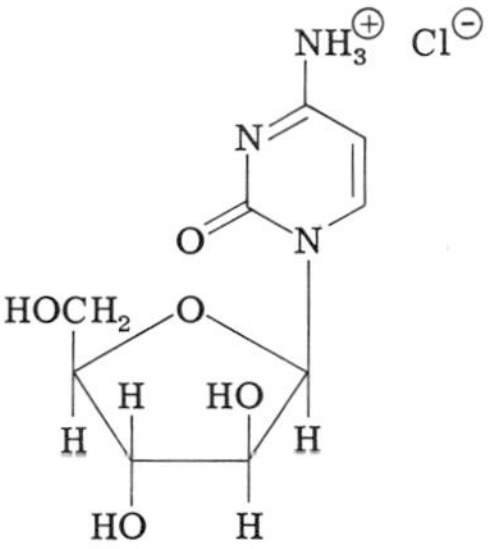

1β-D-Arabinofuranosyl-
cytosine · HCl

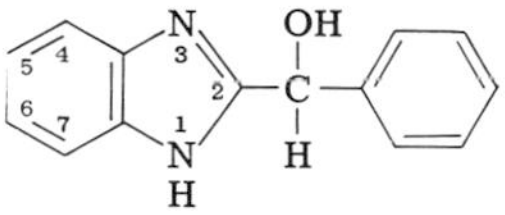

2(α-Hydroxybenzyl)
benzimidazole

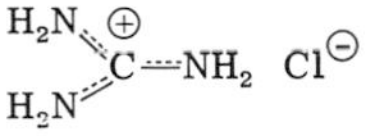

Guanidine · HCl

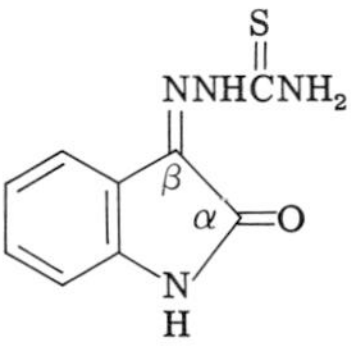

Isatin β-thiosemi-
carbazone

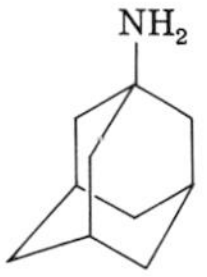

1-Aminoadamantane
hydrochloride

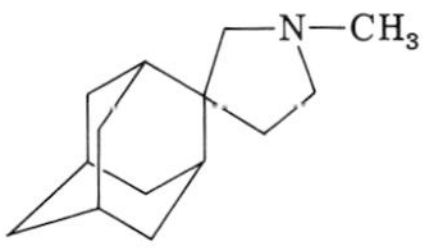

1' - Methyl spiro(adamantane-
2, 3'-pyrrolidine) maleate

FIG. 13-1. *Structural formulas of some antiviral drugs: rifampicin; 5-iodo-2'-deox-yuridine; 1β-D-arabinofuranosylcytosine (cytosine arabinoside, Ara-C); 2(α-hydrox-ybenzyl) benzimidazole (HBB); guanidine·HCl (the positive charge is distributed over the whole molecule); isatin β-thiosemicarbazone; amantadine (1-amino-adamantane hydrochloride); 1'-methyl spiro(adamantane-2,3'-pyrrolidine) maleate.*

THIOSEMICARBAZONES

The multiplication of poxviruses is inhibited by 1-methylisatin β-thiosemicarbazone, better known as methisazone or the trade name Marboran (Fig. 13-1). Methisazone interacts with some product of the first 3 hours of the poxvirus multiplication cycle (perhaps a viral or cellular protein) to prevent the translation of late viral mRNA.

In the face of serious epidemics of smallpox in India and elsewhere, the drug has been administered orally to large groups of people; the incidence of disease appeared to be reduced in some trials but not others. Clearly the drug is no substitute for vaccination. It may have a role in the treatment of complications of smallpox vaccination, such as vaccinia gangrenosa or eczema vaccinatum.

AMANTADINE AND DERIVATIVES

A simple three-ringed symmetrical amine, 1-adamantanamine hydrochloride, or 1-amino adamantane hydrochloride, or amantadine (Symmetrel) (see Fig. 13-1) has been shown to block the multiplication of influenza A2 viruses at the stage of penetration or uncoating. The efficacy and safety of amantadine in man have been the subject of considerable controversy. A number of clinical trials have indicated a twofold reduction in the incidence of influenza A2 (but not B) following oral administration of amantadine prior to exposure to the virus, but no effect if treatment was commenced after infection. Certainly the benefits of the drug are marginal at best and the incidence of side effects upon the central nervous system is disturbing if the recommended dosage is exceeded, as well it might be with a drug freely available for a common ailment like "flu."

The related compound, 1'-methyl spiro(adamantane-2,3'-pyrrolidine) maleate (1:1) (see Fig. 13-1) has been said to show promise in early clinical trials.

SOME OTHER COMPOUNDS OF POTENTIAL VALUE

Rifamycin Derivatives (Ansamycins)

Rifamycin-SV and the derivative, rifampin, otherwise known as rifampicin (Fig. 13-1) inhibit bacterial RNA synthesis by binding to DNA-dependent RNA polymerase, but have no such effect on mammalian polymerases. It was demonstrated that these agents are active in cell culture against poxviruses, which contain a similar DNA-based transcriptase, and the initial assumption was that the mechanism was the same. However, the drug concentrations required were over a thousand times higher than those effective against bacteria, and it transpired that the target was not the transcriptase *per se,* but an early step in assembly of the virion. More recently, however, it has been shown that rifampicin and other rifamycin derivatives inhibit the reverse transcriptase of retroviruses at concentrations slightly lower than those blocking cellular DNA-dependent DNA polymerase.

Benzimidazoles

The benzimidazole derivative, 2(α-hydroxybenzyl) benzimidazole (Fig. 13-1), known as HBB, inhibits the multiplication of many picornaviruses, notably coxsackie B and echoviruses, at noncytocidal concentrations in cultured cells, but is of no value at all *in vivo.* The same is true of the simpler compound, guanidine hydrochloride. Both have been widely used as tools to probe the mechanism of multiplication of picornaviruses. One salutory finding to come out of these studies is that drug-resistant, and even drug-dependent, mutants rapidly emerge when viruses are grown in the presence of these agents.

SUMMARY

Most chemicals that interfere with viral replication are toxic for human cells, but a few show selectivity of action (Table 13-1). Iododeoxyuridine and other nucleoside derivatives such as adenine arabinoside and cytosine arabinoside, all of which affect DNA synthesis, can be successfully employed to treat infections of the eye and perhaps other epithelial surfaces caused by DNA viruses, notably herpes simplex, because they can be applied locally in effective dosage without damage to cells elsewhere in the body. Several other drugs show antiviral activity at nontoxic concentrations in cultured cells, but lack significant activity *in vivo.*

A rational search for new antiviral drugs must be based on our expanding knowledge of those particular biochemical steps in the viral multiplication cycle that are not vital to cell survival and may therefore be vulnerable to selective attack: (a) attachment of the virion to specific receptors, (b) transcription of early viral mRNA by the virion's transcriptase, (c) translation of early proteins from viral mRNA, (d) replication of viral nucleic acid, particularly RNA, by virus-coded polymerases, (e) posttranslational cleavage of proteins and assembly of virions, and (f) regulation of viral gene expression.

However, there is reason to expect that even when such drugs are discovered they may suffer from the disadvantage of being specific for a limited number of viruses, and the rapid emergence of drug-resistant mutants may nullify their effectiveness in practice. For these reasons interferon (or a drug with a similar mode of action) is potentially the most useful of all viral inhibitors, since it is completely nontoxic and can be used against a wide range of viruses, thus circumventing the need for precise diagnosis, which frequently cannot be made until the disease is advanced. There are formidable difficulties in obtaining enough purified interferon for direct administration to man, but the induction of interferon by nonviral inducers (e.g., synthetic polynucleotides) affords the most promising approach in sight for effective antiviral chemotherapy.

FURTHER READING

Adamson, R. H., Levy, H. B., and Baron, S. (1972). The interferon system. In "Search for New Drugs" (A. A. Rubin, ed.), p. 292. Dekker, New York.

Bauer, D. J. (1972). "Chemotherapy of Virus Diseases," Vol. I, Int. Encycl. Pharmacol. Ther., Sect. 61. Pergamon, Oxford.

Carter, J., ed. (1973). "Selective Inhibitors of Viral Functions." Chem. Rubber Publ. Co., Cleveland, Ohio.

Finter, N. B., ed. (1973). "Interferons and Interferon Inducers." North-Holland Publ., Amsterdam.

Herrmann, E. C., and Stinebring, W. R., eds. (1970). "Second Conference on Antiviral Substances," Ann. N.Y. Acad. Sci. No. 173, Art. 1. N.Y. Acad. Sci., New York.

Hill, D. A., Baron, S., Levy, H. B., Bellanti, J., Buckler, C. E., Cannellos, G., Carbone, P., Chanock, R. M., DeVita, V., Guggenheim, M. A., Homan, E., Kapikian, A. Z., Kirschstein, R. L., Mills, J., Perkins, J. C., Van Kirk, J. E., and Worthington, M. (1971). Clinical studies on the induction of interferon by polyinosinic-polycytidylic acid. Perspect. Virol. 7, 197.

Hilleman, M. R. (1970). Double-stranded RNAs (poly I:C) in the prevention of viral infections. Arch. Intern. Med. 126, 109.

Hilleman, M. R., Lampson, G. P., Tytell, A. A., Field, A. K., Nemes, M. M., Krakoff, I. H., and Young, C. W. (1971). Double-stranded RNA's in relation to interferon induction and adjuvant activity. In "Biological Effects of Polynucleotides" (R. F. Beers and W. Braun, eds.), p. 26. Springer-Verlag, Berlin and New York.

Ho, M., and Armstrong, J. A. (1975). Interferon. Annu. Rev. Microbiol. 29, 131.

Luby, J. P., Johnson, M. T., and Jones, S. R. (1974). Antiviral chemotherapy. Annu. Rev. Med. 25, 251.

Merigan, T. C., Jordan, G. W., and Fried, R. P. (1975). Clinical utilization of exogenous human interferon. Perspect. Virol. 9, 249.

Pavan-Langston, D., Buchanan, R. A., and Alford, C. A. (1975). "Adenine Arabinoside: An Antiviral Agent." North-Holland Publ., Amsterdam.

Shugar, D., ed. (1972). "Virus-Cell Interactions and Viral Antimetabolites," FEBS Symp., Vol. 22. Academic Press, New York.

Tilles, J. G. (1974). Antiviral agents. Annu. Rev. Pharmacol. 14, 469.

CHAPTER 14

Laboratory Diagnosis of Viral Disease

INTRODUCTION

Almost all known human viruses can now be isolated in cultured cells, and techniques are continuously being refined to accelerate the process of identification. As a consequence, viral diagnostic facilities are becoming a standard feature of many hospital and public health laboratories. Because of the expense and the delay involved in obtaining a definitive virological diagnosis, physicians should be discriminating in their use of these services.

Indications for Laboratory Diagnosis

Viral diseases that call for laboratory confirmation fall into four main categories.

1. Diseases in which the management of the patient, as well as the prognosis, depend on accurate diagnosis. A good example is suspected rubella in a pregnant woman. When antiviral chemotherapy becomes a practical reality, rapid diagnosis may be required for a much wider range of diseases to assist in the selection of the appropriate drug.

2. Dangerous epidemic diseases like smallpox, yellow fever, poliomyelitis, encephalitis, and influenza, where early identification of "sentinel cases" may be of vital importance in alerting the authorities to the need to instigate programs of immunization, quarantine, or other forms of control or surveillance essential to the maintenance of public health.

3. Epidemiological surveillance, for example, screening of blood donors for hepatitis, or determining the efficacy of an immunization campaign, or studying the prevalence and distribution of particular viruses in the community.

4. Investigation of new syndromes or outbreaks. Viral identification

255

has played a valuable role during the last 25 years in defining the etiology of such previously poorly understood syndromes as "aseptic meningitis," "URTI" (upper respiratory tract infection), "ARD" (acute respiratory disease), "atypical pneumonia," nonbacterial gastroenteritis, and many of the unexplained exanthemata. Presumably many viruses and virus–disease associations remain to be discovered by alert physicians working in concert with cooperative laboratories.

Approaches to Laboratory Diagnosis

As in the diagnosis of bacterial infections there are three major approaches to the identification of viruses in the laboratory.

Microscopy. Virions may be identified by electron microscopy, viral antigens may be located in infected cells by immunofluorescence, or virus-induced histopathology may be recognized by light microscopy in specimens taken directly from the patient. Such methods, while obviously desirable because they provide an answer within an hour or so, can only be applied to specimens containing relatively large numbers of virions or virus-infected cells respectively. Moreover, techniques such as electron microscopy and immunofluorescence demand expensive equipment, carefully controlled standards, and experienced personnel often not available in the average laboratory. Nevertheless, this is the direction in which diagnostic virology is moving.

Virus Isolation. Viruses may be grown in cultured cells (or animals). This is still the method of choice for most viruses. Provisional identification based on the nature of the cytopathic effects produced in cultured cells may take anywhere from a couple of days to a couple of weeks or even longer. Positive identification, which is not always required, requires further time for serological "typing" of the isolate.

Serology. Antibodies specific for a particular virus may be identified in the patient's serum. Since the sensitivity of most of the available tests does not allow detection of the relatively small quantities of antibody synthesized during the first few days of illness, a diagnosis cannot be established by serology for at least a week or two. Such retrospective diagnoses are of value mainly in relatively drawn-out infections, such as infectious mononucleosis, or in instances where the result does not come too late to influence the management of the case, e.g., rubella in pregnancy, or where viral isolation is notoriously unsuccessful, e.g., encephalitis.

RAPID DIAGNOSTIC TECHNIQUES

By and large, laboratory diagnosis of viral infection is too slow to provide information that will influence the management of the case. Isolation of viruses in cultured cells or detection of specific antibody rises in sera rarely provides a definitive answer within less than a week or so, by which time it may be of little more than academic interest to doctor or patient, although useful to the epidemiologist. Attention is therefore being directed to new techniques which yield a diagnosis within hours of the patient's admission to the hospital. It may be anticipated that such techniques will be refined during the next few years to render them applicable to a much wider range of situations.

Electron Microscopy and Immunoelectron Microscopy

For many years direct electron microscopy (EM) of negatively stained vesicle or pustule fluid has been used for the urgent differentiation of the brick-shaped smallpox virion from the enveloped icosohedral virion of varicella (chickenpox). The practicability of the procedure in this situation rests on the high concentration of virions to be found in the readily accessible skin lesions in these two diseases. Recently the approach has been extended to other situations. For example, virions of hepatitis A, or the "Norwalk agent," or the rotavirus responsible for infantile gastroenteritis may be concentrated from feces by ultracentrifugation (following preliminary clarification and/or filtration to remove bacteria and debris) and then negatively stained and examined by electron microscopy. Providing that the morphology of the virions is sufficiently characteristic, the agent can immediately be allocated to the correct family.

A refinement of this approach is to mix the specimen with a specific antiserum and then to spin down the resulting viral aggregate, which is readily identifiable by EM; this powerful technique, known as immunoelectron microscopy has recently led to the discovery of three important new viruses in feces—"rotavirus," hepatitis A and "Norwalk agent" (Plate 14-1). No doubt we shall soon see this technique applied much more widely to identify viruses from respiratory secretions and other sources as well as from feces, and to discover new groups of agents not yet able to be grown in cultured cells.

Immunofluorescence

"Fluorescent antibody" is specific immunoglobulin which has been "tagged" with a dye such as fluorescein or rhodamine that fluoresces

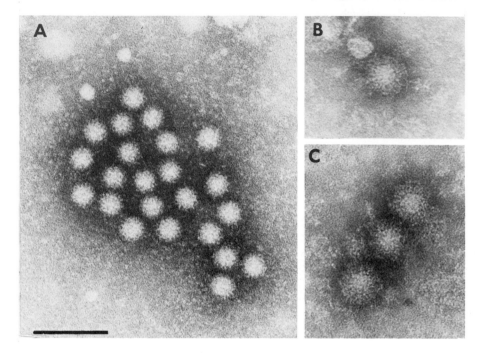

Plate 14-1. *Immunoelectron microscopy. Norwalk agent from the feces of a patient with gastroenteritis; aggregation with convalescent serum containing a low (A) or high (B and C) titer of antibody against the virus. Bar = 100 nm (Courtesy Dr. A. Z. Kapikian.)*

on exposure to ultraviolet or blue light. The chemical process of conjugation can be accomplished as a simple laboratory routine, but fluorescein-conjugated immunoglobulins are now commercially available. There are two main variants of this technique.

Direct Immunofluorescence (A–af). Viral antigen (A) (e.g., in the form of an acetone-fixed, virus-infected, cell monolayer on a coverslip) is exposed to fluorescein-tagged antiviral serum (af). Excess af is washed away and the cells are inspected microscopically using a powerful light source (xenon or mercury vapor lamp) from which all but the light of short wavelength has been filtered out. Additional filters in the eyepieces, in turn, absorb all the blue and ultraviolet light, so that the specimen appears black except for those areas from which fluorescein is emitting greenish-yellow light (Plate 14-2).

Indirect Immunofluorescence (A-a-gf). This system, sometimes known as the "sandwich" technique, differs in that the antiviral an-

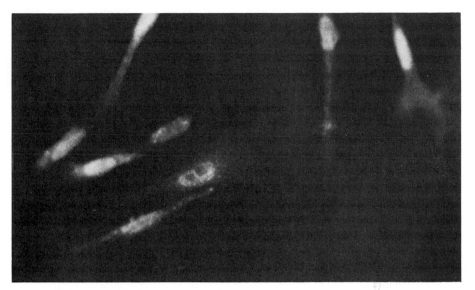

PLATE 14-2. *Immunofluorescence, used here for determining the site of assembly of components of influenza virus. Antibody against the NP (nucleocapsid) antigen shows nuclear accumulation at 4 hours after infection of chick cells. Guinea pig antiserum to NP antigen; fluorescein-conjugated rabbit anti-guinea pig IgG. (Courtesy Dr. N. Dimmock.)*

tibody (*a*) is untagged, but fills the role of the "meat in the sandwich." It binds to antigen (A) and also to a fluorescein-conjugated anti-γ-globulin (*gf*), which is subsequently added. For example, if the antiviral serum were a human convalescent serum, then it would be appropriate for *g* to be antibody globulin made in a goat or a rabbit by injecting normal human globulin. The indirect technique has the advantages of greater sensitivity and, more important in diagnostic virology, of requiring only a single tagged reagent, e.g., fluorescein-conjugated goat antihuman γ-globulin, with which to test the interaction between any antigen and the corresponding human antibody.

Considering that immunofluorescence has now been available for over 30 years it is odd that this attractive technique has not become the standard approach to laboratory diagnosis of viral infection. No doubt this can be attributed to the cost, and to the technical difficulties involved in ensuring that the procedure is specific, as well as to the multitude of distinct serotypes responsible for most respiratory and enteric infections. Nevertheless a considerable future can be predicted for the use of immunofluorescence in the identification of viral antigens in infected cells taken directly from the patient. An obvious prerequisite is that the doctor has ready access to infected cells and that an adequate number of them can be removed without danger or pain to the patient. In fact, there is no real difficulty in removing partially detached infected cells from the mucous membrane of the upper respiratory tract, genital tract, eye, or from the skin simply by swabbing or scraping the infected area with reasonable firmness. Examples of diseases that may be diagnosed satisfactorily from scrapings taken from these various sites are smallpox (skin), herpes simplex (skin, eye, or vagina), and measles (throat). Respiratory syncytial virus, influenza, and parainfluenza infections may be rapidly differentiated by immunofluorescence on fixed smears of cells deposited by centrifugation from mucus aspirated by suction from the nasopharynx of infected infants. The technique is most applicable to diseases with only one or a very small number of possible etiological agents because specific antisera are required to test for every different serotype.

Biopsy is sometimes justifiable to obtain infected tissue from very ill patients. Using such material, immunofluorescence has been applied to the diagnosis of hepatitis (liver biopsy), herpes simplex encephalitis or measles SSPE (brain). Post-mortem (PM), immunofluorescence has for several years been the standard approach to the verification of rabies in the brain of animals trapped after biting man, or to the identification of a number of lethal viral infections of the human brain. The major viral diseases readily identifiable by immunofluorescence applied to fixed smears of cells taken directly from the patient are listed in Table 14-1.

A second application of immunofluorescence to laboratory diagnosis of viral infection is the identification of specific viral antigen in cell cultures inoculated a day or two earlier with material taken from the patient. This very sensitive probe reveals the growth of virus even in a very small number of infected cells, long before pathognomonic cytopathic effects become manifest (Plate 14-2).

Radioimmunoassay and Other Sensitive Probes for Viral Antigen

Under certain circumstances viral antigens may be present in sufficient quantity in vesicle fluid, serum, or even feces, to permit identification using any of a variety of sensitive serological procedures. For instance, hepatitis B surface antigen is routinely identified in serum by such sensitive procedures as radioimmunoassay, reverse passive hemagglutination, or cross-over immunoelectrophoresis. Mention has already been made of the use of immunoelectron microscopy to

TABLE 14-1
Rapid Identification of Viruses by Immunofluorescence

SPECIMEN	VIRUS
Brain (biopsy or PM)	Rabies
	Herpes simplex
	Measles (SSPE)
	Papovavirus (PML)
Corneal scraping	Herpes simplex
Vesicle scraping	Smallpox
	Varicella
	Herpes simplex
Nasopharyngeal aspirate	Respiratory syncytial virus
	Parainfluenza
	Influenza
	Measles
Blood leukocytes	Cytomegalovirus
	Measles
	Arboviruses
	Many others
Heart (PM)	Coxsackieviruses
Liver biopsy	Hepatitis B

identify a number of previously unknown viruses in feces. The first hint of the identity of the hepatitis A virus came from the relatively simple technique of immunodiffusion, which had for years been utilized to differentiate smallpox from chickenpox. These various serological procedures are described in detail later in the chapter in the context of identifying antibodies in the patient's serum, but the point to be emphasized here is that they are also finding increasing application to the rapid identification of viral *antigen* in specimens taken directly from the patient.

VIRUS ISOLATION

Collection and Preparation of Specimens

The chance of isolating a virus critically depends on the attention given by the attending physician to the collection of the specimen. Clearly, such a specimen must be taken from the right place at the right time. The "right time" is always as soon as possible after first seeing the patient, because virus is usually present in maximum titer at about the time symptoms first develop, and then disappears during the ensuing few days. Specimens taken as a last resort when days or weeks of empirically chosen antibacterial chemotherapy have failed are almost invariably a waste of effort.

The site from which the specimen is collected will depend on the pathogenesis of the particular infection. Thus a nasopharyngeal swab or aspirate will be the appropriate specimen to take from a patient with an upper respiratory infection, feces from an intestinal infection, cerebrospinal fluid (CSF) from meningitis, vesicle fluid and scrapings from the base of the lesion from a vesicular rash and a piece of any affected organ at autopsy. (Needless to say, tissue taken at autopsy or biopsy for the purposes of virus isolation must not be placed in formalin or any other preservative.) In the case of many generalized viral diseases it may not be obvious what specimen to take. As a rough working rule it can be said that at a sufficiently early stage in the disease, virus can usually be isolated from a throat swab, or feces, or leukocytes from blood.

Because of the lability of many viruses, specimens must always be kept cold and moist. Immediately after collection the swab should be swirled around in a small screw-capped bottle containing about 2–5 ml of an isotonic, balanced salt solution, such as Hanks' BSS or tryptose phosphate broth or cell culture maintenance medium, buffered at pH 7 and containing 0.25–0.5% gelatin or bovine serum albumin, 200–1000 units of penicillin, 200–1000 μg of streptomycin, and 50–100 units of nystatin per ml—the higher concentrations of antibiotics being used for fecal specimens. (If it is at all probable that the specimen will also be used for attempted isolation of bacteria, rickettsiae, chlamydiae, or mycoplasmas the collection medium must not contain antibiotics—the portion used for virus isolation can be treated with antibiotics later.) The swab stick is then broken off aseptically into the fluid, the cap tightly fastened and secured with adherent tape, and then dispatched immediately to the laboratory (in a thermos flask of crushed ice if a delay or transit time of more than a few minutes is envisaged). It should be accompanied by an informative clinical history and provisional diagnosis (Plate 14-3).

On arrival in the laboratory the specimen is immediately refrigerated and processed as soon as possible. If delays of more than a few hours are anticipated fecal samples are stored at −70°. (CSF must be frozen unless used immediately because some viruses are labile in this "medium.") Before inoculation into cell culture, material is shaken from the swab into the suspending medium by hand or on an appropriate mechanical device, sometimes with the help of glass beads; tissue specimens are homogenized in a high-speed blender or ground with a mortar and pestle. Cell debris and bacteria are deposited from

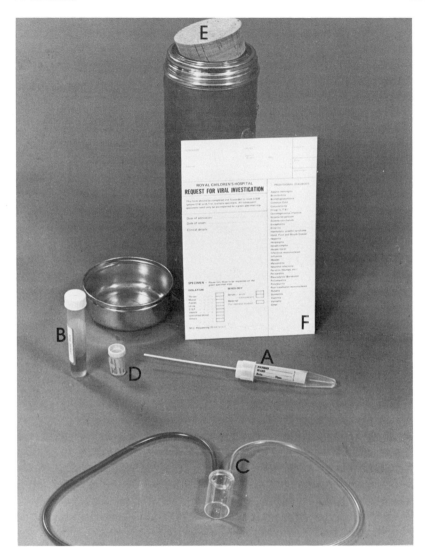

PLATE 14-3. *Basic equipment for collection of specimens for virus isolation.* (*A*) *Sterile swab.* (*B*) *Bottle containing a few ml of buffered isotonic balanced salt solution plus gelatin.* (*C*) *Apparatus for aspirating nasopharyngeal mucus by suction.* (*D*) *Vial containing anticoagulant for collection of blood (leukocytes may be taken for virus isolation and the plasma used for antibody determinations).* (*E*) *Thermos flask containing plastic bag full of ice (for short trips) or dry ice (if long delays are anticipated).* (*F*) *Requisition form. (Courtesy I. Jack.)*

the fluid by centrifugation, after which "dirty" specimens like feces may be passed through a membrane filter. Some of the clarified fluid is than "snap-frozen" in thin-walled ampules and stored at −70° for future reference. The remainder is inoculated into cell cultures, and rarely into chick embryos or newborn mice. Methods of inoculation and cultivation of viruses in these hosts have been discussed in Chapter 2.

Inoculation and Maintenance of Cultures

The choice of the most suitable host cell system will depend partly on availability, but mainly on the virus one expects to find, in the light of the patient's clinical history and the source of the specimen. The precise routine differs from one laboratory to another. There are essentially three different types of cell culture available. (1) Primary cultures of human embryonic kidney or monkey kidney cells have a "broad viral spectrum" because they contain a variety of types of differentiated cells; many RNA viruses, particularly paramyxoviruses and enteroviruses, are most easily isolated in such cultures. Monkey kidneys are expensive in some countries and almost unobtainable in others, and they are frequently contaminated with adventitious simian viruses; meanwhile, human embryos are becoming less readily available in many countries as a result of changes in abortion laws. (2) Continuous malignant human epithelial cell lines, such as HEp-2 or HeLa, are most useful for the growth of adenoviruses, rhinoviruses, and respiratory syncytial viruses. (3) Diploid cell strains of human embryonic fibroblasts have a broad spectrum that overlaps with the other two; they are particularly valuable for the isolation of rhinoviruses, cytomegalovirus, varicella virus, and several enteroviruses and indeed are the most versatile and useful of all the cultured cells available for viral isolation today. Some laboratories rely on Hayflick's well-known strain WI-38, while others have derived their own diploid strain of fibroblasts from human embryonic lung (HEL); the sensitivity of such strains to infection by individual viruses varies somewhat. It should also be appreciated that the cell strain (or line) that is more sensitive to a particular virus (i.e., which will "pick up" virus in the highest percentage of specimens) is usually but not always the strain in which CPE can be visualized earliest; both considerations are important.

In order to cover themselves against most contingencies many laboratories have a policy of inoculating a given specimen simultaneously into one type of primary culture, one diploid strain of human fibroblasts, and perhaps, in the case of throat but not fecal specimens, one malignant cell line. Even so, some viruses are more specialized in their requirements. For example, togaviruses, including rubella virus,

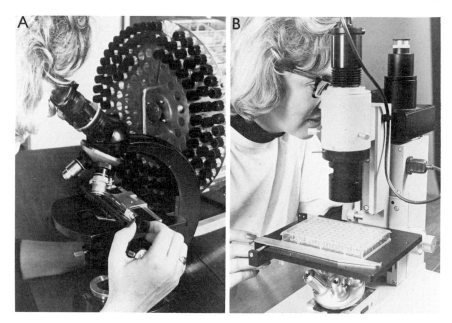

PLATE 14-4. *Inspection of cell cultures. (A) Note: Screw-capped culture tube containing monolayer of cells over a limited area of the glass near the bottom of the tube; detachable rack for supporting tube on microscope stage; binocular microscope, set on low power (×30–60), with condenser down and iris diaphragm partly shut; roller drum for rotating tubes in warm room (36°). (B) Note: Plastic tray containing numerous "minicultures"; inverted microscope for reading CPE from below. (Courtesy I. Jack.)*

grow best in continuous lines derived originally from baby hamster kidney (BHK-21), rabbit kidney (RK-13), or green monkey kidney (Vero).

Conventionally, monolayer cultures for viral diagnostic purposes are grown in screw-capped glass tubes, and this is still perhaps the surest way of handling relatively small numbers of specimens and minimizing the chance of error. There is an increasing trend towards the substitution of "miniature cell cultures" growing as monolayers in very small volumes of medium in the wells of disposable sterile plastic trays ("plates") (Plate 14-4, B). In this way large numbers of specimens may be processed rapidly with considerable economy of cells and media. To maintain the desired pH in the presence of conventional media, which are based on a CO_2-$NaHCO_3$ buffering system, the plates must be maintained in a special incubator which is fully humidified and supplied continuously with 5% CO_2 in air; alternatively, organic buffers such as HEPES may be substituted.

The inoculated cultures are held at the temperature and pH of the human body (37°, pH 7.4), except in the case of the rhinoviruses and coronaviruses, which grow best at 33° (the temperature encountered in the nasal mucosa). They may be slowly rotated on a roller drum, or kept stationary. If the pH of the maintenance medium falls below 7.0, it may be aseptically replaced with fresh medium, but this practice should be reduced to a minimum because of the risk of cross-contamination. Meanwhile, the cultures are observed at intervals of 1–2 days for the development of CPE (cytopathic effects) (Plate 14-4).

Recognition of Viral Growth

The time at which cytopathic changes first become detectable will depend, to some extent, on the number of virions that the specimen happens to contain, but, far more important, on the growth rate of the virus in question. Poliovirus or herpes simplex virus, for example, have a short latent period and a high yield, hence they will often show detectable CPE after 24 hours and destroy the monolayer completely within about 3 days. On the other hand, cytomegaloviruses, rubella, and some of the more slowly growing adenoviruses may not produce detectable CPE for 1–4 weeks. By this time, the uninoculated control cultures will often be showing nonspecific degeneration, so robbing the virologist of his standard of comparison. Accordingly, it may be necessary to subinoculate the cells and supernatant fluid from the infected culture into fresh monolayers ("blind passage").

When, at any stage, degenerative changes suggestive of viral multiplication become evident, the virologist has a number of courses open to him. The CPE is often sufficiently characteristic, even in the living unstained culture viewed *in situ,* for the trained observer to be able to tender a provisional diagnosis to the physician immediately (Table 2-1). Alternatively, he may need to fix and stain the infected monolayer to identify the CPE, for which purpose it is usual to include a coverslip in one of the culture tubes. Inclusion bodies and multinucleated giant cells (syncytia) can be identified by staining with hematoxylin and eosin, or Giemsa stain; if present, these changes are usually sufficiently characteristic to enable the virologist to place the isolate at least within its correct family (Table 14-2). Coverslip cultures can also be stained with fluorescent antibody for positive identification. If none of these histological methods is diagnostic, virus must be extracted from the culture for further examination.

Some viruses are relatively noncytocidal for cultured cells (see Chapter 5 and Table 2-1). Their growth in monolayer culture may sometimes be recognized by means of hemadsorption or interference

TABLE 14-2
Inclusion Bodies

SITE	STAINING	VIRUS[a]
Nuclear	Basophilic	Adenoviruses
Nuclear	Acidophilic[b]	Herpetoviruses
		Papovaviruses
Cytoplasmic	Acidophilic	Paramyxoviruses
		Reoviruses
		Rhabdoviruses
		Poxviruses
		(Togaviruses)
Nuclear and cytoplasmic	Acidophilic	Measles virus
		Cytomegalovirus
		(Orthomyxoviruses)

[a] The viruses in parentheses do not produce conspicuous inclusion bodies in many cell systems.
[b] Occasionally basophilic.

(see Chapter 2 for details). Viruses that hemagglutinate will be amenable to test by hemadsorption; the growth of paramyxoviruses or orthomyxoviruses and, to a lesser extent, the togaviruses is routinely recognized in this way (Plate 2-2). Interference, on the other hand, is really a research tool rather than a routine diagnostic aid. The growth of the rhinoviruses and rubella virus in cultured cells was first recognized in this way, but both groups of viruses can now be grown quite satisfactorily in other cell types in which they give adequate CPE.

Organ Cultures, Eggs, and Animals

Some viruses, not readily recoverable in conventional monolayer cultures, can be grown in organ cultures. For example, certain strains of coronavirus can be cultivated only in rafts of intact fully-differentiated human nasal or laryngeal epithelium (see Chapter 2). The agents responsible for certain persistent viral infections are demonstrable only from explants of infected human tissue, e.g., adenovirus from adenoids explanted *in vitro;* "cocultivation" of explanted tissue with a "permissive" cell line or even fusion of the cells growing out from the explant with such permissive cells may sometimes be necessary, e.g., measles virus from SSPE brain biopsies (see Chapter 8).

Many viruses will grow satisfactorily in chick embryos or newborn mice (see Chapter 2) but neither animal is now commonly used in the diagnostic laboratory, so neither will be discussed in any detail here.

The use of mice can be limited to the isolation of arboviruses, rabies virus, and some of the group A coxsackieviruses; suckling mice less than 24 hours old are injected with approximately 0.02 ml of virus suspension intracerebrally and 0.03 ml intraperitoneally, then observed for up to 14 days for the development of pathognomonic signs before sacrificing for histological examination (for details, see "Laboratory Diagnosis" section of the relevant chapters in Part II). Chick embryos are still used, together with primate kidney cell cultures, for the isolation of influenza A viruses from man. Three to four days after amniotic and allantoic inoculation of 10-day-old embryonated hen's eggs, the fluids are tested from hemagglutination (see Chapter 20 for details). Smallpox virus is isolated somewhat more readily on the chorioallantoic membrane than in cultured cells and may be differentiated from vaccinia and other poxviruses by pock morphology (Chapter 17).

Larger animals are used very infrequently for routine viral isolation, although monkeys, especially chimpanzees, are employed routinely today, by laboratories that can afford them, for attempted isolation of human viruses not yet cultivable in any other nonhuman host, e.g., hepatitis viruses and various agents responsible for slow infections of the human brain (Chapter 8), as well as for experimental studies of viral pathogenesis and immunity, for testing new viral vaccines (Chapter 12), and for production of standard antisera.

Typing of Viral Isolates

A virus newly isolated in cell culture, eggs, or suckling mice can usually be provisionally allocated to a particular family on the basis of the patient's clinical history, the laboratory host in which the virus was grown, and the visible result of this growth (CPE, hemadsorption, pock type, hemagglutination, etc.). Final identification, however, rests on serological procedures (described below). By using the new isolate as antigen against known antisera, e.g., in a complement fixation test, the virus can be placed into its correct family or genus. Having allocated it to a particular family (e.g., Adenoviridae), one can then go on to determine the serotype (e.g., adenovirus type 5) by the more specific serological procedures of neutralization or hemagglutination inhibition. This approach is of course applicable only to families that share a common family antigen, hence is of no help with the picornaviruses.

Neutralization tests can be exceedingly tedious in the case of viral genera that contain large numbers of different serotypes, e.g., the enteroviruses, rhinoviruses and adenoviruses. To avoid the necessity to test the viral isolate against every single type-specific antiserum, "intersecting serum pools" can be employed. Up to a dozen antisera are combined to give one pool, while a second pool comprises a few of those antisera together with several additional ones, and so on. With appropriate advice from a mathematician one can so construct the pools that an isolate may be positively identified by observing which particular pools neutralize it and which do not.

By far the quickest and simplest way of identifying a newly isolated virus is by fluorescent antibody staining of the infected cell monolayer, or cells freed from the monolayer, which may give a definitive answer within an hour or so of recognizing early CPE. Fluorescent antibody staining is best suited to the identification of monotypic viruses, but can be more widely applied if suitable antisera are available. However, the reliability of diagnosis by fluorescent antibody staining is greatly influenced by the experience that a particular laboratory has had with the technique, which must always be adequately controlled.

Interpretation

The isolation and identification of a particular virus from a patient with a given disease is not necessarily meaningful in itself. Fortuitous subclinical infection with a virus unrelated to the illness in question is not uncommon. Koch's postulates are as apposite here as in any other microbiological context, but are not always easy to fulfill. In attempting to interpret the significance of any virus isolation one must be guided by the following considerations: (a) the site from which the virus was isolated, e.g., one would be quite confident about the etiological significance of rubella virus isolated from any organ of a congenitally deformed infant, or of mumps virus isolated from the CSF of a patient with meningitis, because these sites are usually sterile, i.e., they have no normal bacterial or viral flora. On the other hand, recovery of an echovirus from the feces or herpes simplex from the throat may not necessarily be significant, because such viruses are often associated with inapparent infections. Interpretation of the significance of the isolation in such instances will be facilitated by (b) isolation of the same virus from several cases of the same illness during an epidemic, and (c) knowledge that the virus and the disease in question are often causally associated.

MEASUREMENT OF SERUM ANTIBODIES

Serological techniques may also be employed in reverse, as it were, to identify an unknown antibody using known antigens. "Paired" sera are taken from the patient, the first ("acute") specimen as early as pos-

TABLE 14-3
Serological Procedures Used in Virology

PROCEDURE	PRINCIPLE	SENSI-TIVITY[a]	SPECI-FICITY[b]
Virus neutralization	Antibody neutralizes infectivity	High	High
Hemagglutination inhibition	Antibody inhibits viral hemagglutination by coating the virus	High	High
Immunodiffusion	Antibodies and soluble antigens diffuse toward one another through agar and produce visible lines of precipitate where homologous antigens and antibodies are present in optimal proportions	Low	High
Complement fixation	Antigen–antibody complex binds complement, which is thereafter unavailable for the lysis of sheep RBC by hemolysin	Low	Low
Immunofluorescence	Antibody (usually) or antigen can be "tagged" by conjugation to a dye (e.g., fluorescein) which fluoresces on excitation by ultraviolet or blue light	Low[c]	Low
Radioimmunoassay	Antibody or antigen can be radioactively labeled and precipitation of antibody–antigen can be monitored by radioisotopic counting or autoradiography	High	High
Immunoelectron microscopy	Antibody aggregates virions into clumps visible by negative staining	Low	High

[a] The sensitivity of a serological test refers to its ability to detect small quantities of antibody.

[b] The specificity of a serological test refers to its ability to discriminate between serotypes within a genus.

[c] "Low" in the sense that the test is negative if the serum is greatly diluted, but high in that small numbers of infected cells can be readily detected.

sible in the illness, the second ("convalescent") specimen at least 1–2 weeks later. Blood should be collected in the absence of anticoagulants, given time to clot, and serum separated before freezing for storage. A simple method of collecting a known volume of blood from a fingerprick is to fill a circle of fixed diameter on a disc of filterpaper by capillarity; such discs may be dried and stored conveniently without appreciable denaturation of IgG (IgM is less stable).

When both the acute and the convalescent specimen are available they are heated at 56° for 30 minutes and sometimes treated by additional methods to destroy various types of nonspecific inhibitors, then they are titrated simultaneously for antibodies by any of a number of serological techniques (summarized in Table 14-3).

A rise in antibody titer between the two serum samples is indicative of recent infection, an increase of fourfold or greater being statistically significant. Since conventional serological techniques cannot detect a rise during the first few days of the illness, serology provides a diagnosis only in retrospect. Nevertheless, it is of value if (a) virus isolation is not practicable, in the case of viruses which are notoriously difficult to isolate (e.g., some togaviruses), or (b) it is too late to expect attempts at virus isolation to be successful (e.g., rubella in a pregnant woman belatedly reporting a rash). Perhaps the most important application of serology today is in epidemiological surveys to determine a community's experience of a given virus (see Chapter 10).

The serological techniques applicable to viruses include many of those with which the reader will already be familiar from bacteriology. In practice, the most useful are complement fixation, hemagglutination inhibition, neutralization, immunodiffusion, immunofluorescence, and radioimmunoassay. It should be noted that agglutination is not a practicable diagnostic procedure, because viruses are too small to yield a visible aggregate. Such aggregation can nevertheless be demonstrated by highly sensitive methods such as immunoelectron microscopy, and precipitation of radioactively labeled virus; both these methods are beginning to be exploited by the better equipped laboratories. Virtually all of the serological procedures described below are susceptible to "miniaturization" and automation. Small volumes of concentrated reagents are dispensed as single drops from calibrated pipettes into "microtrays" ("plastic plates") and dilutions are made with calibrated spiral wire "loops."

Complement Fixation

Antigen–antibody complexes will "fix" complement, and virus–antibody complexes are no exception to the rule. Indeed, following attachment of antibody to enveloped viruses, the phospholipase that comprises an integral component of the complement system may actually puncture the viral envelope.

For the complement fixation (CF) test the acute and convalescent sera are heated (56° for 30 minutes) to inactivate complement, then serially diluted. Usually small plastic trays are used, in order to conserve reagents. Two to four units of antigen (e.g., a crude preparation of live or inactivated virus) are then added to each serum dilution together with two units of complement, derived from (viral antibody-free) guinea pig serum. The reagents are allowed to interact at 4° overnight (or at ambient temperature for a shorter interval) to allow complement to become "fixed." Sheep erythrocytes, "sensitized" by the

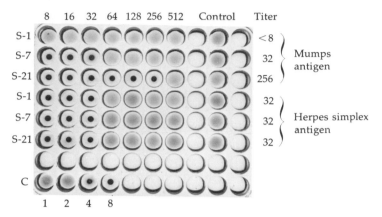

PLATE 14-5. *Complement fixation test. (Titers are expressed as reciprocals of dilutions.) The example illustrates· the results of examination of three consecutive serum samples from a patient with aseptic meningitis complicated by herpes simplex infection of the lip (S-1, S-7, and S-21, serum taken, respectively on day 1, 7, and 21 following admission to hospital). Following heating at 56° for 30 minutes to inactivate complement, each serum was diluted in twofold steps from 1/8 to 1/512. A standard dose of antigen (inactivated mumps or herpes simplex virus) and complement (two hemolytic units) was added to each cup, and allowed to stand at 4° overnight. Sheep erythrocytes, "sensitized" by addition of "hemolysin" (rabbit antiserum against sheep erythrocytes), were then added, and the tray incubated at 37° for 45 minutes. Where complement has been fixed, there is no lysis, and the red blood cells have sedimented to the bottom of the cup. Titration of the complemtnt used in the test is shown in the lowest row (C). Interpretation: (1) The rising titer of antibody against mumps antigen confirms the diagnosis of mumps meningitis. (2) The unchanged titer of antibody against herpes simplex antigen indicates that the herpes labialis represented a recrudescence of a previously existing infection. (Courtesy I. Jack.)*

addition of rabbit antiserum against them ("hemolysin"), are then added and the trays are incubated at 37° for about 45 minutes. In those cups where complement has been fixed by the virus–antibody complex, the hemolysin fails to lyse the RBC; where complement is still available, the RBC's are lysed (Plate 14-5).

To ensure that the test is working properly, controls must be set up to exclude the following:

1. The serum may be "anticomplementary," i.e., it may fix complement even in the absence of viral antigen. This is most commonly due to hemolysis or bacterial contamination in serum that has been improperly collected or stored, or to the high lipid content of serum taken too soon after a fatty meal. The anticomplementary activity can sometimes be removed by heating, by absorption with kaolin, or by absorption with excess complement followed by heating.

2. The viral antigen preparation may be anticomplementary, i.e., it fixes complement even in the absence of antiviral serum. This does not commonly occur, except with mouse brain extracts (arboviruses and coxsackieviruses). The offending lipids can be removed with acetone or fluorocarbon.

3. The viral antigen preparation may fix complement "nonspecifically" as a result of interaction between cellular antigens and anticellular antibodies in the serum. Such problems can be avoided by proper care in the selection of host species for the growth of virus and the preparation of antisera, respectively.

4. On the other hand, anticomplementary activity of serum may indicate the presence of immune complexes in the circulation, which may bind complement *in vivo*. Since such antigen–antibody complexes may play an important role in pathogenesis (see Chapter 8) it is necessary to differentiate their presence from nonspecific anticomplementary activity of the types just mentioned.

The crude preparations of virus (e.g., cell culture supernatants) employed as antigen in most CF tests contain, beside virions, "soluble" antigens corresponding to unassembled components of the virion and also virus-coded nonvirion proteins. Antibodies to all these antigens may be present in convalescent sera and will register by CF. Accordingly, the CF test, as usually performed, will demonstrate substantial "crossing" between serotypes within certain genera or families of viruses if those serotypes share common antigens (e.g., adenoviruses). In this sense, it is not a highly specific test, since it does not always allow one to discriminate between antibodies to different serotypes. It is also not a particularly sensitive test, in that antibody titers determined by CF are lower than those obtained in neutralization or hemagglutination-inhibition (HI) tests on the same serum. Yet, this is the test of choice for the preliminary screening of a serum for antibody, when one has little idea of even the group to which the causative virus belongs. For example, a single CF test using any human adenovirus serotype as antigen will detect serum antibodies provoked by any other human adenovirus serotype. Once "group-specific" antibodies have been determined in this way, the neutralization or HI test can be invoked to identify "type-specific" antibodies, should that information be required.

Hemagglutination Inhibition

Antibody can inhibit virus-mediated hemagglutination (HA) by blocking the particular antigens on the surface of the virion that are responsible for this phenomenon. The hemagglutination-inhibition

(HI) test is conducted as follows. The serum being tested is first treated to remove or destroy nonspecific inhibitors of HA. For example, inhibitors of influenza virus HA are of two varieties: serum glycoproteins, which can be destroyed by periodate, trypsin, or the receptor-destroying enzyme of *Vibrio cholerae;* and heat-labile inhibitors, destroyed by heating at 56° for 30 minutes. A more complex protocol is necessary to prepare sera for HI tests for rubella and other togaviruses (see Chapters 19 and 23). Nonspecific inhibitors of hemagglutination by many viruses can be removed from serum by absorption with bentonite, kaolin, or rivanol (for details see appropriate chapters in Part II). The treated sera are then diluted serially in a plastic tray using a calibrated wire loop. About four or five "agglutinating doses" of the virus being tested are added to each cup from a calibrated dropping pipette, and erythrocytes of the appropriate species are added in the same way. The tray is allowed to stand at a suitable temperature and pH for enough time for the erythrocytes to settle (about 45 minutes). The HI titer is taken as the highest dilution of serum inhibiting hemagglutination (Plate 14-6).

In order to increase the hemagglutination titer of certain viruses, the virions are dissociated with detergents, e.g., Tween 80 plus ether, to release individual peplomers which then reaggregate by their hydrophobic "feet" to form numerous hemagglutinating "rosettes." [See, for example, measles (Chapter 21) and rubella (Chapter 23).]

The HI test is highly sensitive and, except in the case of the togaviruses, highly specific. It measures only those antibodies that bind directly to viral hemagglutinin (i.e., to the projecting tip of the peplomers of most enveloped viruses), and possibly also those capable of attaching to other antigens so closely adjacent to the hemagglutinin that HA is inhibited by steric hindrance. Such superficial antigens are usually type-specific. Moreover, the HI test is simple, inexpensive, and rapid. It is the serological procedure of choice for assaying antibodies to any virus that causes hemagglutination.

A number of less commonly used serological tests are also based upon the property of hemagglutination. For instance, the *hemadsorption inhibition* test is a rapid method of identifying a hemadsorbing virus newly isolated in cell culture; the infected culture is treated with antiviral serum which binds to the hemagglutinin present in exposed plasma membranes and inhibits hemadsorption when erythrocytes are subsequently added. The *immune adherence hemagglutination* test for hepatitis antigen is described in Chapter 23. Even viruses that do not hemagglutinate, e.g., herpetoviruses, can be attached to tannic acid-treated red cells, or coupled to them with chromium chloride, and such

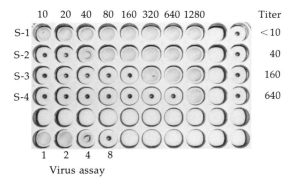

PLATE 14-6. *Hemagglutination inhibition test, used for titrating antibodies to the viral hemagglutinin. (Titers are expressed as reciprocals of dilutions.) In the example illustrated, a cardiac invalid was immunized against the influenza strain, A2/Hongkong/68. Serum samples, S-1, S-2, S-3, and S-4 were taken, respectively, before immunization, 1 week after the first injection, 4 weeks after the first injection, and 4 weeks after the second injection. The sera were treated with periodate and heated at 56° for 30 minutes to inactivate nonspecific inhibitors of hemagglutination, then diluted in two fold steps from $\frac{1}{10}$ to $\frac{1}{1280}$. Each cup then received four hemagglutinating (HA) units of influenza A2/Hongkong/68 virus, and a drop of red blood cells. Where enough antibody is present to coat the virions, hemagglutination has been inhibited, hence the erythrocytes settle to form a button on the bottom of the cup. On the other hand, where insufficient antibody is present, erythrocytes are agglutinated by virus and form a shield. The virus assay (bottom line) indicates that the viral hemagglutinin used gave partial agglutination (the endpoint) when diluted $\frac{1}{4}$. Interpretation: The patient originally had no hemagglutinin-inhibiting antibodies. One injection of vaccine produced some antibody; the second injection provided a useful booster response. (Courtesy I. Jack.)*

erythrocytes are then agglutinable by specific antiviral serum—*indirect (or passive) hemagglutination.*

Virus Neutralization

Certain antibodies interact with virions and neutralize their infectivity. The neutralization test is usually conducted as follows. Serum is heated at 56° for 30 minutes to destroy nonspecific inhibitors of viral infectivity. Dilutions of the serum are then mixed with a constant dose of virus, say 100 $TCID_{50}$. The mixtures are allowed to stand for a time, e.g., 60 minutes at ambient temperature, and then assayed for residual infectivity by inoculation into cultured cells, embryonated eggs, or laboratory animals. The endpoint of the titration is taken as the highest dilution of antiserum that inhibits the development of CPE in

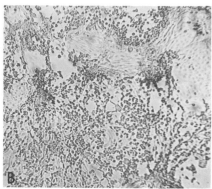

PLATE 14-7. *Virus neutralization test, a specific and sensitive measure of antiviral antibodies. "Acute" and "convalescent" sera from a patient with aseptic meningitis was heated at 56° for 30 minutes to inactivate nonspecific inhibitors of infectivity, then serially diluted. Aliquots of each dilution were separately mixed with 100 $TCID_{50}$ of coxsackievirus type B1, B2, B3, B4, B5, or B6. Following brief standing, each mixture was inoculated into monolayers of monkey kidney cells, and incubated at 36° for several days. All cultures were inspected daily for development of CPE in comparison with controls receiving virus only. Only two key cultures are shown (unstained, ×23). Tube A received a 10^{-2} dilution of convalescent serum plus 10^2 $TCID_{50}$ of coxsackievirus B2. Tube B received a 10^{-2} dilution of convalescent serum plus 10^2 $TCID_{50}$ of coxsackievirus B1, B3, B4, B5, or B6 (all giving the same picture, which corresponded also to that obtained when the "acute" serum was inoculated together with any one of the six serotypes). Interpretation: The meningitis was probably attributable to coxsackievirus B2. (Courtesy I. Jack.)*

cultured cells (Plate 14-7), or the multiplication of virus in the animal used.

The interaction of virions and antibodies is to some extent reversible, so that the apparent titer of antiserum will be influenced by factors such as the time and temperature of incubation of the virus-antiserum mixture, the volume of fluid in which the test is conducted, the susceptibility of the cell type used, and the time at which the CPE is read. Since all these factors may affect the final equilibrium between bound and free antibodies, strict comparisons of titers are only valid within a laboratory where a particular protocol has become standard. For example, all tests should be terminated as soon as the "virus only" controls (no antiserum) show complete CPE; otherwise, the apparent neutralizing titer of the serum drops the later the test is read, as the virus "breaks through" the higher dilutions of antiserum.

If a newly isolated virus proves to be "untypable," i.e., not neutralizable by antisera against any of the known serotypes, it is often because the preparation contains aggregated virions, perhaps associated with cell debris, which are not all accessible to immunoglobulin molecules. These should then be removed by ultrafiltration, or in the case of nonenveloped viruses, e.g., picornaviruses, by treatment with sodium deoxycholate.

The sensitivity of neutralization tests can be increased by the addition of complement, e.g., in the form of fresh serum, which facilitates the lytic destruction of enveloped virions by complement-fixing classes of immunoglobulin. Similarly, the addition of anti-immunoglobulin increases the neutralizing titer of an antiserum. However, neither of these technical tricks are employed in the routine diagnostic laboratory.

There are a number of technical variations on this theme. In keeping with the general trend towards miniaturization, most neutralization tests today are conducted in disposable, nontoxic, sterile, plastic "microtrays" with flat-bottomed wells in which a cell monolayer can be established (Plate 14-4, B). Virus-antiserum dilutions can be added either to established monolayers or simultaneously with the cell suspension. An example of the trend towards automation is the "piggyback" carrier—a plate containing U-shaped wells with a pin-point hole at the bottom; virus-antiserum mixtures prepared in this tray are retained in the wells by surface tension but may be transferred into a tray of mini-cultures simply by placing the first tray on top of the second such that the pointed bottoms touch the meniscus in the tray below.

With many viruses the endpoint can be read simply by color—the so-called *metabolic inhibition test,* in which antibody, by neutralizing the infectivity of the virus, protects the cells against viral destruction and, hence, allows cellular metabolism to continue and the resulting acid turns the indicator yellow. In the *plaque reduction test,* which may also be conducted in microtrays, cell monolayers inoculated with virus-antiserum mixtures are overlaid with agar and incubated until countable plaques develop (Chapter 2); the endpoint is taken to be the highest dilution of antiserum reducing the number of plaques by at least 50%. The *kinetic neutralization test* is an even more sensitive technique used to differentiate closely related strains of virus, e.g., serologically similar arboviruses from different geographical localities (Chapter 19), or vaccine strains of poliovirus from "wild" strains (Chapter 18), or herpes simplex type 1 from type 2 (Chapter 16); the test determines the rate of neutralization of virus by a given concentration of serum by measuring the number of plaque-forming units (pfu) surviving after incubating the virus-antiserum mixture for various lengths of time.

The virus neutralization test is the most sensitive and the most specific serological procedure available. Only antibodies directed against surface antigens of the virion, particularly those antigens involved in adsorption of the virion to the host cell, will register in this test. These superficial antigens are those that have been most subject to the selective pressures of evolution (e.g., by antigenic drift, see Chapter 11), and they are therefore specific to the viral type or strain. Hence, neutralizing antibodies against a given serotype usually show little or no cross-reaction with other viruses within the same genus.

Immunodiffusion

Antigen–antibody interactions can be detected by observing immunoprecipitation reactions in semisolid gels. Antigen and antibody are placed in separate wells cut in a thin layer of agar or agarose on a glass slide or petri dish. The reactants diffuse through the agar at a rate inversely related to their molecular weights. Where antigen and antibody meet in "optimal proportions" a sharp line of precipitate forms in the agar. If several antigens and their corresponding antibodies are present, as in the case of most unpurified preparations of virus and of unabsorbed antiviral sera, each antigen–antibody complex forms a discrete line. If two qualitatively identical preparations of antigen (or antiserum) are placed in adjacent wells and allowed to diffuse toward a common well of antiserum (or antigen), equidistant from them both, the corresponding pairs of precipitin lines will join exactly. This is called the *reaction of identity* (Plate 14-8).

Immunodiffusion is a powerful technique because it permits the simultaneous recognition of all the antibody (or antigen) specificities present in the test material. Furthermore, it is capable of providing a definitive answer overnight or sometimes within a few hours. On the other hand, the method does not lend itself readily to quantitation, i.e., it gives no accurate idea of the "titer" of the unknown antibody or antigen. Moreover, rather high concentrations of both reagents are necessary to ensure the formation of a visible line. It should be appreciated that as a rule purified viruses cannot be used as antigen in gel diffusion because most virions are too large to diffuse through the agar at a satisfactory rate; crude preparations, on the other hand, usually contain most of the antigens of the virion in "soluble" or small particulate form. This can present something of a technical problem, because if a particular antigen is present in particles of different size, several precipitin lines may result.

Many of these drawbacks are overcome by an important new development whereby gel diffusion is rendered much more sensitive and

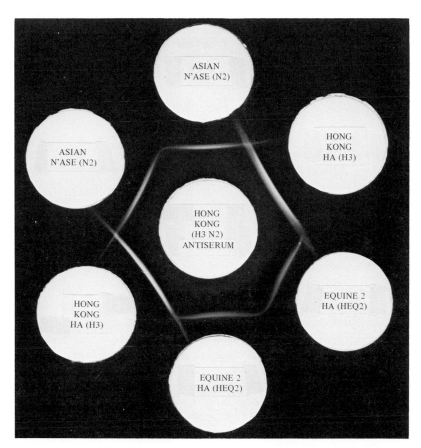

PLATE 14-8. *Immunodiffusion test. Example illustrates its use to analyze relationships between envelope antigens of influenza A virus. Center well: antiserum to Hong Kong influenza virus (H3 N2). Peripheral wells, purified antigens: Hong Kong hemagglutinin (H3), Asian neuraminidase (N2), and Equine 2 hemagglutinin (HEq 2). Antiserum to Hong Kong virus contains antibodies to all the antigens tested. Note (1) two pairs of antigens (N2 and HEq 2) each show fusion of precipitin lines ("reaction of identity"), (2) neuraminidase N2 and hemagglutinin H3 show complete crossing over of precipitin lines ("reaction of complete nonidentity"), and (3) equine (HEq 2) and Hong Kong (H3) hemagglutinins show partial fusion of lines ("reaction of partial identity") indicating serological cross-reactivity. (Courtesy Dr. R. G. Webster.)*

can be quantitated. Antigen, which can be whole virions but is preferably "soluble," is incorporated throughout the agar; the lower the antigen concentration the higher the sensitivity of the test. Antisera placed in wells diffuses out and forms a precipitate with the antigen. The test is quantitated by measuring the diameter of the circle of precipitate. Sensitivity is increased by staining with thiazine red after washing out unprecipitated antigen (if "soluble"). Even greater sensitivity can be achieved, if the antigen is radioactively labeled with ^{125}I by autoradiography or by counting a punched out circle of agar in a spectrometer.

A further refinement of the immunodiffusion technique is a procedure known variously as immunoelectroosmophoresis, or counterimmunoelectrophoresis, or crossover-immunoelectrophoresis. Antigen and antibody can be induced to migrate towards one another much more rapidly by the application of an electric current, so increasing both the speed and the sensitivity of the test. It has proved useful in the detection of hepatitis B carriers by blood banks (see Chapter 23).

Radioimmunoassay

The availability of carrier-free preparations of radioisotopes, especially of iodine (^{125}I or ^{131}I), which can be readily coupled to the tyrosine residues of proteins, has led to the development of radioimmunoassays for analyzing either antigens or antibodies. These procedures are very sensitive, e.g., the technique of autoradiography of cells labeled with radioactive ^{125}I antigen or antibody is over a thousand times more sensitive than is the use of fluorescent antigen or antibody. Radioimmunoassays are also very accurate and simple to perform. One of the two major reagents, either antigen or antibody, is labeled with isotope so that it can be followed by standard procedures, such as autoradiography or counting in a spectrometer. There are three major requirements: (a) One of the reagents, antigen or antibody, should be available in pure form. (b) The extent of substitution of the protein with iodide should be kept low, otherwise the antigenic properties of the protein may be modified. (c) The process of iodination should not change these properties.

Two commonly used types of radioimmunoassay are radioimmunoelectrophoresis and radioimmunodiffusion. In both procedures use is made of the knowledge that globulins with antibody activity may react with antigen even after their precipitation by antisera previously prepared against the globulin. For example, if one wishes to know what class of antibody molecules which specifically react with a given antigen are present in serum, the following approach could be

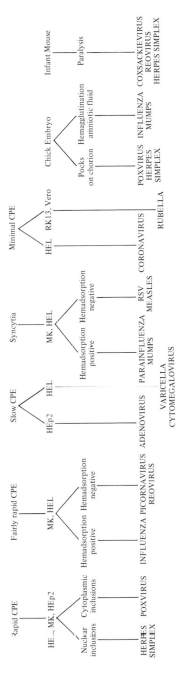

Fig. 14-1. *Isolation of virus from a throat swab.*

TABLE 14-4
Laboratory Diagnosis of Viral Disease[a]

VIRUS	SPECIMEN	HOST[b]	RECOGNITION[b]
Papovaviridae			
PML	PM (brain), urine	CC	CPE
Warts	Biopsy	Nil	Nucl incl, EM (virions)
Adenoviridae, types 1–31	Throat, feces, conjunctiva	CC hu	Slow CPE, nucl incl (B)
Herpetoviridae			
Herpes simplex	Vesicle, throat, cornea, brain	CC hu	Rapid focal CPE, nucl incl (A), fluor
Varicella	Vesicle, throat, PM (lung)	CC hu fibro	Slow focal CPE, syncytia, nucl incl, fluor
Cytomegalovirus	Throat, leukocytes, urine, PM	CC hu fibro	Slow focal CPE, giant cells, nucl cyt incl
Infectious mononucleosis (EBV)	Serum	Nil	Heterophil agglutinins
	Lymphocytes	Cocultivation	Fluor
Poxviridae			
Smallpox, Vaccinia	Vesicle fluid	Nil	EM (virus), CF, gel diff
		Egg CAM	Pocks
		CC primate	Focal CPE, fluor
	Vesicle scrapings	Nil	Cyt incl (A), fluor
Picornaviridae			
Poliovirus, types 1–3	Feces, throat, PM (CNS)	CC primate	Rapid CPE
Coxsackie A, types 1–24	Feces, throat, vesicle, CSF	Newborn mouse	Generalized myositis
		CC monk kid, hu fibro	CPE (some types)
Coxsackie B, types 1–6	Feces, throat, CSF, PM (heart)	CC primate kid	Rapid CPE
		Newborn mouse	Myositis, encephalitis
Echovirus, types 1–33 and Enterovirus 68–71	Feces, throat, CSF	CC primate kid, hu fibro	CPE (often incomplete)
Rhinovirus, >100 types	Nose	CC hu kid or fibro or HeLa	Slow CPE (often incomplete)

Virus	Specimen	Host/culture	Detection
Togaviridae and Bunyaviridae, >200 types	Blood, PM (brain)	Newborn mouse	Encephalitis
		CC Vero, BHK21	CPE, hemadsorption
	Serum	Nil	Serology (HI, neut)
Reoviridae, types 1–3	Feces, throat	CC primate kid	Slow CPE, cyt incl (A)
		Newborn mouse	Steatorrhea, hepatitis, encephalitis
Rotavirus	Feces	Nil	EM (virus)
Orthomyxoviridae			
Influenza A, B	Throat, PM (lung)	CC primate kid	Hemadsorption
		Egg amnion	Hemagglutination
Paramyxoviridae			
Measles	Throat, urine	CC primate kid	Syncytia, cyt nucl incl (A), fluor
	Leukocytes	Nil	Fluor
Mumps	Throat, saliva, CSF, urine	CC primate kid	Hemadsorption, syncytia, cyt incl, fluor
Parainfluenza, types 1–4	Throat	CC primate kid	Hemadsorption, cyt incl (A), fluor
Respiratory syncytial (RS)	Throat, PM (lung)	CC HEp2, hu fibro	Syncytia, cyt incl (A), fluor
Coronaviridae	Throat, nose	CC hu fibro	Incomplete CPE
Arenaviridae	Blood, CSF, PM (organs)	Newborn mouse	Encephalitis
Rhabdoviridae			
Rabies	PM (brain)	Nil	Fluor
Unclassified			
Rubella	Serum	Nil	HI
	Throat, urine, PM (organs)	RK13, Vero, SRIC	Slow focal CPE, fluor
Hepatitis A	Feces	Nil	Immuno-EM (virus)
Hepatitis B	Serum	Nil	RIA, reverse passive HA, CIEP
Norwalk agent	Feces	Nil	Immuno-EM (virus)

[a] Tests for antibody in the patient's serum have been omitted, except where the available techniques of viral isolation are unreliable.

[b] Abbreviations: A, acidophilic; B, basophilic; cyt, cytoplasmic; CC, cell culture; CAM, chorioallantoic membrane; CF, complement fixation; CIEP, counterimmunoelectrophoresis; CPE, cytopathic effect; CSF, cerebrospinal fluid; egg, embryonated hen's egg; EM, electron microscopy; emb, embryonic; fibro, fibroblastic; fluor, immunofluorescence; gel diff, gel diffusion; HI, hemagglutination inhibition; hu, human; immuno-EM, immunoelectron microscopy; incl, inclusion body; kid, kidney; monk, monkey; nucl, nuclear; neut, neutralization; PM, post-mortem; RIA, radioimmunoassay.

used. A sample of the serum is subjected in the normal way to immunoelectrophoresis in agar gels on glass slides. Antiglobulin serum is then diffused into the gel so that arcs of precipitation occur, corresponding to the different classes of immunoglobulin. The slides containing the agar and reagents are thoroughly washed and the radiolabeled antigen is diffused into the gel. The slides are again thoroughly washed, stained, and subjected to autoradiography. Identification of radioactive arcs (and hence specific antibody) is made by superimposing the film bearing the autoradiograph over the stained slide.

While useful, this procedure does not readily allow a quantitative estimate of the amount of specific antibody (or antigen). This may be achieved in several ways, one of which is as follows. Aliquots of ^{125}I-labeled antigen (or antibody) are mixed with serial dilutions of the antiserum (or antigen). An appropriate amount of antiglobulin antibody is added, the mixture incubated for a few hours at 37°, and allowed to stand for 16 hours at 4°. The mixtures are then centrifuged so that complexes of antigen, antibody and antiglobulin antibody, but not free antigen or antibody, are sedimented. The radioactivity in the washed sediments is then counted. Several variants of this procedure are currently used as the most sensitive assays for hepatitis B antigens in sera.

SUMMARY

Table 14-4 summarizes the procedures best suited to the isolation and identification in the laboratory of each of the viruses of man. Much fuller details are given in Chapters 15–23.

Figure 14-1 is a highly stylized flow-sheet intended to provide an example of how a virus diagnostic laboratory might process a throat swab, and reach a provisional diagnosis.

FURTHER READING

Grist, N. R., Ross, C. A. C., and Bell, E. J. (1975). "Diagnostic Methods in Clinical Virology," 2nd ed. Blackwell, Oxford.

Hsiung, G. D. (1973). "Diagnostic Virology." Yale Univ. Press, New Haven, Connecticut.

Kapikian, A. Z., Feinstone, S. M., Purcell, R. H., Wyatt, R. G., Thornhill, T. S., Kalica, A. R., and Chanock, R. M. (1975). Detection and identification by immune electron microscopy of fastidious agents associated with respiratory illness, acute nonbacterial gastroenteritis and hepatitis A. *Perspect. Virol.* **9**, 9.

Kurstak, E., and Morisset, R., eds. (1974). "Viral Immunodiagnosis." Academic Press, New York.

Lennette, E. H., and Schmidt, N. J., eds. (1969). "Diagnostic Procedures for Viral and Rickettsial Infections," 4th ed. Amer. Pub. Health Ass., New York.

Lennette, E. H., Spaulding, E. H., and Truant, J. P. (1974). "Manual of Clinical Microbiology," 2nd ed. Amer. Soc. Microbiol., Washington, D.C.

McCracken, A. W., and Newman, J. T. (1975). The current status of the laboratory diagnosis of viral diseases of man. *Critical Rev. Lab. Clin. Med.* **5,** 331.

Rapp, F., and Melnick, J. L. (1964). Applications of tissue culture methods in the virus laboratory. *Progr. Med. Virol.* **6,** 268.

Schmidt, N. J., and Lennette, E. H. (1973). Advances in the serodiagnosis of viral infections. *Progr. Med. Virol.* **15,** 244.

Symposium. (1972). Laboratory diagnosis of viral infections: Recent advances and their clinical application. *Amer. J. Clin. Pathol.* **57,** 731.

PART II

Viruses of Man

CHAPTER 15

Adenoviridae

INTRODUCTION

In 1953, Rowe and his colleagues made an accidental discovery of far-reaching importance. Searching for a suitable type of cultured cell in which to isolate the elusive "common cold virus," they embedded fragments of adenoids in clotted plasma and waited for a sheet of cells to grow out. Healthy monolayers did indeed develop, but many subsequently degenerated with a "viral" type of cytopathic effect. The cultures yielded a new agent, fittingly named the "adenovirus" (see Table 15-1). It soon became evident that these viruses are present as a latent infection of the tonsils and adenoids of most children. Hilleman was quick to discover that they are responsible for a proportion of the respiratory ailments of man (see Table 15-2). Their role in diseases of the alimentary tract, from which they are also frequently isolated, is still being assessed.

During the 1960's attention was focused on the startling finding that some human adenoviruses produce cancer following inoculation into baby rodents, and are capable of transforming cells to the malignant state *in vitro* (see Chapter 9). However, the evidence now available indicates that adenoviruses are not naturally carcinogenic in man.

PROPERTIES OF THE VIRUS

The virion consists of an outer icosahedral capsid, 70–80 nm in diameter, and an inner core, 40–45 nm in diameter, composed of histone-like polypeptides in association with a double-stranded DNA molecule with a molecular weight of 20–25 million daltons. The capsid is composed of 252 capsomers, of which 240 are *hexons* (hollow hexagons, each adjoining 6 neighboring capsomers), while the 12 capsomers situated at the vertices of the icosahedron are *pentons* (hollow pentagons composed of a different type of polypeptide, and bounded

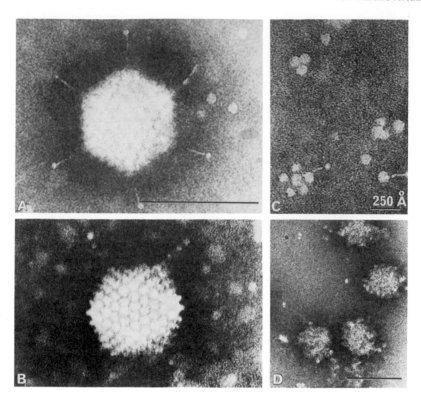

Plate 15-1. *Adenovirus. Negatively stained preparations of adenovirus type 5 (bars = 100 nm). (A) Virion showing the fibers projecting from the vertices of the icosahedron. (B) Virion showing the icosahedral array of capsomers. The capsomers at the vertices are surrounded by five nearest neighbors, all the others by six. (C) Clusters of capsomers from disrupted virion. Penton base with attached fiber is surrounded by a group of five hexons. (D) Purified cores released from acetone-disrupted virions with some contaminating groups of nine hexons and single capsomers. [A, B, and C from R. C. Valentine and H. G. Pereira, J. Mol. Biol.* **13,** *13 (1965); D from W. G. Laver, N. G. Wrigley, and H. G. Pereira, Virology* **39,** *599 (1969).]*

TABLE 15-1
Properties of Adenoviruses

Icosahedron, diameter 70–80 nm
252 capsomers, 12 fibers
Inner protein core, diameter 40–45 nm
DNA, double-stranded, mol. wt. 20–25 million

by 5 neighbors). From each penton projects a "fiber," with a terminal knob (Plate 15-1). The fibers, as well as the virions themselves agglutinate rat, monkey, and/or human erythrocytes.

Adenoviruses are relatively stable to heat and acid, and since they contain no lipid are absolutely resistant to lipid solvents, including bile salts. For this reason the adenoviruses survive the conditions prevailing in the gastrointestinal tract and can often be isolated from the feces, as well as from the throat.

There are 33 human serotypes as well as many more infecting other animals. All human serotypes share at least two major antigenic determinants, so that they cross-react almost completely in complement fixation tests and show lines of identity in gel diffusion. On the other hand, there are type-specific determinants on the hexon and fiber polypeptides, which enable the serotypes to be readily distinguished in other tests. Type-specific antibodies binding to fibers inhibit hemagglutination; those binding to hexons neutralize infectivity. Thus, serotypes whose hexon antigens are clearly distinguishable by neutralization tests may nevertheless share a common fiber antigen and, hence, appear identical by hemagglutination inhibition tests.

A large number of adenoviruses occur naturally in most animal species that have been adequately studied. Most are host-specific, but human serotypes 12, 18, and 31 are highly oncogenic when injected into baby rodents, while other human serotypes have more limited oncogenic potential. Sarcomas develop a few months after inoculation of newborn hamsters, while rodent or human cells may be transformed to malignancy in culture (Chapter 9).

The description of the multiplication cycle of a "typical" DNA virus given in Chapter 3 can be taken, in most of its basic essentials, to represent that of an adenovirus. Accordingly, little need be added here, except to say that the replication of viral DNA and assembly of virions is completed in the nucleus, whereas viral proteins must first be synthesized on cytoplasmic polyribosomes. Viral DNA and structural proteins are made in considerable excess; most of the newly synthesized protein is quickly polymerized to form capsomers, but the majority of these are never assembled into virions. Host cell macromolecular synthesis is not shut down until relatively late in the cycle.

HUMAN INFECTIONS

Clinical Features

Perhaps the best known clinical expression of adenovirus infection is the syndrome known as "acute respiratory disease" (ARD), caused

TABLE 15-2
Diseases Produced by Adenoviruses

COMMON	LESS COMMON
"Acute respiratory disease"	Epidemic keratoconjunctivitis
Pharyngitis	Meningoencephalitis
Conjunctivitis	Hemorrhagic cystitis
Pharyngoconjunctival fever	? Intussusception
Pneumonia in infants	
Mesenteric adenitis	

mainly by types 4 and 7, but also 3, 14, 21, and others, as classically described in outbreaks among military recruits. Pharyngitis, with an accompanying cough, headache, fever, and lymphadenopathy, are the dominant features; a rash is rarely seen. In the community at large, on the other hand, types 3 and 7 more commonly lead to a milder pharyngitis or conjunctivitis, or to a syndrome known as pharyngoconjunctival fever, often occurring as summer epidemics of "swimming pool conjunctivitis" among children. In young children, these viruses may also involve the lower respiratory tract, presenting as bronchitis, bronchiolitis, croup, or bronchopneumonia, which is occasionally fatal. A serious type of keratoconjunctivitis, caused by type 8, is seen mainly in workers exposed to dust and trauma in industry ("shipyard eye"). By contrast, types 1, 2, 5, and 6 produce very mild upper respiratory infections of a sporadic nature in young children.

The relationship of adenoviruses to disease in the alimentary tract is rather more tenuous. Adenoviruses have been recovered from several outbreaks of gastroenteritis. However they are also frequently seen by electron microscopy in feces from normal people. There also appears to be an indirect association between adenoviruses and intussusception. This surgical emergency of infancy arises when excessive peristaltic activity leads to invagination of one segment of the small intestine into another with consequent obstruction. The Peyer's patches and the draining mesenteric lymph nodes are often found to be inflamed and yield an adenovirus on culture.

Recently adenoviruses have been incriminated in two serious diseases, acute hemorrhagic cystitis (types 2, 11, 21) and meningoencephalitis (type 7).

Pathogenesis and Immunity

Adenoviruses multiply initially in the pharynx, conjunctiva, or small intestine, and rarely spread beyond the draining cervical, preauricular, or mesenteric lymph nodes. The disease process remains relatively localized, and the incubation period is correspondingly short (5–7 days), but excretion of virus in the feces may be quite prolonged. Very few adenoviral infections come to autopsy, but the lungs from infants who have died from pneumonia due to adenovirus contain the typical basophilic nuclear inclusions.

Tonsils and adenoids removed from otherwise normal children more often than not carry an adenovirus, usually type 1, 2, or 5. The virus is not demonstrable when the tissue is macerated and inoculated into human epithelial cell lines in the conventional way; the tissue itself must be cultured for several weeks before viral CPE becomes apparent in the resulting outgrowth. The state of the virus during the persistent infection *in vivo* is not certain, but its multiplication may be held in check by the considerable quantities of antibody synthesized by these lymphoid organs.

In contrast to most other respiratory viral infections, acquired immunity to adenovirus infections is long-lasting, and second attacks by the same serotype are rare. Perhaps this reflects the extent of involvement of lymphoid cells in the alimentary tract and the regional lymph nodes.

Laboratory Diagnosis

Virus may be recovered from the throat, feces, or conjunctiva by inoculation of human epithelial cell cultures such as primary embryonic kidney, or more conveniently, continuous malignant lines such as HEp-2, KB or HeLa (which must, however, be free of *Mycoplasma* because this common contaminant of animal sera depletes the medium of arginine, which is an absolute requirement for the growth of adenoviruses). The characteristic rounding, swelling, and aggregation of cells into "grape-like clusters" usually takes several days to develop; indeed with some of the higher numbered serotypes recovered from feces, CPE develops so slowly that one blind passage may be required to reveal it after 2–4 weeks. Fixation and staining is not usually necessary but does reveal highly characteristic intranuclear inclusions; initially small and acidophilic, they enlarge and turn basophilic, sometimes connected to the nuclear periphery by strands of chromatin (Plate 15-2).

The isolate may be identified as an adenovirus by complement fixation, then typed by HI or neutralization, perhaps after preliminary allocation into one of several HA groups defined by the virus' ability to agglutinate erythrocytes from rats, monkeys, or man. Some of the higher numbered serotypes from feces grow poorly and others (un-

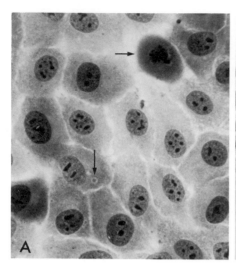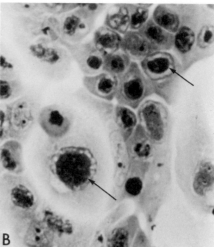

PLATE 15-2. *Cytopathic effects induced by adenoviruses (hematoxylin and eosin; ×400). (A) Normal monolayer of HEp-2 cells. Horizontal arrow, cell in mitosis. Vertical arrow, phagocytosed cell debris, not to be confused with viral inclusion. (B) Cytopathic effects produced by adenovirus in HEp-2 cells. Note distended cells containing basophilic intranuclear inclusions (arrows), from which threads of chromatin sometimes radiate to the periphery of the nucleus. (Courtesy I. Jack.)*

typed), not at all. The latter are often seen when fecal specimens are examined directly by electron microscopy. As these probably represent new serotypes of unknown pathogenicity, efforts should be directed toward their cultivation and serological differentiation by immunoelectron microscopy.

Epidemiology

Since the human adenoviruses seem to be species-specific, spread occurs only from man to man, via the respiratory and alimentary routes. However, the virus is stable enough to survive in sewage, swimming pools, and dust, which may sometimes be the source of infection in gastroenteritis, conjunctivitis, and epidemic keratoconjunctivitis, respectively.

Types 1, 2, 5, and 6 are endemic, infecting most children by the age of three, and often persisting for long periods thereafter as a latent infection of tonsils and adenoids. By contrast, 3, 4, 7, and, to a lesser extent, types 14 and 21 are usually associated with epidemics of acute respiratory and ocular disease in closed communities like military camps and boarding schools. The propensity of type 4 to cause ARD in military recruits, but rarely to produce infection in civilians, has few parallels in epidemiology. Overall, the adenoviruses account for something less than 5% of the total spectrum of viral respiratory disease.

Prevention and Control

There is no pressing need for adenovirus vaccines, except perhaps in military recruits, for whom oral vaccines have proved satisfactory. By enclosing virus in enteric-coated capsules, live unattenuated adenoviruses 4 and 7 are introduced directly into the intestine without running any risk of the virus multiplying in the throat and causing a respiratory disease, nor of their being inactivated by the acid environment of the stomach. The resulting immunity to challenge is of the order of 50–75%, and excreted virus spreads only rarely to contacts, probably via the alimentary route. Needless to say, the adenoviruses employed in these vaccines are grown in human embryonic fibroblasts, not monkey kidney, because of the danger of formation of adenovirus-SV40 hybrids (see below).

ADENOVIRUSES AND DEFECTIVENESS

A virus is said to be "defective," in a particular host cell, when its replication in that cell falls short of the production of infectious virus. The adenoviruses are associated with two types of defectiveness, discussed in detail elsewhere in this book; here we wish only to draw the reader's attention to these accounts.

Adeno-Associated Viruses (AAV) (see Chapters 1, 4, and 22)

These are very small single-stranded DNA parvoviruses (Plate 15-3) of 4 serotypes that are defective in human cells (and perhaps in all cells), but can be complemented by adenoviruses with which they cohabit in the human throat. They have not been causally connected with human disease.

Adenovirus-SV40 Hybrids (see Chapter 4)

Human adenoviruses will not multiply in simian cells unless they are complemented by a simian adenovirus or by the papovavirus SV40. When simultaneously infected with SV40 and an adenovirus, the simian cells yield some hybrid particles consisting of adenovirus capsids enclosing either SV40 DNA, or SV40 DNA covalently linked with adenovirus DNA. A great variety of situations have been recorded involving various combinations of entire or partial genomes (see Chapter 4). Discovered as a laboratory artifact, these mixed

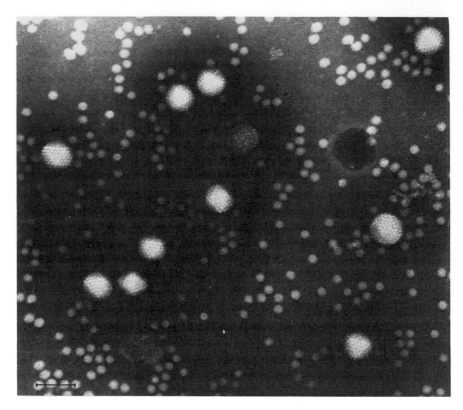

PLATE 15-3. *Adeno-associated virus (AAV) particles together with their helper adenovirus (bar = 100 nm). (Courtesy Dr. H. D. Mayor.)*

virions nevertheless have considerable theoretical interest, and practical relevance to both vaccine production and oncogenesis.

FURTHER READING

Fenner, F., McAuslan, B. R., Mims, C. A., Sambrook, J. F., and White, D. O. (1974). "The Biology of Animal Viruses," 2nd ed., pp. 16, 79–85 and 191–198. Academic Press, New York.

Green, M. (1970). Oncogenic adenoviruses. *Mod. Trends Med. Virol.* **2**, 164.

Jackson, G. G., and Muldoon, R. L. (1973). Viruses causing common respiratory infection in man. IV. Reoviruses and adenoviruses. *J. Infec. Dis.* **128**, 811.

Norrby, E. (1971). Adenoviruses. *In* "Comparative Virology" (K. Maramorosch and E. Kurstak, eds.), Academic Press, New York.

Philipson, L., and Lindenberg, U. (1974). Reproduction of adenoviruses. *In* "Comprehensive Virology" (H. Fraenkel-Conrat and R. R. Wagner, eds.), Vol. 3, p. 143. Plenum, New York.

Philipson, L., Pettersson, U., and Lindberg, U. (1975). Molecular biology of adenoviruses. *Virol. Monogr.* **41**, 1.

CHAPTER 16

Herpetoviridae

INTRODUCTION

These large intranuclear DNA viruses (Table 16-1) display a remarkable propensity for establishing latent infections which may persist for the life of the host. The clinical manifestations of primary infection are set out in Table 16-2. Thereafter, the virus remains quiescent, being demonstrable only sporadically or not at all, until it is reactivated by one of several known types of stimulus, such as irradiation or immunosuppression. Such exacerbations of endogenous disease may take the form of a crop of vesicles on the skin in the case of herpes simplex or zoster, or more generalized effects in the case of cytomegalovirus or EB virus. The capacity to persist indefinitely as a latent infection equips these viruses uniquely well for long-term survival in nature. During the last few years, attention has turned to the possibility that herpetoviruses such as EB virus and herpes simplex type 2 may play an etiological role in human cancer (see Chapter 9).

PROPERTIES OF THE VIRUS

The herpetovirion has an icosahedral capsid composed of 162 capsomers, generally described as elongated hollow hexagonal and pentagonal prisms with a central axial hole. Surrounding the particle is a loose lipoprotein envelope, giving the virus an overall diameter of 150 nm. "Naked" (nonenveloped) particles (100 nm) are often also seen and thin sections reveal that there are at least two additional protein shells within the outer capsid (Plate 16-1). The DNA is double-stranded, with a molecular weight of 100 million daltons and an unusually high guanine plus cytosine content.

The nucleocapsids of the various herpetoviruses share a common group antigen demonstrable by immunodiffusion or complement fixa-

TABLE 16-1

Properties of Herpetoviruses

Spherical virion, diameter 150 nm
Icosahedral capsid, 100 nm, 162 capsomers
Multiple shells
Envelope
DNA, double-stranded, mol. wt. 100 million

tion. The human herpetoviruses, listed in Table 16-2, differ in respect to numerous proteins. There is only a single serotype of varicella, EB virus, and the human cytomegalovirus, and only two of herpes simplex virus. Other members of the family Herpetoviridae infect a wide variety of animals and birds. Of particular interest are the simian herpetovirus B, lethal for man infected by monkey-bite, and certain oncogenic herpetoviruses, such as Marek's virus of fowls, the Lucké renal carcinoma virus of frogs, and *Herpesvirus saimiri* and *Herpesvirus ateles* of monkeys, which were discussed in Chapter 9.

In cultured cells herpetoviruses produce intranuclear inclusions and often multinucleate giant cells which are diagnostic (Plate 16-2).

TABLE 16-2

Diseases Produced by Herpetoviruses

	DISEASE	
VIRUS	COMMON	LESS COMMON
Herpes simplex	Gingivostomatitis	Encephalitis
	Pharyngitis, tonsillitis	Eczema herpeticum
	Herpes labialis ("fever blisters")	Neonatal herpes (type 2)
	Genital herpes (type 2)	Traumatic herpes
	Keratoconjunctivitis	Hepatitis
Varicella–Zoster	Varicella (chickenpox)	Pneumonitis
	Herpes zoster (shingles)	Encephalitis
Cytomegalovirus	Cytomegalic inclusion disease	Post-transfusion mononucleosis
		Hepatitis
		Pneumonitis
EB virus	Infectious mononucleosis	Post-transfusion mononucleosis
	Burkitt's lymphoma?	Guillain–Barré syndrome
	Nasopharyngeal carcinoma?	Bell's palsy, transverse myelitis

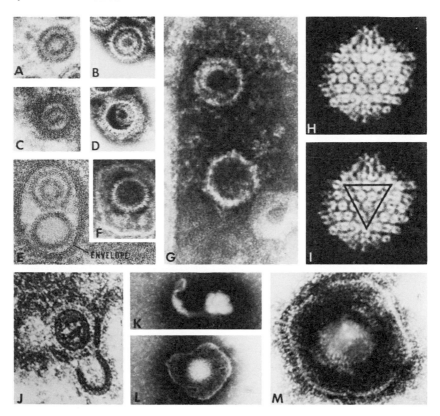

PLATE 16-1. *Herpetoviridae.* (*A–F*) *EB virus from Burkitt lymphoma,* (*G*) *Virus of Lucké adenocarcinoma of the frog,* (*H–M*) *Herpes simplex virus.* (*A, C, E, and J*) *thin sections; others, negatively stained whole mounts.* (*A and B*) *Particles consisting of a core and three concentric capsids.* (*C and D*) *The same as* (*A*) *and* (*B*) *but coated by amorphous material (inner membrane).* (*E and F*) *Enveloped nucleocapsids.* (*G*) *Naked nucleocapsids from Lucké adenocarcinoma, the upper nucleocapsid showing details of an internal structure.* (*H and I*) *Naked nucleocapsid of herpes simplex virus with a triangular face of the icosahedron outlined.* (*J*) *Thin section of a particle, apparently coated with an inner membrane, in the process of acquiring its outer envelope from the nuclear membrane.* (*K and L*) *Enveloped particles at the same magnification, one with an intact envelope impermeable to negative stain* (*K*), *and one into which the stain has penetrated* (*L*). (*M*) *Enveloped particle penetrated by stain and showing details of the envelope.* [*From B. Roizman and P. G. Spear, Herpesvirus In "Ultrastructure of Animal Viruses and Bacteriophages: An Atlas* (*A. J. Dalton and F. Haguenau, eds.*), *p. 83. Academic Press, New York, 1973. Courtesy Dr. B. Roizman.*]

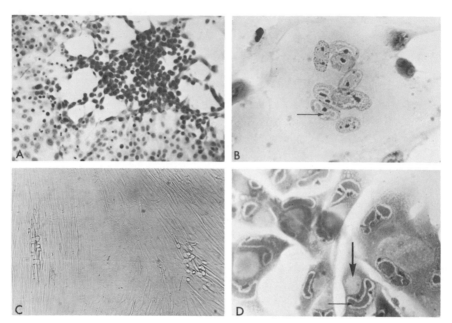

PLATE 16-2. *Cytopathic effects (CPE) induced by herpetoviruses. (A) Herpes simplex virus in HEp-2 (Hemotoxylin and Eosin stain; ×57). Note early focal CPE. (B) Varicella-zoster virus in human kidney (Hemotoxylin and Eosin stain; ×228). Note multinucleated giant cell containing acidophilic intranuclear inclusions (arrow). (C) Cytomegalovirus in human fibroblasts (unstained; ×35). Note two foci of slowly developing CPE. (D) Cytomegalovirus in human fibroblasts (Hematoxylin and Eosin stain; ×228). Note giant cells with acidophilic inclusions in the nuclei (small arrow) and cytoplasm (large arrow) the latter being characteristically large and round. (Courtesy I. Jack.)*

HERPES SIMPLEX

Introduction

Among the most common of all human viral diseases are the "cold sores" or "fever blisters" that strike most of us at frequent intervals throughout life. We carry the causative virus, herpes simplex, as a latent infection, localized to certain sensory nerve ganglia. On occasions of stress due to fever, menstruation, sunburn, and so forth, the virus is induced to produce vesicular lesions of the skin, especially on the face, lips, and nose.

Occasionally infection of the cornea causes keratoconjunctivitis leading to dendritic ulceration and even blindness. Fortunately, this is the one viral disease so far accessible to specific antiviral chemotherapy, as inhibitors of DNA synthesis can be safely instilled into the eye without running any risk of systemic toxicity.

Herpes simplex type 2 is a common venereal infection which is associated (possibly causally) with carcinoma of the cervix in women.

Clinical Features

Most primary infections occur during childhood and are inapparent. Of the clinical conditions, by far the most common are *vesicular pharyngitis* and *tonsillitis*, being seen increasingly in adolescents experiencing the virus for the first time, and *gingivostomatitis*, which is the classical presentation in younger children. The mouth and gums become covered with vesicles which rupture to become ulcers (Plate 16-3). Though febrile and irritable, the child invariably recovers rapidly. Not so lucky are those with infantile eczema who contract *eczema herpeticum*. Occasionally, *disseminated neonatal herpes* overwhelms a newborn (often premature) baby, hepatitis being a particularly conspicuous feature. Most of these fatal infections are acquired at the time

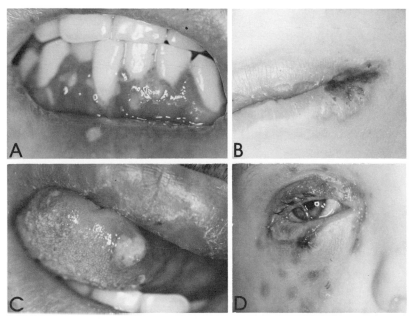

PLATE 16-3. *Herpes simplex. Note vesicles, rupturing to become ulcers, on gums (A), lips (B), tongue (C), and eye (D). (Courtesy Dr. J. Forbes, Fairfield Hospital for Infectious Diseases, Melbourne.)*

of birth, when a baby encounters genital herpes (type 2) in the mother's vagina; others are contracted from attendants with "cold sores." Disseminated herpes with hepatitis may also complicate immunosuppressive therapy in patients of any age. Meningitis, encephalitis, or *meningoencephalitis* are occasional serious complications of the primary infection in all age groups; herpes simplex virus is the most common cause of sporadic necrotizing encephalitis in most western countries. Herpetic *keratoconjunctivitis* is not uncommon; classically, the keratitis progresses to a dendritic ulcer, or less commonly, a geographic ulcer of the cornea, which may, though rarely, go on to disciform keratitis or to corneal scarring with resultant blindness. A rather exotic presentation occasionally seen in dentists, doctors, or nurses is the herpetic whitlow of the finger; another example of infection acquired through a break in integrity of the skin (*traumatic herpes*) is "herpes gladiatorum" of wrestlers. *Vulvovaginitis*, occurring in children as well as young women, is caused by type 2 virus, as are most infections of the penis. Indeed, herpes simplex type 2 has now replaced syphilis as the second most common venereal disease, and is on the increase in the permissive society. The possible role of the virus in congenital malformations is still a matter of debate.

The classical form of recurrent infection is known variously as herpes simplex, herpes labialis, herpes facialis, or herpes febrilis. Attacks are sometimes predictable, being heralded by prodromal hyperesthesia lasting a few hours before a crop of vesicles breaks out, often around the mucocutaneous junction of the lips or nose. Less commonly, recurrent herpes occurs elsewhere on the skin or mucous membranes, e.g., as genital herpes or recurrent keratoconjunctivitis.

Pathogenesis and Immunity

Herpes simplex virus acquired in childhood is usually retained in the body for life, and the individual is thenceforth subject to recurrent attacks of "fever blisters." Provocative stimuli include menstruation, excessive exposure to sunlight or cold wind, pituitary or adrenal hormones, allergic reactions, or classically, fever. In this connection it is intriguing that herpes simplex is more commonly provoked by pneumococcal pneumonia, malaria, meningococcal meningitis, and influenza than by other fevers. The blisters always tend to break out on the same part of the body of any given individual. Between attacks the virus can only occasionally be isolated from the saliva (2% of carriers) and not at all from biopsies of skin taken from the area customarily affected, nor from organ cultures of that skin. Furthermore, if such skin is transplanted to another part of the body and the patient subjected to a provocative stimulus, herpetic vesicles appear at the usual site but not on the graft. This would suggest that virus persists in the body as an asymptomatic infection of some area other than that in which the lesions subsequently appear. Recent investigations in animals and man have demonstrated conclusively that the virus persists in the neurons of sensory ganglia, commonly the trigeminal ganglion. The state of the virus in the neurons during the years of latency is unknown. Suffice to say that infectious virions cannot be demonstrated until they have been "induced" to multiply by some trick such as explantation or cocultivation of the ganglion *in vitro* (see Chapter 8).

A likely mechanism for latency and reactivation of herpes simplex is as follows. Following primary infection of the mucous membrane of the mouth, the virus ascends the sensory nerves of the maxillary and mandibular branches by growing along the nerve sheath. On reaching the ganglion it infects some of the ganglion cells and sets up a latent noncytocidal infection of these cells. After the appropriate type of stimulus, which may be general (perhaps hormonal) or local, the virus spreads down the nerve fiber, again probably by contiguous growth along the sheath. When it reaches the sensory nerve endings, virus spreads to the cells of the epidermis and multiplies freely to produce a blister (see Figure 8-2).

Superficially, recurrent herpes seems to represent an immunological paradox. Adults with high levels of antibody are subject to recurrent attacks of herpes simplex, while those with no antibody have no such attacks because they have never acquired the primary infection. The explanation of these apparent anomalies is clear enough. Antibody, though present in high titer in the serum following the primary infection, cannot penetrate cells to eliminate the latent virus from the body during remissions. By contrast, individuals without antibody have never acquired the primary infection and probably never will, once they have reached adulthood because for some reason those who get through childhood without contracting the infection rarely contract it later.

Laboratory Diagnosis

Vesicle fluid, saliva, throat or vaginal swab, or brain or liver biopsy can be taken for virus isolation. A corneal scraping from the base of a dendritic ulcer yields virus much more consistently than a conjuctival swab, but must be taken during the first few days of the infection.

Cultures of human fetal fibroblasts, human or rabbit kidney, or continuous cell lines are all suitable for inoculation. Cytopathic changes
develop so rapidly that a tentative diagnosis can often be provided
within 24–48 hours on the basis of the distinctive foci of swollen,
rounded cells (Plate 16-2A). Positive identification can be confirmed
by immunofluorescence or neutralization.

The virus kills infant mice and produces pocks on the chorioallantoic membrane of embryonated eggs; type 2 pocks are larger than
those of type 1. As herpetic vesicles are usually readily accessible, it is
possible to establish a diagnosis even more rapidly by scraping infected cells from the base of the lesion for direct demonstration of viral
antigen by immunofluorescence or conventional staining. The typical
giant cell is multinucleated with Cowdry type A inclusions in each
nucleus (Plate 5-1A and Fig. 5-1B). Where facilities are available, the
virions themselves may be demonstrated by direct electron microscopy on vesicle fluid.

Epidemiology

Saliva from patients with active herpetic lesions or from asymptomatic carriers is a ready source of infection for infants being fondled
by well-meaning parents, or sharing contaminated eating utensils.
Most cases of primary herpetic gingivostomatitis used to be acquired
in this way during the months following the disappearance of maternal antibody. In this sense, herpetic infections were spread "vertically," i.e., familially. Children from poorer socioeconomic environments still tend to pick up their infection in the second year of life, but
those from more affluent surroundings nowadays escape often until
adolescense. A high proportion of adults have antibody and most of
these suffer from current episodes of herpes labialis.

Herpes simplex type 2, by contrast, is a venereal disease, seen
mainly in young adults and steadily increasing in incidence.

Prevention and Control

Prevention of herpes simplex infection demands education of
mothers in the relevant aspects of hygiene. Eczematous children are
particularly vulnerable. In order to lower the risk of neonatal type 2
herpes simplex in infants born of mothers with genital infection, a
strong case can be made for Caesarean section in the case of women
with overt vulvovaginitis at the time of parturition.

Herpetic keratoconjunctivitis is one of the very few viral diseases
amenable to specific chemotherapy (see Chapter 13). Inhibitors of

DNA synthesis, instilled topically every hour or two as drops or ointment, can halt the infection, provided that treatment is commenced in
the first few days, before the lesion has progressed to a deep metaherpetic ulcer. Such drugs inhibit cellular as well as viral DNA synthesis,
but are harmless to nondividing cells. The high doses of drug administered topically to the eye remain localized to this relatively avascular environment for long periods. When it does enter the blood
stream, the agent is diluted out to the point of being quite innocuous
to other cells in the body. Most of the clinical experience so far has
been with 5-iodo-2'-deoxyuridine (IUdR, idoxuridine, Stoxil), but
other substituted nucleosides, notably cytosine arabinoside (ara-C)
and adenine arabinoside (ara-A) are also effective. The same agents
embodied in a dimethyl sulfoxide-based ointment are also widely
used outside the USA for the treatment of herpes labialis. Intravenous
administration of these highly toxic drugs to patients with the lethal
disease, herpetic meningoencephalitis, is currently controversial; the
severe myelosuppression these drugs inflict cannot readily be justified
in the light of the benefit they bestow, which is marginal at best (see
Chapter 13).

VARICELLA–ZOSTER

Introduction

Varicella (chickenpox) is well known to every young mother as one
of the "Big Four" viral diseases of childhood. Herpes zoster (shingles),
on the other hand, means more to the grandparent, who experiences a
painful crop of blisters on his trunk or face. Yet both diseases are
caused by the same virus. The child can catch chickenpox from the
grandfather's zoster; the old man is subject to zoster only if he has suffered from chickenpox earlier in his life, following which the virus has
remained latent in dorsal root ganglia.

Clinical Features

Varicella begins with fever, followed within a day by the sudden
eruption of crops of skin lesions distributed mainly "centripetally"
(trunk), by contrast with the "centrifugal" distribution (face and
limbs) in smallpox. Successive crops of vesicles progress to pustules
then scabs, so that (again by contrast with smallpox) all stages of
lesion development may be seen simultaneously. Intense itching
tempts the child to scratch the lesions, leading to bacterial superinfection and permanent scarring. Complications are unusual, except in

adults, in whom a severe *pneumonitis,* sometimes accompanied by hepatitis, may occasionally supervene a few days after the rash appears. *Encephalitis* is rare (less than 1 in 1,000).

Chickenpox is a dangerous disease in nonimmune neonates who can contract the disease transplacentally or postnatally from mothers suffering from the disease at the time of delivery. Another particularly susceptible group are immunologically compromised children with leukemia, nephrotic syndrome, splenectomy, etc., or indeed patients of any age undergoing immunosuppressive therapy with corticosteroids, antimetabolites, or radiation; in such individuals varicella pneumonitis can be lethal (Plate 24-2).

Herpes zoster (zoster = "girdle") is characterized by the abrupt development in an adult of pain and then blisters over an area of skin supplied by a particular sensory nerve. Accordingly, the lesions are unilateral, terminating abruptly at the midline. The trunk and face are most commonly affected. Paralysis is an occasional development if motor roots become involved. Pain is extremely severe for 1–4 weeks, often persisting for months as post-zoster neuralgia. When the ophthalmic division of the trigeminal nerve is involved, corneal ulceration may complicate the facial rash (Plate 16-4). Disseminated herpes zoster is sometimes seen in teminal cancer patients or the immunosuppressed.

Pathogenesis and Immunity

Varicella virus is thought to enter the body by the respiratory route. An incubation period of 2 weeks (occasionally up to 3) follows before the generalized vesicular rash breaks out on the skin. Infected prickle cells show ballooning and intranuclear inclusions; virus is plentiful in the vesicle fluid. In those rare instances when the patient dies of disseminated varicella, giant cells are found in many organs of the body.

Herpes zoster occurs only in those with a history of varicella. Crops of vesicles appear on the skin supplied by a particular sensory nerve, usually in one of the areas affected by the original attack of chickenpox, namely the thorax, abdomen or face. Second attacks are unusual (~4%) and do not necessarily occur on the same dermatome. Zoster affects almost 1% of the population annually, the incidence increasing strikingly with age, such that more than half of all cases occur in those over 50 and the majority of 80-year-olds have suffered from the disease. The condition is particularly common in patients suffering from Hodgkin's disease, lymphatic leukemia, or other malignancies, or following treatment with immunosuppressive drugs or, in

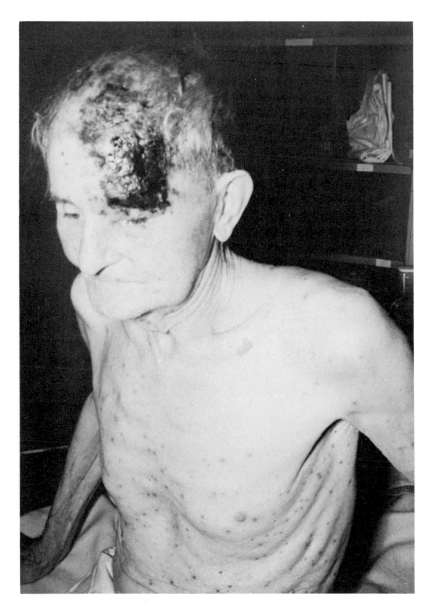

PLATE 16-4. *Ophthalmic zoster. Note distribution of lesions corresponding to distribution of ophthalmic division of the trigeminal nerve. In addition there are scattered lesions elsewhere on the skin (disseminated zoster), not an uncommon complication of the disease. (Courtesy Dr. J. Forbes.)*

particular, following injury to or irradiation of the spine. Under the influence of these or other unknown inducing agents, the virus is believed to travel down a sensory nerve from the corresponding dorsal root ganglion or cranial nerve ganglion, where it has been residing in some occult form between the attacks (see Chapter 8).

Following varicella, immunity to exogenous reinfection is lifelong, so that second attacks of the disease are virtually unknown. Nevertheless, the high levels of antibody that result do not succeed in eliminating the virus from the dorsal root ganglia of those subject to subsequent episodes of herpes zoster. Indeed, zoster triggers an anamnestic antibody response (IgG not IgM) which may help to bring that particular recrudescent attack to a halt, but still fails to abolish the latent infection.

Laboratory Diagnosis

The clinical picture is so distinctive that the laboratory is rarely called upon for assistance. Differential diagnosis from smallpox may be obtained rapidly by direct electron microscopic examination of vesicle fluid for virions. Alternatively, scrapings taken from the base of a lesion may be stained for the presence of inclusions, which are nuclear, not cytoplasmic. A definitive diagnosis may also be sought postmortem following neonatal generalized varicella or pneumonia. Cultured human fetal fibroblasts or thyroid cells slowly develop foci of rounded refractile cells which on staining are seen to include giant cells containing intranuclear inclusions. The CPE is not easy to distinguish from that produced by other herpetoviruses or measles; immunofluorescence or neutralization confirms the diagnosis.

Epidemiology and Control

Varicella is endemic the year round but is most prevalent during late winter and spring. Cases are customarily excluded from school until the last lesion has disappeared, but the evidence indicates that, whereas vesicles shed virus, pustules and scabs are not a source of infection. Varicella patients are contagious to contacts for up to 5 days after the rash first appears. Transmission is so effective (secondary attack rate in susceptible siblings = 75%) that most children contract chickenpox during the first 6–8 years of life. This highly infectious, monotypic virus, restricted to man, may depend for its ultimate survival on its ability to persist indefinitely in dorsal root ganglia of immune hosts (see Chapters 8 and 10).

Zoster immune globulin, obtained from patients convalescing

from herpes zoster, is effective in preventing varicella in newborn infants of mothers with varicella, or in immunosuppressed children exposed to the risk of hospital cross-infection. An experimental live varicella vaccine passaged in cultured human embryonic lung fibroblasts has been tested in man and found to be insufficiently attenuated to be generally acceptable. Zoster may be treated by continuous application to the affected skin of 5–40% iododeoxyuridine in dimethyl sulfoxide.

CYTOMEGALOVIRUS

Introduction

Cultured for the first time in 1956 in the laboratories of Rowe, Smith, and Weller, all working independently, the human cytomegaloviruses are now acknowledged to be among the most common parasites of man. Yet the virus rarely causes disease, but persists for long periods as a latent or chronic infection of salivary and other glands. The characteristic cells that give the virus its name can be found in the salivary glands or renal tubules of many asymptomatic children. Transplacental infection may lead to "cytomegalic inclusion disease," which, though responsible for only 1% of all neonatal deaths, is the most important cause of microcephaly and is today an even more frequent cause of congenital abnormalities than rubella.

Clinical Features

Fortunately, most cytomegalovirus infections are inapparent. On the rare occasions when recognizable disease occurs, the outcome is frequently fatal. A number of situations should be distinguished.

Congenital infections, acquired transplacentally, present as *cytomegalic inclusion disease* (CMID) of the newborn, which is often fatal within days or weeks (Plate 16-5). The infant has hepatosplenomegaly with jaundice, thrombocytopenic purpura, and hemolytic anemia. Should the child survive, it is usually left permanently mentally retarded with microcephaly and other cerebral abnormalities. Chorioretinitis, which is commonly present, invites confusion with congenital toxoplasmosis. Recently, it has become clear that an even larger number of babies (1%) are infected *in utero* than previously recognized. Many of these may survive without any obvious sequelae, but others sustain a variety of significant defects such as minor loss of

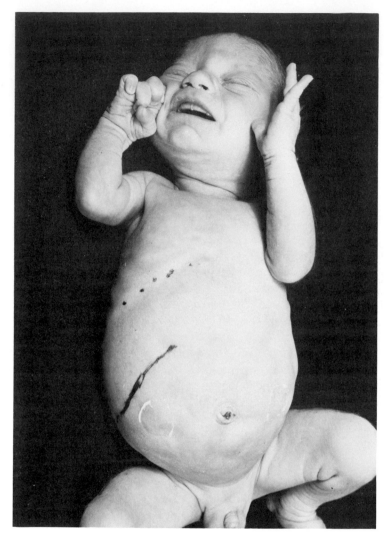

PLATE 16-5. *Prenatal cytomegalovirus infection: cytomegalic inclusion disease in newborn premature baby. Features are hepatosplenomegaly, thrombocytopenia, and microcephaly with severe mental retardation. (Courtesy Dr. K. Hayes.)*

hearing, mental dysfunction, growth retardation, and bone abnormalities, which may escape detection unless carefully sought.

Childhood infections, acquired by the respiratory route, lead to chronic carrier conditions, which are usually inapparent, but occasionally present as an insidious hepatitis or pneumonitis. Such infections may be overwhelming, when they complicate debilitating diseases like fibrocystic disease of the pancreas.

"*Reactivation*" of a latent cytomegalovirus infection commonly occurs in children or adults undergoing prolonged immunosuppressive therapy for kidney transplants, leukemia, or cancer. In these instances the disease is usually widely disseminated throughout the body, but often presents as an interstitial pneumonia, or less commonly, as hepatitis.

Post-transfusion mononucleosis ("postperfusion syndrome") is a febrile condition characterized by an increase of atypical mononuclear cells in recipients of recent blood transfusions. It is particularly frequent in patients receiving massive transfusion, e.g., for open-heart surgery. Cytomegalovirus can be isolated from the leukocytes. The evidence suggests that the disease results from transmission of the virus via the donor's blood more commonly than from activation of a latent infection preexisting in the recipient.

Pathogenesis and Immunity

The typical cytomegalic cell is greatly swollen (up to 40μm in diameter) with an enlarged nucleus distended by a huge acidophilic (or occasionally basophilic) inclusion, up to 15μm across, separated by a nonstaining halo from the nuclear membrane; inclusions may also be seen in the cytoplasm (Plate 16-2). Such "owl's eye" cells are demonstrable in the salivary glands and/or kidney tubules of about 10–15% of normal children under the age of four (Plate 16-6A). The same children excrete virus in their saliva or urine for months or years on end, or may yield virus when their tonsils or adenoids are explanted in culture. Older children and adults rarely excrete the virus, even though primary subclinical infections continue to occur at these ages.

Primary maternal infection during pregnancy can lead to transplacental transmission of the virus to the baby with the production of cytomegalic inclusion disease. Many of these children die shortly before or after birth; others survive for a year or two; most recover with or without permanent neurological sequelae attributable to microcephaly, microgyria, and periventricular calcification, which is diagnostic. Up to 1% of all stillborn infants contain the telltale cytomegalic cells in most organs of the body. Cytomegalic cells are also found in the placental chorionic villi, and the virus itself is found in the mother's urine. The fact that subsequent babies are only very rarely affected seems to indicate that viremia in the mother does not persist for more than a few months.

Congenitally infected infants do not show immunological tolerance.

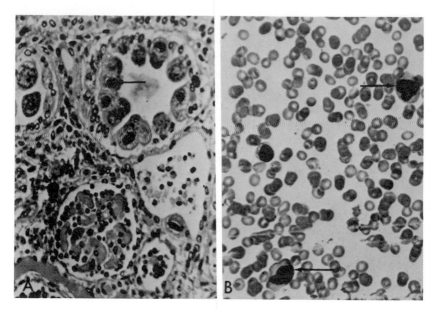

PLATE 16-6. *Histopathology of some diseases induced by herpetoviruses. (A) Cytomegalic inclusion disease. Section of kidney, (Hematoxylin and Eosin stain; ×250). Arrow indicates swollen ("cytomegalic") cells with intranuclear inclusions, inside a renal tubule. (B) Infectious mononucleosis. Smear of peripheral blood. (Leishman; ×400). Note large "atypical lymphocytes" (arrows). (Courtesy I. Jack.)*

Even *in utero* they develop very high titers of IgM, as well as IgG, the latter persisting for life. In this regard the immune response to infection resembles that of congenital rubella.

Laboratory Diagnosis

Cytomegalovirus may be grown from urine, saliva, milk, liver biopsy, various organs taken at autopsy, or leukocytes (from cases of post-transfusion mononucleosis), by inoculating cultures of human fetal fibroblasts. Foci of swollen, rounded cells usually appear within a week but take up to 4–6 weeks to involve the whole cell sheet, probably because most of the newly synthesized virus is noninfectious and much of it remains cell-associated. At this stage the monolayer should be embedded in collodion and stripped off the tube prior to staining for "cytomegalic cells." These swollen cells contain a number of nuclei, with large acidophilic (or amphophilic) nuclear inclusions, as well as strikingly regular, smooth, round acidophilic masses in the cytoplasm that are composed of aggregated virions and lysosomes. Serial passage of the virus to fresh cultures can be achieved satisfactorily only by transferring the infected cells themselves.

A somewhat less certain way of detecting cytomegalovirus infection is to identify exfoliated cytomegalic cells in a centrifuged deposit from urine or saliva. A thick smear stained by the Giemsa method may reveal the pathognomonic cytomegalic cells with their large acidophilic nuclear inclusions. Similar cells are plentiful in the salivary glands, liver, kidneys, lungs, and brain postmortem. "Exfoliative cytology" should really be regarded as obsolete now that reliable techniques of viral isolation are commonplace.

Suspected cases of cytomegalic inclusion disease can be diagnosed by virus isolation from the urine or throat, or by demonstration of anti-CMV IgM in cord serum using indirect immunofluorescence.

Epidemiology and Control

Transplacental (prenatal) infection accounts for the CMV infections that are demonstrable in approximately 1% of neonates. It is clear that most people must become infected postnatally, because the proportion of the community with demonstrable antibody increases steadily with age, to reach 50–80% in adults. Very probably this alternative route of transmission is via saliva, as in the case of the related herpes simplex virus. By direct analogy, cytomegalovirus infection has been found to be (a) readily transmitted to siblings in the same family, (b) 10–20 times more common in children living under institutional conditions than in the general population, and (c) acquired at an earlier age in backward countries than in those with higher standards of hygiene. The vast majority of such natural postnatal CMV infections are subclinical. *Iatrogenic* infections, on the other hand, quite commonly give rise to manifest disease, e.g., post-transfusion mononucleosis and pneumonitis in immunosuppressed patients.

Ara-A, ara-C and iododeoxyuridine have been used to treat newborn infants with CMID. It is too early to claim that the results are permanently beneficial. Experimental vaccines have also been produced but there are great problems with this seemingly attractive approach — inactivated vaccines are impractical because of the obstacles to obtaining high yields of cell-free virions from cultured cells; live attenuated vaccines, on the other hand, may be extremely dangerous for any of the herpetoviruses because of their known tendency to establish prolonged, and perhaps oncogenic, persistent infections.

EB VIRUS

Introduction

In 1958 Burkitt described an unusual malignant lymphoma of African children (Plate 16-7) that appeared to be restricted to the hot and wet areas of the continent below an altitude of 4500 feet. Reasoning that this distribution corresponded to the "mosquito belt," Burkitt postulated that the cancer might be caused by a mosquito-borne virus. There followed an intensive search for viruses in lymphomas and in cultured cell lines derived from them, which unearthed not an arbovirus but a new herpetovirus, now known as EB virus after its discoverers, Epstein and Barr. The genome of EB virus is integrated with that of the cells of all African Burkitt lymphomas and all nasopharyngeal carcinomas of a particular type encountered in people of Southern Chinese extraction. The key question currently engaging the attention of cancer virologists around the world is whether this regular association is etiological or fortuitous (see Chapter 9).

If indeed EB virus does turn out to be oncogenic this is certainly a rare complication of its interaction with man. The great majority of EBV infections are inapparent. Most people throughout the world

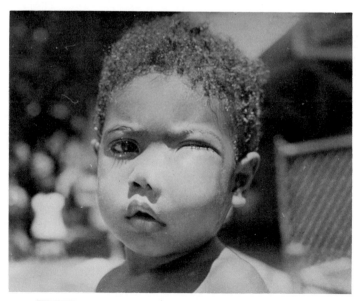

PLATE 16-7. *Burkitt lymphoma. Characteristic facial tumor in a child from New Guinea. (Courtesy Dr. J. Biddulph.)*

develop antibodies to the agent during childhood. When primary infection is delayed until adolescence the individual may develop infectious mononucleosis (glandular fever). Henle has unequivocally shown that EBV is the etiological agent of this important disease.

Clinical Features

Infectious mononucleosis in the young adult presents typically as fever, pharyngitis, lymphadenitis, and splenomegaly, together with disturbed liver function and characteristic "atypical large lymphocytes" in peripheral blood (Plate 16-6,B). The onset is often insidious, with general malaise and lethargy. Then the patient develops a high fluctuating fever, chills, and headache. In about half the cases the pharyngitis is severe, with a white, or grey malodorous, exudate covering the tonsils. Differential diagnosis from diphtheria or Vincent's angina may be assisted by the later development of petechiae on the palate. Subsequent enlargement of the posterior cervical lymph nodes is usual; axillary and other glands also frequently become involved as the disease develops. The spleen is often enlarged, the liver occasionally. Jaundice is detectable in only about 5% of cases, but some degree of hepatitis, as evidenced by abnormal liver function tests, is almost invariable. Rashes are uncommon, being usually "morbilliform" (maculopapular and blotchy) and concentrated on the trunk. The disease usually lasts about 1–2 weeks, but convalescence can be prolonged. Relapses are rare, and death extremely rare, usually from traumatic rupture of the spleen.

Neurological complications occasionally occur, particularly the Guillain–Barré syndrome, Bell's palsy, meningoencephalitis and transverse myelitis, and it has recently been reported that EBV can be associated with these diseases even in the absence of infectious mononucleosis.

Pathogenesis and Immunity

Once infected with EBV, usually in childhood, one develops neutralizing antibodies which confer a lifelong immunity to exogenous reinfection, but one also becomes a lifelong carrier. Virus continues to be secreted from the nasopharynx of up to 20% of normal people and up to 50% of those on immunosuppressive therapy. In addition, nonproductive infection of peripheral blood lymphocytes may continue for years. The precise nature of the relationship between the viral genome and the lymphocyte that carries it is not fully understood. Virions are not demonstrable in lymphocytes *in vivo* but are inducible by treatment with bromodeoxyuridine or by *in vitro* cultivation. Lymphoblas-

toid cell lines can be established from peripheral leukocytes taken
from normal EBV carriers, as well as from patients with infectious
mononucleosis; in most respects these immortal cell lines are indistinguishable from those derived from Burkitt's lymphoma or nasopharyngeal carcinoma (see Chapter 9). Some factor, presumably immunological, must usually inhibit the progression of these EBV-in-
fected lymphocytes to malignancy *in vivo*. Biopsy of lymph nodes,
liver, or tonsils during an acute attack of infectious mononucleosis
does reveal massive nodular infiltration of atypical large lymphocytes,
but lymphomas do not develop. The relationship if any, between infectious mononucleosis, Burkitt's lymphoma and nasopharyngeal carcinoma is currently the subject of intensive study.

Laboratory Diagnosis

Definitive diagnosis of infectious mononucleosis demands help
from the laboratory. Although methods have recently developed for
culturing the virus they are too elaborate to be practised as a routine.
Currently therefore, diagnosis rests on a number of less direct approaches.

A blood count during the first few days of the illness may reveal a
leukopenia due to a drop in the number of polymorphs. This is soon
followed by a striking mononuclear leukocytosis (10,000–50,000
cells/mm^3), some of which are characteristic "large atypical lymphocytes" (lymphoblasts) with deeply basophilic vacuolated cytoplasm
and kidney-shaped nuclei criss-crossed with a lattice of fenestrated
chromatin (Plate 16-6,B).

The Paul–Bunnell test for heterophil agglutinins has been the standard diagnostic test until now. It is an empirically based serological
procedure, which detects the ephemeral appearance of an IgM that
agglutinates sheep erythrocytes. The patient's serum is inactivated (45°
for 30 minutes), diluted serially, and incubated with 1% fresh washed
sheep RBC at 37° for 2 hours, then 4° overnight. Demonstration of a
rising antibody titer would be the ideal, but heterophil agglutinins
develop quite early in this protracted disease, so that levels are usually
high by the time the first specimen is taken, and have usually disappeared again within 1–2 months. Titers of over 1:100 are considered to
be diagnostic. Sera from patients with serum sickness, and from
normal individuals, can give false positives. However, the agglutinins
from those with hypersensitivity reactions may be absorbed out of the
serum by incubation at 37° for 30 minutes with bovine erythrocytes or
minced guinea pig kidney; those from normal individuals are ab-

sorbed by guinea pig kidney only; and those from infectious mononucleosis patients by bovine RBC only. If absorption with guinea pig kidney is properly conducted, heterophil agglutinin titers as low as 1:10 are diagnostic for infectious mononucleosis. The test may also be rendered more specific by treating the sheep RBC with a protease. The fact that only 60–80% of clinically and hematologically diagnosed cases of infectious mononucleosis show a positive Paul–Bunnell test (the negative ones being chiefly mild cases occurring in children) has led to some controversy about whether or not two different agents are involved. Certainly cytomegalovirus occasionally produces the syndrome and infectious mononucleosis is sometimes confused clinically with infectious lymphocytosis, but many heterophil antibody-negative cases show definite evidence of anti-EBV antibodies using one or more of the specific serological procedures described below.

While detection of heterophil agglutinins is still the simplest and most widely employed serological technique for diagnosing infectious mononucleosis, it can now be supplemented with any of a large number of more specific assays for anti-EBV antibodies. Perhaps the most discerning indicator of EBV infection available today is the demonstration of a rising titer of specific neutralizing antibody; this is based on the neutralization by serum of the ability of EBV to infect the Raji line of Burkitt lymphoma lymphoblasts, so preventing the cells from forming colonies in plastic trays. Technically simpler is the indirect immunofluorescence test for serum antibodies against the various antigens synthesized by EBV-induced lymphoblastoid cell lines, namely VCA (viral capsid antigen), D and R ("early antigens") and "nuclear" antigen. Anti-VCA antibodies rise rapidly following infection and persist, at a slightly lower level, for many years. Anti-D antibodies rise more slowly and fall off rapidly as does EBV-specific IgM, hence a positive result with either of these tests on a single serum specimen is diagnostic of recent infection.

The only way in which EBV can currently be "grown" is in human umbilical cord leukocytes; the lymphocytes become transformed into blast cells with the capacity for indefinite propagation, and EBV antigen is demonstrable by immunofluorescence in their nuclei.

Epidemiology and Control

EB virus is ubiquitous, infecting most children subclinically before school-age; 80–90% of normal adults have antibody. The virus is thought to be transmitted mainly via saliva, e.g., by kissing; parenteral transmission via blood transfusion also occurs. Antibody is acquired at an earlier age in lower socioeconomic groups and in countries with lower standards of living. When primary infection occurs in adolescence there is a much higher probability (50%) that the individual will develop infectious mononucleosis. Infectious mononucleosis is principally a disease of priviliged adolescents and young adults (e.g., college students) exchanging EBV by "intimate osculatory contact."

The etiological relationship with Burkitt's lymphoma is still uncertain, as is the nature of the presumed climate-dependent cofactor (hyperendemic malaria?) postulated by Burkitt to account for the striking geographical localization of this disease despite the ubiquity of the virus. With nasopharyngeal carcinoma genetic predisposition appears to be more important because the disease is virtually confined to certain ethnic groups of Chinese men.

Advocates of immunization against EB virus make the point that effective prevention of Burkitt lymphoma by this means may be the only way of proving whether the two are etiologically connected or not. There are precedents which indicate that immunization with an inactivated vaccine might be successful—vaccines are now employed routinely to prevent Marek's disease in fowls and an experimental vaccine against *Herpesvirus saimiri* has been reported to retard the development of malignant lymphomas in marmosets and monkeys.

SUBACUTE MYELOOPTICONEUROPATHY

A new agent, morphologically resembling a herpetovirus and serologically related to the herpetovirus of avian infectious laryngotracheitis, has been recovered from the CSF and feces of patients with the disease subacute myeloopticoneuropathy. The disease is a type of polyneuritis which is widespread in Japan, and is particularly seen in immunosuppressed patients. The virus grows in human diploid cell strains and on the chorioallantoic membrane of eggs.

FURTHER READING

Epstein, M. A., and Achong, B. G. (1973). The EB virus. *Annu. Rev. Microbiol.* **27,** 413.
Fenner, F., McAuslan, B. R., Mims, C. A., Sambrook, J. F., and White, D. O. (1974). "The Biology of Animal Viruses," 2nd ed. Academic Press, New York.
Henle, W., and Henle, G. (1972). Epstein-Barr virus: The cause of infectious mononucleosis—a review. *In* "Oncogenesis and Herpesviruses" (P. M. Biggs, G. de-The; and L. N. Payne, eds.) Publ. No. 2. Int. Agency Cancer Res., Lyons, France.
Juel-Jensen, B. E., and MacCallum, F. O. (1972). "Herpes Simplex Varicella and Zoster: Clinical Manifestations and Treatment." Lippincott, Philadelphia, Pennsylvania.
Kaplan, A. S., ed. (1973). "The Herpesviruses." Academic Press, New York.

Nahmias, A. J., and Roizman, B. (1973). Infection with herpes-simplex viruses 1 and 2. *N. Engl. J. Med.* **289.** 667, 719, and 781.

Plummer, G. (1973). Cytomegaloviruses of man and animals. *Progr. Med. Virol.* **15,** 92.

Roizman, B., and Furlong, D. (1974). The replication of herpesviruses. *In* "Comprehensive Virology" (H. Fraenkel-Conrat and R. R. Wagner, eds.), Vol. 3, p. 229. Plenum, New York.

Roizman, B., Spear, P. G., and Kieff, E. D. (1973). Herpes simplex viruses I and II: A biochemical definition. *Perspect. Virol.* **8,** 129.

Stevens, J. G., and Cook, M. L. (1973). Latent herpes simplex virus in sensory ganglia. *Perspect. Virol.* **8,** 171.

Taylor-Robinson, D., and Caunt, A. E. (1972). Varicella virus. *Virol. Monogr.* **12,** 1.

Weller, T. H. (1971). The cytomegaloviruses: Ubiquitous agents with protean clinical manifestations. *N. Engl. J. Med.* **285,** 203 and 267.

CHAPTER 17

Poxviridae

The poxviruses are the largest and most complex of all viruses (Table 17-1). The family, Poxviridae, is divided into six genera on the basis of antigenic and morphological differences. Several poxviruses cause infections of man: smallpox, molluscum contagiosum, cowpox, milkers' nodes, and orf (Table 17-2). Others cause economically important infections of domestic animals, and myxoma virus has been used with dramatic success in the biological control of a pest animal, the rabbit.

Most diseases due to poxviruses are associated with pustular skin lesions, which may be localized or may be part of a generalized rash, as in the case of smallpox. Smallpox, as one of the great plagues of mankind, has played an important role in human history and a central role in the development of virology, from the time of Jenner to the present day.

PROPERTIES OF THE VIRUS

Poxviruses are just large enough to be seen under a light microscope using critical illumination, but knowledge of their structure comes from electron microscopic studies. Plate 17-1A, C, D illustrates the structure of the virion of vaccinia (or smallpox which appears identical). There is no nucleocapsid conforming to either of the two types of symmetry found in most other viruses (Chapter 1), hence it is sometimes called a "complex" virion. An outer "membrane" of tubular-shaped lipoprotein subunits, arranged rather irregularly, encloses a dumbbell-shaped core and two "lateral bodies" of unknown nature. The core contains the viral DNA together with protein, but the detailed arrangement of these components has not been determined.

TABLE 17-1

Properties of Poxviruses

Brick-shaped virion, 300 × 240 × 100 nm
Complex structure with core, lateral bodies, and outer membrane
DNA, double-stranded, mol. wt. 160 million
Transcriptase in virion

The nucleic acid is double-stranded DNA, with a molecular weight of 160 million daltons, and an unusually low G + C content (35%). There are at least 30 different polypeptides in the virion, and probably many more. The core proteins include a transcriptase and several other enzymes. The lipoprotein "membrane" of the virion is synthesized *de novo*, not derived by budding from cellular membranes.

As would be expected from its multiplicity of polypeptides, the virion contains numerous antigens recognizable by gel diffusion. Most of these are common to all members of any one genus, and one is shared by all poxviruses. There is extensive cross-neutralization between viruses belonging to the same genus, but none between viruses of different genera.

TABLE 17-2

Diseases Produced by Poxviruses

GENUS	DISEASE	CLINICAL FEATURES
Orthopoxvirus	Smallpox	
	Variola major	Generalized infection with pustular rash; mortality 15%
	Variola minor	Generalized infection with pustular rash; mortality less than 1%
	Complications of vaccination (rare)	Postvaccinial encephalitis, high mortality
		Vaccinia gangrenosa; high mortality
		Eczema vaccinatum; low mortality
		Autoinoculation and generalized vaccinia; nonlethal
	Cowpox	Localized ulcerating infection of skin, acquired from cows
Parapoxvirus	Milkers' nodes	Localized nodular infection of skin acquired from cows
	Orf	Localized ulcerating infection of skin acquired from sheep
Unclassified	Molluscum contagiosum	Multiple benign nodules in skin
	Yaba and Tanapox	Localized skin tumors acquired from monkeys

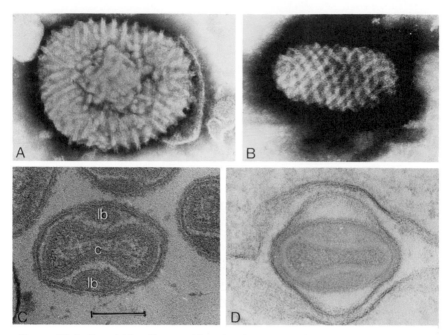

PLATE 17-1. *Poxviridae (bar = 100 nm). (A) Negatively stained vaccinia virion, showing surface structure of rodlets or tubules characteristic of the Orthopoxvirus genus. (B) Negatively stained orf virion, showing characteristic surface structure of the Parapoxvirus genus. (C) Thin section of vaccinia virion in its narrow aspect, showing the biconcave core (c) and the two lateral bodies (lb). (D) Thin section of mature extracellular vaccinia virion lying between two cells. The virion is enclosed by a single membrane originating from the inner membrane of the cisterna. (A and D, from S. Dales, J. Cell Biol.* **18,** *51 (1963); B, from J. Nagington et al., Virology* **23,** *461 (1964); C, from B. G. T. Pogo and S. Dales, Proc. Nat. Acad. Sci. U.S.* **63,** *820 (1969).*

SMALLPOX

Smallpox was the first of the great scourges of man to yield to preventive measures, viz., vaccination with cowpox (vaccinia) virus which was introduced by Edward Jenner in 1798. As we go to press it has just been announced that the disease has apparently been eradicated from the last remaining pockets of infection: the Indian subcontinent and Ethiopia. The success of the WHO Smallpox Eradication Campaign is a tribute to persistence in the face of great obstacles. If thorough surveillance over the next few years confirms that the virus of smallpox no longer exists anywhere on earth then this chapter will disappear from future editions of this book. In the meantime we shall retain it in the expectation that governments will retain a measure of caution for at least a year or two before totally abandoning the requirement for vaccination, or at least surveillance, of people traveling through the Indian subcontinent and Ethiopia.

Clinical Features

Smallpox is a horrifying disease (Plate 17-2). The case mortality rate is about 15% for *variola major* (classical smallpox) and 1% for *alastrim* (a much milder disease caused by a different strain of virus); the survivors bear permanent scars.

Classical smallpox begins with high fever which may drop temporarily after 2–4 days, coincident with the eruption of lesions. These appear first on the mucous membrane of the mouth and throat, then the skin of the face and extremities, spreading "centripetally" to the trunk. The lesions progress synchronously from vesicle through pustule to scab. In about 5% of cases the lesions take the form of flat soft vesicles which do not progress; this variant of the disease has a high mortality. The most lethal syndrome of all, however, is the rare (<1%) hemorrhagic smallpox, which is characterized by fulminating hemorrhages in the skin and mucous membranes, prostration and death usually occurring even before the typical vesicular rash appears. Doctors should remain alert to the possibility of smallpox in patients who have traveled through recently endemic areas within the preceding two weeks. In particular, they should be suspicious of mild attacks of "modified" smallpox in vaccinated individuals, which may present as "chickenpox." Doubtful cases must be promptly referred to a public health expert, for the clinical and laboratory diagnosis of such a dangerous disease cannot be left to chance.

Pathogenesis and Immunity

The pathogenesis of smallpox has been described in detail in Chapter 6.

Recovery from smallpox gives lifelong immunity. The immunity following vaccination with vaccinia virus is less prolonged, hence quarantine authorities in most countries insisted until recently on vaccination or revaccination within the 3 years prior to entry of travelers from endemic areas.

Laboratory Diagnosis

Because of the great importance from a public health viewpoint of rapid and accurate diagnosis of smallpox, a good deal of attention has

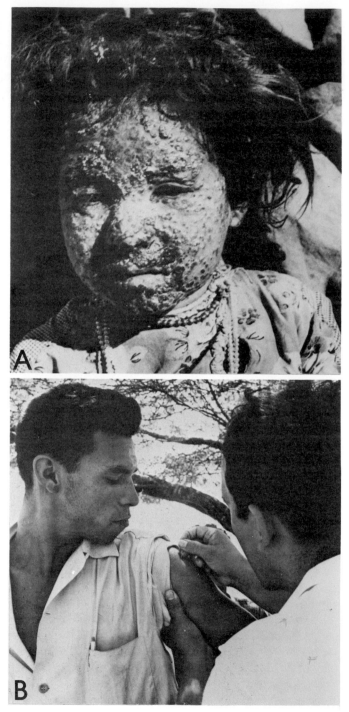

PLATE 17-2. Smallpox. (A) A five-year-old Afghan girl whose face will be pitted for life. (B) The simple answer, Jennerian vaccination. (Courtesy of WHO and Cambridge University Press. From F. M. Burnet and D. O. White, "Natural History of Infectious Disease." Cambridge Univ. Press, London and New York, 1972.)

been devoted to this problem and techniques are constantly being refined. It is a fortunate coincidence that more and better methods are available for the prompt differentiation of smallpox from chickenpox (the principal cause of confusion) than for any other viral diseases (see Chapter 14). The method of choice is the direct examination in the electron microscope of negatively stained smears from vesicle fluid; a diagnosis can be made within a few hours. The brick-shaped smallpox virus is clearly distinguishable from the enveloped icosahedron of chickenpox. The other recommended method of rapid diagnosis is immunodiffusion using vesicle fluid as antigen against a hyperimmune vaccinia antiserum; here again a definitive answer is available in a few hours.

The identification of the poxvirus as variola and not vaccinia can be determined by cultivation on the chorioallantoic membrane of the chick embryo (variola producing greyish-white pocks that are smaller than those of vaccinia virus), testing for the "ceiling temperature" of growth (variola being unable to multiply above 39°), or inoculation of chick embryo fibroblasts, in which variola virus, with its narrow host range, does not produce plaques while vaccinia virus, which grows in a wide range of cultured cell types, does.

Epidemiology

Contrary to popular belief smallpox is not highly infectious (compared with influenza, measles, or chickenpox, for example), but the virus is sufficiently resistant to persist in an infectious state in scabs or on bedding and clothing for prolonged periods. Infection probably occurs via the respiratory route. In the early stage of the disease, when the focal eruption appears, infection is spread from the mouth and nose; later the skin lesions assume importance. Patients are not infectious during the incubation period.

The relatively low infectiousness of smallpox was illustrated by careful analysis of 958 cases of smallpox that occurred in Europe in the 20 years following the second World War which indicated that relatively few of the 49 separate episodes of importation gave rise to more than a handful of cases and that most of the dissemination occurred inside hospitals. Spread within the largely unvaccinated general community almost never went beyond the first round of immediate contacts and was relatively easily contained by prompt detection and "ring vaccination."

Smallpox is a specifically human disease in which a previous attack (or vaccination with vaccinia virus) gives a high degree of protection and recurrent disease does not occur. Intensive investigations over the past decade have shown that there is no animal reservoir; monkeypox, which may cause sporadic cases of a smallpox-like disease in man, has never caused secondary human cases. It was for these reasons that WHO recognized the disease was amenable to eradication by vaccination.

Prevention and Control

Public Health Responsibilities. On the discovery of a suspected case of smallpox the appropriate Public Health Department should be notified by phone immediately. The patient should be isolated in an appropriate special hospital, and his house, bedding and clothing disinfected by formaldehyde gas. Other suspect artifacts should be boiled or burnt. All human contacts of the patient should be traced and vaccinated or revaccinated with vaccinia virus and kept under surveillance for 16 days. The controversial question of methisazone prophylaxis was discussed in Chapter 13; here let us merely reiterate that it is no substitute for vaccination.

Vaccination. Vaccination against smallpox, originally but dangerously practised with the virulent virus itself ("variolation," in China and the Middle East) but converted to a safe procedure by Jenner's use of cowpox virus, constitutes one of the epoch-making advances of medical science. Not only has it protected countless millions of people from a dreaded disease but it provided the seeds for two fields of science, virology and immunology, and the example was used by microbiologists from Pasteur onward in their dramatically successful campaign to reduce mortality and morbidity due to infectious diseases.

Smallpox vaccine virus is still produced in the skins of calves, sheep, and buffaloes, and WHO regulations stipulate that the number of contaminating bacteria per milliliter of vaccine shall not exceed 500—a permissive requirement that would never be tolerated in a modern vaccine! It is typical of the conservative approach to viral vaccines that alternative methods of producing the virus have not been widely adopted, though many have been described, including chick embryo- and cell culture-grown vaccines.

Some technical improvements have been made however. High titer virus is being produced to compensate for the loss of infectivity that occurs in the humid heat of the tropics after frozen vaccine has been thawed, or after lyophilized vaccine has been rehydrated. A recent development involves the incorporation of a dried vaccine into a paste dispensed in a disposable three-pronged hollow needle which can be stored without refrigeration and used directly for inoculating the subject. This procedure of "multiple puncture" into the superficial layers of the epidermis covering an area of skin about ½ cm in diameter over the deltoid (Plate 17-2) is quite as satisfactory as the previously recommended technique of "multiple pressure" through a droplet of vaccine with a straight needle held parallel to the skin. Mass vaccination campaigns often involve the use of a foot-pedal-operated jet injector which fires a fine spray of fluid into the skin under pressure.

Response to Vaccination. Three types of response are recognized: primary, in previously unvaccinated persons, and accelerated and immediate in persons undergoing revaccination.

In the *primary reaction* (Plate 17-3) a papule appears three to five days after inoculation, and rapidly becomes a vesicle, which enlarges in size and develops a secondary erythema. The center of the vesicle usually becomes depressed and by the eighth or ninth day the contents become turbid and there may be some axillary lymphadenitis and fever. After the tenth day the pustule dries up and a scab forms which takes a week to separate, leaving a typical vaccination scar. Immunity to vaccinia and variola appears about the tenth day and persists, to a varying degree in different individuals, for several years.

The *accelerated reaction* is seen in individuals with a limited degree of residual immunity from previous vaccination. It is less pronounced and more rapid than a primary reaction, but vesiculation nevertheless occurs. The maximum intensity is reached between the third and seventh day after vaccination. This type of reaction enhances waning immunity.

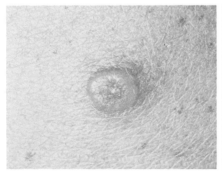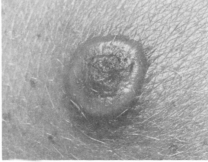

PLATE 17-3. *Primary response to vaccination in man: typical vesiculo-pustular response, maximal at 7–10 days. Left, at 7 days; right, at 11 days. (Courtesy Dr. J. R. L. Forsyth.)*

The *immediate reaction* is usually seen in solidly immune individuals. It is most marked on the second or third day, usually as a papular but occasionally as a vesicular reaction, and is in effect a hypersensitivity response. However, an immediate reaction does not always indicate immunity, nor does it necessarily lead to immunity. If an individual giving this response is likely to be exposed to smallpox, revaccination should be carried out with fresh virus.

Because of difficulties in interpretation, WHO experts classify the responses to revaccination as either a "major reaction" or an "equivocal reaction." The former is characterized by the appearance about a week after vaccination of a pustular lesion or an area of definite induration or congestion surrounding a central lesion, which may be a scab or an ulcer. All other responses are termed "equivocal reactions." In mass vaccination campaigns it is assumed that if freeze-dried vaccine meeting WHO requirements is used and if more than 95% of primary vaccinations are successful the technique is satisfactory and an adequate response can be expected in revaccinated persons.

Complications of Primary Vaccination. Whereas revaccination is very rarely accompanied by adverse effects, in primary vaccination three types of serious complication may occur, all of them being rare.

Recent surveys indicate that there is about one complication in every thousand primary vaccinations, and one death in every million. *Autoinoculation,* especially of the eyes and lids, accounts for nearly one-half these occurrences and *generalized vaccinia* for about another one-quarter. The much more serious *postvaccinial encephalitis* has an incidence of something less than one in every thousand primary vaccinations. *Progressive vaccinia (vaccinia necrosum)* is even rarer (10^{-6}) but is generally fatal; in individuals with a defect in cellular immunity the primary lesion slowly spreads to involve the whole arm. *Abortion* may rarely occur in pregnant women, due to intrauterine infection of the fetus. *Eczema vaccinatum* has decreased markedly in incidence since physicians became aware of the dangers of vaccinating children with eczema, but also occasionally arises by contact infection of an eczematous child from a vaccinated parent. If children with eczema are ever to be vaccinated, it should be done in hospital under cover of anti-vaccinia immunoglobulin injected into the other arm. Vaccination is contraindicated in eczema or other chronic skin conditions, pregnancy, immunological deficiency states, or in those taking immunosuppressive drugs.

Treatment of Complications. Human hyperimmune antivaccinial immunoglobulin should be administered promptly following the development of any of the complications of smallpox vaccination, except encephalitis, for which there is no treatment known. Methisazone may also be of some use in vaccinia gangrenosa although results reported so far are not encouraging.

Vaccination Policy. Routine compulsory vaccination was abandoned some years ago in Western countries in favor of a policy of quarantine, surveillance, and ring vaccination in the event of an introduction. The only people routinely vaccinated today in countries such as the U.S.A., U.K., and Australia are medical and immigration personnel. Nor is vaccination any longer demanded of travelers between such nonendemic, low-risk countries. However, people entering or returning from endemic areas of Asia or Africa must present certification of successful vaccination or revaccination within the preceding 3 years. It may be anticipated that this requirement will also disappear as soon as we can be assured that the disease itself has indeed passed into history.

MOLLUSCUM CONTAGIOSUM

This specifically human disease consists of multiple discrete nodules 2–5 mm in diameter, limited to the epidermis, and occurring anywhere on the body except on the soles and palms. They are pearly white in color and painless. At the top of each lesion there is an opening through which a small white core can be seen. The disease may last for several months before recovery occurs.

Cells in the nodule are greatly hypertrophied and contain large hyaline acidophilic cytoplasmic masses called molluscum bodies. These consist of a spongy matrix divided into cavities in each of which are clustered masses of viral particles which have the same general structure as those of vaccinia virus.

The incubation period in human volunteers has varied between 14 and 50 days. Attempts to transmit the infection to experimental animals have failed, and reported growth in cultured human cells has been hard to reproduce.

The disease is world-wide, but is much more common in some localities than others. It is probably transmitted through minor abrasions, and swimming pools may be a source of infection.

COWPOX AND MILKERS' NODES

Man may acquire two different poxvirus infections from cows, usually as lesions on the hands after milking. Cowpox occurs in cattle as ulcers of the teats and the contiguous parts of the udder, and it is spread through herds by the process of milking. The lesions in man usually appear on the hands and develop just like primary vaccinia. There may be some fever and constitutional symptoms. Cowpox is very like vaccinia virus in most respects, for vaccinia was originally derived from cowpox virus. However, cowpox produces much more hemorrhagic lesions on the chorioallantoic membrane and in the skin of rabbits.

Milkers' nodes also occur on the hands of man, derived from lesions on the teats and udder. In man the lesions are small nonulcerating nodules. They are caused by a poxvirus of the genus *Parapoxvirus* (Plate 17-1,B); the same virus causes bovine pustular stomatitis. Homologous immunity following infection of man does not last long, and second attacks may occur at intervals of a few years. The disease is trivial and no measures for prevention or treatment are warranted. Vaccination with vaccinia virus protects man against cowpox but not against milkers' nodes.

ORF

Orf is an old Saxon term applied to the infection of man with the virus of contagious pustular dermatitis ("scabby mouth") of sheep. The disease of sheep occurs particularly in lambs during spring and summer and consists of a papulovesicular eruption that is usually confined to the lips and surrounding skin. Infection of man usually occurs as a single lesion on the hand or forearm or occasionally on the face; a slowly developing papule becomes a flat vesicle and eventually heals without scarring. Orf is an occupational disease associated with handling of sheep; it has caused industrial trouble among shearers in Australia.

The virus of orf (Plate 17-1B) belongs to the same genus (*Parapoxvirus*) as that of milkers' nodes, but is a distinct species and occurs naturally among sheep and goats rather than cattle.

YABA AND TANAPOX

The Yaba virus, which produces benign tumors under natural conditions in African monkeys, has on rare occasions given rise to similar lesions in laboratory workers handling the animals.

Tanapox virus has been isolated from solitary skin lesions in Kenyans, who may become quite ill. There is speculation that it may also be derived from monkeys, perhaps by insect transmission.

FURTHER READING

Bedson, H. S., and Dumbell, K. R., (1967). Smallpox and vaccinia. *Brit. Med. Bull.* **23,** 119.

Cho, C. T., and Werner, H. A. (1973). Monkeypox virus. *Bacteriol. Rev.* **37,** 1.

Dick, G. (1966). Smallpox: A reconsideration of public health policies. *Progr. Med. Virol.* **8,** 1.

Dumbell, K. R. (1968). Laboratory aids to the control of smallpox in countries where the disease is not endemic. *Progr. Med. Virol.* **10,** 388.

Fenner, F., McAuslan, B. R., Mims, C. A., Sambrook, J. F., and White, D. O. (1974). "The Biology of Animal Viruses," 2nd ed., pp. 18–19, 93–99, and 207–220. Academic Press, New York.

Joklik, W. K. (1968). The poxviruses. *Annu. Rev. Microbiol.* **22,** 359.

Moss, B. (1974). Reproduction of poxviruses. *In* "Comprehensive Virology" (H. Fraenkel-Conrat and R. R. Wagner, eds.), Vol. 3, p. 405. Plenum, New York.

Postlethwaite, R. (1970). Molluscum contagiosum. *Arch. Environ. Health.* **21,** 432.

World Health Organization. (1972). WHO Expert Committee on Smallpox Eradication, Second Report. *World Health Organ., Tech. Rep. Ser.* **493.**

World Health Organization. (1975). Smallpox: Point of no return. *World Health.* Feb.-March.

CHAPTER 18

Picornaviridae

INTRODUCTION

The Picornaviridae comprise one of the largest and most important families of human pathogens. They are also the smallest RNA viruses (Table 18-1), hence the name *"pico* (small) *RNA* virus." Structurally similar viruses are found in other animals and in plants and bacteria. The family is divided into two large genera: *Enterovirus* (found primarily in the enteric tract) and *Rhinovirus* (found primarily in the nose). The enteroviruses, in turn, are subdivided into polioviruses, coxsackieviruses, and echoviruses (Table 18-2). Distinction between the coxsackieviruses and echoviruses, (*e*nteric *c*ytopathogenic *h*uman *o*rphan viruses) many of which were discovered by accident during poliovirus surveillance in the early days of Salk vaccine development, was based on the somewhat artificial criterion that coxsackieviruses are pathogenic for baby mice and echoviruses are not. Both usually cause inapparent infections of the intestinal tract but are also responsible for a wide variety of human illness (see Table 18-3).

PROPERTIES OF THE VIRUS

The fine structure of the picornaviruses is still the subject of some controversy. Being only 25–30 nm in diameter, their capsomers are difficult to discern and count (Plate 18-1). It is clear that the capsid has cubic symmetry, and X-ray diffraction studies indicate 60 structural units. The viral RNA is in the form of a continuous single strand of molecular weight 2.5 million daltons.

The picornaviruses are remarkably stable to inactivation by most agents. The enteroviruses, in particular, are resistant to the acid, proteolytic enzymes, and bile of the intestinal contents, and may survive for long periods in sewage, or even in chlorinated water, providing

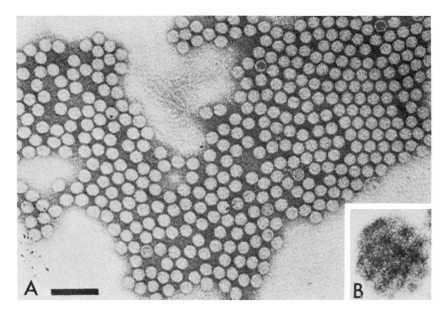

PLATE 18-1. (A) Electron micrograph of poliovirions, negatively stained (bar = 100 nm). (B) Viral particle at fivefold magnification, printed in reverse contrast. Cubic symmetry of virions is evident, but detailed arrangement of capsomers is difficult to see. (Courtesy Dr. H. D. Mayor.)

that adequate organic matter is present to protect them. The rhinoviruses, by contrast, are acid-labile but relatively more stable to heat. Molar $MgCl_2$ protects all picornaviruses against heat inactivation and has been used to stabilize live attenuated poliovaccines against loss of infectivity in the tropics.

There are no antigenic relationships between most of the picornaviruses, although some serotypes do share complement fixing antigens. Definitive identification is based on neutralization tests. Certain serotypes from different subgenera have turned out to be identical, e.g., coxsackie A23 has been found to be indistinguishable from echovirus 9, which had long been considered anomalous because

TABLE 18-1
Properties of Picornaviruses

Spherical virion, diameter 25–30 nm
Icosahedral symmetry, 60 capsomers
RNA, single-stranded, mol. wt. 2.5 million

TABLE 18-2

Classification of Human Picornaviruses

GENUS	SUBGENUS	TYPES
Enterovirus	*Poliovirus*	1–3
	Coxsackievirus	A1–24
		B1–6
	Echovirus	1–33
	Unspecified	68–71
Rhinovirus		1–89

of its pathogenicity for infant mice. It is clear that the division of enteroviruses into coxsackieviruses and echoviruses is a rather arbitrary matter, based largely on the history of their discovery. Eventually we can expect that a more rational numbering system will be introduced. Present policy is something of a compromise; the 67 serotypes allocated to the coxsackievirus or echovirus subgenera prior to 1969 will, for the moment at least, retain their original nomenclature, but enteroviruses discovered after that date will simply be known as "enterovirus 68," etc. It is a reflection of the explosion of discovery in the decade 1955–1965 that only four "new" enteroviruses pathogenic for man have been identified since this convention was introduced.

POLIOVIRUSES

Poliomyelitis was once one of the most feared of all viral diseases; its tragic legacy of paralysis and deformity was a familiar sight a generation ago. Today, by contrast, many medical students have not seen a case, such has been the impact of the Salk and Sabin vaccines. The foundation for these great developments was laid by Enders, Weller, and Robbins in 1949, when they demonstrated the growth of poliovirus in cultures of nonneural cells. From this fundamental discovery flowed all subsequent work involving the recognition of viral growth in cultured cell monolayers by virtue of the cytopathic effects they produce. Enders and his colleagues were rewarded with the Nobel Prize. A new era in virology had begun.

There are three antigenic types of poliovirus (1, 2, and 3), all of which have been recognized for many years. They share common complement-fixing antigens, but are quite distinct in neutralization tests and show almost no cross-protection. Avirulent mutants are now widely distributed in man and his sewage as a result of Sabin vaccination.

Man is the only natural host for the polioviruses. Nevertheless, chimpanzees and other monkeys are highly susceptible to paralysis following intraspinal or intracerebral inoculation, hence, they are extensively used for vaccine testing.

Clinical Features

It is important to realize that paralysis is a relatively infrequent complication of an otherwise trivial infection. Of those infections that become clinically manifest at all, most take the form of a minor illness ("abortive poliomyelitis"), characterized by fever, malaise, and sore throat, with or without headache and vomiting that may indicate some degree of aseptic meningitis. In about 1% of cases muscle pain and stiffness commences 1–3 days later, and is followed by flaccid paralysis, which develops rapidly to maximal involvement. Death from respiratory or cardiac failure may occur with bulbar poliomyelitis (Plate 18-2). Otherwise, some degree of recovery of motor function may occur over the next few months, but paralysis remaining at the end of that time is permanent.

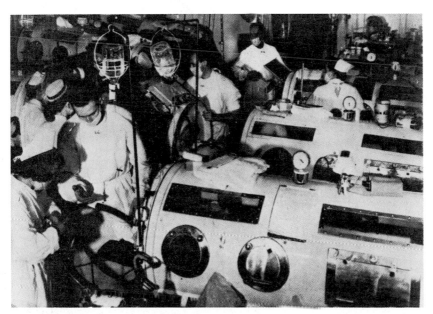

PLATE 18-2. *Totally paralyzed poliomyelitis patients in mechanical respirators during the last great epidemic in the U.S. before the advent of universal immunization (1955).* [*From L. Weinstein, J. Infec. Dis.* **129,** *480 (1974).*]

Pathogenesis and Immunity

Following ingestion, poliovirus multiplies first in the pharynx and/or small intestine. It is not clear whether the mucosa itself is involved, but the lymphoid tissue (tonsils and Peyer's patches) certainly is. Spread to the draining lymph nodes leads to a viremia, enabling the virus to become widely disseminated throughout the body. It is only in the occasional case that the central nervous system becomes involved. Virus is carried via the bloodstream to the anterior horn cells of the spinal cord and motor cortex of the brain. The resulting lesions are widely distributed, but variation in severity gives rise to a spectrum of clinical presentations with spinal poliomyelitis the most common and the bulbar form less so.

The probability of neural involvement with consequent paralysis is influenced by certain well-defined factors, notably age, pregnancy, tonsillectomy, trauma, fatigue, and inoculations (see Chapter 7). Most striking is the influence of age. Among those not protected by immunization or prior infection, paralysis is more frequent and severe in adults than in children. Injections, or trauma of any sort, predispose to paralysis in that limb, while tonsillectomy increases the chance of bulbar poliomyelitis. These observations have been invoked as evidence for spread of virus up peripheral nerves to reach the corresponding segment of the spinal cord or brain stem, but have with equal plausibility, been interpreted in terms of a reflex increase in permeability of the capillaries supplying those areas of the CNS.

The incubation period of paralytic poliomyelitis averages 1–2 weeks, with outer limits of 3 days to 1 month. Acquired immunity is permanent but monotypic.

Laboratory Diagnosis

Virus may be recovered from the throat in the very early stages, but after paralysis has become apparent, it is more readily isolated from the feces. It is virtually never recoverable from the CSF. Any type of human or simian cell culture is satisfactory for isolation, the virus growing so rapidly that cell destruction is usually complete within a few days. Early changes include cell retraction, increased refractivity, cytoplasmic granularity and nuclear pyknosis (Plate 18–3). The isolate is identified by neutralization tests. However, the ubiquity of attenuated Sabin strains following universal adoption of oral immunization poses a very difficult problem for diagnostic laboratories today. Since genetic markers for virulence are not altogether reliable, animal pathogenicity tests must be invoked to distinguish wild strains from

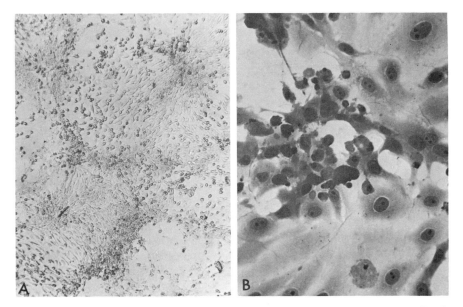

PLATE 18-3. *CPE induced by picornaviruses.* (A) *Enterovirus in primary cell culture derived from monkey kidney, unstained,* ×60. *Note rapidly developing generalized cell destruction.* (B) *Enterovirus in monkey kidney, (hematoxylin and eosin stain). Note disintegrating pyknotic cells.* (*Courtesy I. Jack.*)

vaccine strains. In practice this is only done when a case of paralytic poliomyelitis occurring in a recently immunized individual raises the remote possibility that the disease may be vaccine-induced (see below).

The immune status of individuals in the community is assessed by measuring their serum titers of neutralizing antibody against each of the three serotypes. The "Salk color test" is a technically simple variant of this procedure (see Chapter 14).

Epidemiology

Being enteric in their habitat, polioviruses spread mainly via the fecal–oral route. Direct fecal contamination of hands, thence to food or eating utensils, is probably responsible for most case-to-case spread. Explosive epidemics have been known to result from contamination of water supplies by sewage. In the tropics the disease is endemic throughout the year; in temperate countries before the introduction of vaccination it classically occurred in summer epidemics. The fact that sporadic winter cases nevertheless occur raises the possibility that the virus may also spread by the respiratory route. The chain of infection is rarely obvious, because most infections are inapparent. Dissemination is most effective between unimmunized school children and family contacts. Historically, type 1 has been the most common serotype but type 3 has become prominent since the introduction of Sabin vaccine; cases of poliomyelitis are now very rare in most countries.

Two major changes have influenced the epidemiology of poliomyelitis over the years. The first of these, paradoxically, was the introduction of modern standards of hygiene to the more advanced countries of the world, which led to a change in age incidence of the disease. The consequent reduction in the spread of viruses by the fecal–oral route limited the circulation of polioviruses and the incidence of infection in the community as a whole. As a result, most people had no acquired immunity by the time they reached adolescence or even adulthood. Primary infection of adults is, for reasons still unknown, very much more likely to result in severe paralytic disease than is primary infection of young children. The consequence has been a shift in the age incidence of paralytic poliomyelitis to these young adults, so rendering the old term "infantile paralysis" an obsolete misnomer. The greater susceptibility of young adults to paralysis has been exemplified most strikingly in "virgin soil" epidemics occurring in isolated communities with no prior experience of the virus, and the generality of this age-related severity in viral diseases is illustrated by recent experience with hepatitis.

The other major influence on the epidemiology of poliomyelitis has only become apparent during the last two decades. In those countries where a vigorous policy of immunization with Sabin vaccine has been successfully pursued, not only has the disease been abolished, but the virus itself has almost disappeared. In North America, Europe, and Australasia the incidence of paralytic poliomyelitis has been reduced well over a hundredfold since 1955, and wild polioviruses are rarely found in human sewage (though vaccine strains are now ubiquitous). The few remaining cases tend to occur in unimmunized pockets of poverty or ignorance. In the tropical countries of Africa, Asia, and Central and South America immunization has not met with the success anticipated and in some countries the incidence of poliomyelitis actually seems to be increasing. To some extent this disturbing trend may be an inevitable consequence of the improving living standards of the developing countries. In large part however, it is attributable to practical problems associated with the administration of oral vaccine in tropical climates. The problems, and current attempts to overcome them, are discussed below.

Prevention and Control

In the late 1940's, the U.S. National Foundation for Infantile Paralysis organized a nationwide doorknock appeal, imaginatively labeled "The March of Dimes." The response was overwhelming and the Foundation set about sponsoring a massive research drive on several fronts, in an attempt to turn Enders' recent discovery to advantage by developing an antipoliomyelitis vaccine. Salk was commissioned to work toward an inactivated vaccine; Sabin, Koprowski, and others toward a living attenuated one. The formalin-inactivated Salk vaccine was the first to be licensed (1954) and was enthusiastically embraced in North America, Europe, and Australia during the mid-1950's. Sweden, the country in which paralytic poliomyelitis first became apparent in epidemic form and which for many years continued to have the highest rate of poliomyelitis in the world, eradicated the disease by the use of Salk vaccine.

Meanwhile the living attenuated oral vaccine of Sabin was adopted in the USSR and Eastern Europe where it was demonstrated to be so successful that the whole world, except some Scandinavian countries now uses Sabin vaccine exclusively. The several advantages of such living vaccines over their inactivated counterparts were set out in detail in Chapter 12.

Sabin vaccine is delivered by mouth, in the form of flavored drops—easily administered on a large scale by unqualified personnel in remote regions of developing countries, and highly acceptable to children with a fear of needles. Being live, the virus multiplies in the gastrointestinal tract, hence the immunogenic stimulus approximates that which follows a natural subclinical infection. Sabin vaccination induces the production not only of serum IgG, as does the Salk vaccine, but also of local IgA, synthesized by plasma cells present in the wall of the gut (see Chapter 7). In consequence, the child is protected not only against spread of natural polioviruses through the blood stream to the spinal cord but also against the initial multiplication of such viruses in the bowel.

Ideally, vaccination should be commenced at about 6 months of age, after breast-feeding has ceased, because poliovirus IgA antibodies are present in milk. In some African countries where women habitually breast-feed their infants for up to 2 years, it is stipulated that the child not be breast-fed for about 6 hours before and for the same period after administration of the vaccine. To minimize the possibility that one of the three attenuated serotypes in the vaccine will fail to "take" as a result of interference from one of the others, or from some unrelated enterovirus that happens to be present in the intestine at the time, the trivalent vaccine is usually administered on three occasions about 6–8 weeks apart. In tropical countries with low standards of hygiene such inapparent infections can be a real problem, as anything up to 80–90% of young children may be carrying an enterovirus at any given time, and recalling infants from inaccessible villages on three consecutive occasions is not always practicable. Care must also be taken to keep the vaccine refrigerated; even under these conditions its shelf-life is limited to a few months.

Some concern has been felt about the possibility that the type 3 vaccine strain (Leon) may be insufficiently attenuated. It is certainly less stable genetically than the other two and occasionally back-mutates on passage in man to regain neurovirulence for the monkey. Statistical analysis of "vaccine-associated" cases of poliomyelitis (defined as those instances in which paralysis develops within a month after Sabin vaccination) indicates a higher than chance incidence of virulent type 3 isolations and suggests that this component of the vaccine may indeed be causing paralysis in something less than one in a million recipients. It is also notable that a much higher proportion of all cases of paralytic poliomyelitis today are attributable to type 3 rather than type 1, which used to be the most common; this is probably a result of the relatively poor immunogenicity of the present type 3 attenuated strain. A new type 3 strain is needed for the vaccine, perhaps the relatively stable and immunopotent Czechoslovakian USOL-D strain currently under test.

In many parts of the world, poliovaccine is still grown in monkey kidney cells, despite the clear warnings provided by the Marburg, B virus, and SV40 incidents (see Chapters 9, 16, 22) and the discovery of other cryptic simian viruses. The World Health Organization Committee on Biological Standardization has laid down very strict recommendations on the quarantine and surveillance of monkeys in closed colonies, screening of cell cultures, and safety-testing of the finished product, but it is difficult to justify the continued use of primary monkey kidney cells. Particularly for an oral vaccine, the human diploid cell strain WI-38 would seem to be in all ways superior (for discussion see Chapter 12).

There is no reason in theory why poliomyelitis should not be eradicated by the conscientious administration of Sabin vaccine on a global scale. Considerable problems, not insuperable, are impeding progress in some of the less advanced nations, but they too should conquer poliomyelitis in due course. Nevertheless, we must end with a warning against complacency. Already most of the public in the developed nations have lost their fear of poliomyelitis, because the disease is no longer part of their experience. The incentive to immunize one's children is no longer there. But as long as there remains even the smallest pocket of infection in some remote corner of the world, no country can ever afford to relax its policy of universal immunization. For it must be realized that, by eradicating all wild strains of poliovirus, we are removing nature's way of immunizing children by inapparent infection. If, under these circumstances, we were to relax our insistence on Sabin vaccination to the point where substantial numbers of people were unprotected, an unexpected outbreak would be as devastating to the adults in the community as were the classical virgin soil epidemics of former times.

COXSACKIEVIRUSES, ECHOVIRUSES, AND OTHER ENTEROVIRUSES

In 1948 Dalldorf and Sickles were investigating an outbreak of poliomyelitis in the village of Coxsackie in New York State. Anxious to throw the diagnostic net as wide as possible, they chose to inoculate specimens into the brains of newborn mice. This led to the discovery of the coxsackieviruses, some of which still defy cultivation *in vitro*. In man, they produce pathological effects in target organs as varied as brain, heart, muscle, respiratory epithelium, and skin (Table 18-3). Like the coxsackieviruses, the echoviruses were discovered quite by accident in the course of investigations on poliomyelitis. Because they were originally isolated in cell culture from the feces of apparently normal individuals, they were designated "orphans," i.e., viruses without a parent disease. Echovirus, the name that finally stuck, is a sigla derived from the words *e*nteric *c*ytopathogenic *h*uman *o*rphan virus. Many of the orphans have found their parents (see Table 18-3). In the great variety of diseases with which they are now known to be associated, they closely resemble the coxsackieviruses, which can, indeed, be regarded as echoviruses with a special capacity to grow in baby mice. Indeed, as indicated in the Introduction to this chapter, the distinction between echoviruses and coxsackieviruses is no longer drawn and newly discovered enteroviruses are simply allocated a number.

The human enteroviruses are grouped together solely on the basis of morphological and chemical similarity of the virions. There is no group antigen common to all the echoviruses or the coxsackieviruses, although complement fixation tests do detect cross-reactions between some serotypes, e.g., between all 6 coxsackie B serotypes and coxsackie A9. Neutralization tests are completely type-specific except insofar as some serotypes may be further subdivided into "prime strains" which are incompletely neutralized by antiserum, e.g., partial neutralization of echovirus 6' and 6" by type 6 antiserum.

Some of the serotypes originally allocated to the coxsackievirus or echovirus groups have subsequently been found by more careful antigenic analysis to be closely related or identical to other types. Coxsackie A23 is now recognized as being identical with echo 9; coxsackie A24, of which there are several antigenic variants, is closely related to "echo 34"; coxsackie A18 is indistinguishable from coxsackie A13; echo 1 and 8 are very closely related to one another; coxsackie A9 is nonpathogenic for infant mice, therefore more properly regarded as an echovirus, whereas two strains of echo 6 paralyze infant mice and could therefore be deemed to be coxsackieviruses; echo 10 and 28 no longer exist, having been reclassified years ago as reovirus 1 and rhinovirus 1, respectively. To suggest that the taxonomy of human enteroviruses is in need of revision could be regarded as an understatement.

The enteroviruses are host-specific. Just as the ECHOviruses are confined to man, so monkeys have their ECMO viruses, swine ECSO viruses, bovines ECBO viruses, etc. Parenteral inoculation of monkeys with human enteroviruses regularly produces a subclinical infection, and occasionally paralysis or carditis. The newborn mouse still represents the only laboratory host available for the isolation of many coxsackie A serotypes, and also supports the growth of coxsackie B viruses. Indeed, the division of coxsackieviruses into these two subgroups can be seen in terms of the difference in response of baby mice (Plate 18-4). Susceptibility drops sharply with age, especially to coxsackie B, which does not readily infect suckling mice more than 1 day old. All enteroviruses except some coxsackie A serotypes grow well in primary cell cultures of human or monkey kidney, or human amnion. Many of the echoviruses grow better in diploid strains of human fibroblasts, and several of the coxsackie A viruses grow only in such cell strains.

Clinical Features

Once again it should be stressed that most enteroviral infections are subclinical, particularly in young children. Nevertheless they can cause a wide range of clinical syndromes involving any of the major body systems (Table 18-3). In general, coxsackieviruses tend to be more pathogenic than echoviruses; they are commonly associated with diseases such as exanthemata and aseptic meningitis, and are also responsible for several syndromes not seen with echoviruses, e.g.,

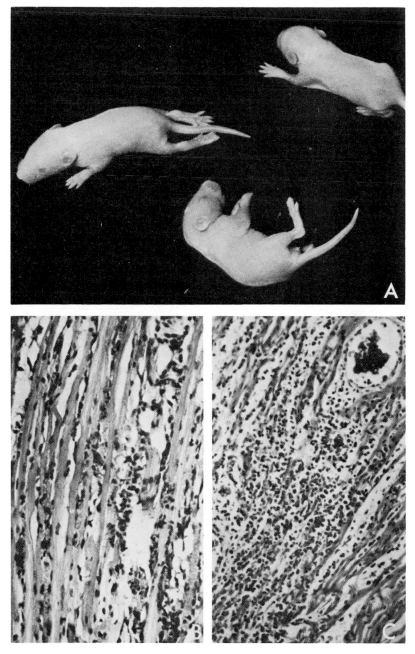

PLATE 18 4. Coxsackieviruses (A) Infant mice 4 days after inoculation with coxsackievirus type A. The mice are moribund, with flaccid paralysis, particularly noticeable in the hind legs. (B) Section of muscle from infant mouse infected with coxsackievirus type B. (H and E stain, ×250). Note focal myositis. (C) Section of heart
from fatal case of coxsackievirus type B-induced myocarditis in a newborn baby
(hematoxylin and eosin stain, ×100). (Courtesy I. Jack.)

TABLE 18-3
Diseases Caused by Enteroviruses

| SYNDROME | VIRUSES | |
	COMMON	LESS COMMON
Paralysis	Poliovirus 1, 2, 3	Coxsackie A7
Meningitis	Echovirus 4, 6, 9, 11, 14, 16, 30	
	Coxsackie A7, 9, 23; B1–6	Other enteroviruses
Encephalitis	Enterovirus 71	
Rash (macular)	Echovirus 4, 6, 9, 16	Echovirus 2, 11 and others
	Coxsackie A9, 16, 23	Coxsackie A2, 4; B1, 3, 5 and others
		Enterovirus 71
Rash (vesicular)	Coxsackie A16	Coxsackie A4, 5, 9, 10
		Enterovirus 71
"Summer febrile illness"	Many enteroviruses	
Vesicular pharyngitis (herpangina)	Coxsackie A1–6, 8, 10	Other enteroviruses
Upper respiratory tract infection (URTI)	Coxsackie A21 Echovirus 11, 20	Coxsackie A10, 24; B2–5 Echovirus 4, 8, 9, 22, 25
Pneumonitis and bronchiolitis	Enterovirus 68	
Epidemic myalgia (Bornholm disease)	Coxsackie B1–5	Other enteroviruses
Carditis	Coxsackie B1–5	Other enteroviruses
Acute hemorrhagic conjunctivitis and radiculomyelitis	Enterovirus 70	
Hemolytic uremia	Coxsackie B	
Pancreatitis	Coxsackie B	
Hepatitis		Echovirus 19
Gastroenteritis		?Coxsackie A and echoviruses

vesicular pharyngitis, Bornholm disease, hand-food-and-mouth disease, pancreatitis, and hemolytic uremia.

Vesicular Pharyngitis (Herpangina). This is a severe febrile pharyngitis (having nothing to do with herpetovirus) sometimes accompanied by vomiting and abdominal pain. Pathognomonic small, grey vesicles, rupturing to become ulcers, are present on the mucous membrane of the throat, palate, or tongue. It is the most common clinical manifestation of coxsackie A infection in young children. A syndrome known as lymphonodular pharyngitis has been reported with coxsackie A10.

"Summer Minor Illnesses." These episodes, frequently observed during coxsackievirus and echovirus epidemics in the summer or autumn, often take the form of a brief febrile illness of no great consequence.

Common Colds. Pharyngitis, and other mild upper respiratory tract infections (URTI) are not caused exclusively by the rhinoviruses. Coxsackie A21, echovirus 11 and 20, and less frequently coxsackie A10, A24, B3, and B5 have been associated with colds.

Pneumonitis and Bronchiolitis. In children these ailments have been associated with the new enterovirus type 68.

Epidemic Myalgia (Bornholm Disease). Bornholm is one of the islands forming part of Denmark where holidaying Danes first encountered this disease which is perhaps the most common substantial syndrome attributable to coxsackie B. The victim suffers a rather severe constitutional disturbance with fever, headache, and generalized muscle aches and pains, sometimes extending to stabbing pains in the muscles of the chest and abdomen. The diagnostic confusion with pleurisy (or an acute abdomen) has been perpetuated in the misleading pseudonym "epidemic pleurodynia." For all its unpleasantness, recovery is invariable.

Exanthemata. Macular or maculopapular ("rubelliform") rashes, with accompanying fever and pharyngitis, are seen, particularly in children, during epidemics of coxsackie A9, 16, or 23, and echovirus 4, 6, 9, or 16, as well as several others including the new enterovirus 71 (see Chapter 24). The syndrome may occasionally include abdominal pain, and rarely meningitis.

Vesicular rashes are rarer, but coxsackie A16, and occasionally A4, 5, 9, or 10, as well as enterovirus 71, can produce a syndrome characterized by vesicular lesions on the palms, soles, and mouth, with the rather quaint name of "hand, foot, and mouth disease" (unconnected with the more notorious foot and mouth disease of cattle which is also caused by a picornavirus but is not transmissible to man).

Aseptic Meningitis. This can follow infection by any of several enteroviruses, notably echovirus 3, 4, 6, 9, 11, 14, 16, and 30 and coxsackie B1–6, and A7, 9, or 23. The syndrome is much more common than bacterial meningitis, but fortunately is not nearly as serious (see Chapter 24). Muscle weakness occasionally develops (e.g., with echovirus 6 or 9) but permanent paralysis is extremely rare, except with coxsackie A7. A rubelliform rash occasionally accompanies the meningitis, e.g., in epidemics of coxsackie A9, echovirus 6, 9, 16, 18, and enterovirus 71, and others.

Meningoencephalitis. The most recently discovered enterovirus, type 71, has been recovered from the feces of patients with encephalitis, meningitis, or both.

Myocarditis and Pericarditis. This condition, caused by any of the coxsackie B types, or less commonly by coxsackie A or echoviruses, is uncommon but grave, with a high mortality in newborn infants. Classically, a neonate within the first week of life becomes suddenly dyspneic and cyanosed and on examination reveals a marked tachycardia with electrocardiographic abnormalities. Recovery may be followed in later life by cardiac malfunction. Milder cases probably go undiagnosed or are passed off as "failure to thrive" in lethargic infants with feeding difficulties. Some cases of sudden, unexplained "cot deaths" in infants also yield a coxsackie B virus at autopsy.

Recently it has become apparent that a substantial proportion of the myocarditis and pericarditis in older children and adults has the same etiology.

Acute Hemorrhagic Conjunctivitis and Radiculomyelitis. During 1969–1971 there occurred a pandemic of a highly contagious "new" disease characterized by subconjunctival hemorrhage with frequent corneal involvement and occasional neurological complications (see Chapter 24). The causal agent has been designated enterovirus type 70. Other enteroviruses have been associated only with very mild cases of conjunctivitis.

Hemolytic Uremia. This is a relatively uncommon but dangerous syndrome of which the most common viral cause appears to be coxsackie B. As the name implies the disease is characterized by hemolytic anemia with thrombocytopenia and intravascular coagulation and uremia. It is seen among young children, principally in late summer.

Acute Pancreatitis. This is sometimes associated with coxsackie B infection.

The possible role of coxsackieviruses in the etiology of diabetes mellitus is a much more controversial issue. Reports that juvenile diabetes tends to occur in epidemic clusters and that these children show serum antibodies to coxsackievirus B4 more often than would be expected by chance provoked a prospective study of children infected with coxsackie B3 or B4 which failed to support a causal relationship. Mice infected with coxsackie B viruses certainly develop pancreatitis, but it is the acinar cells, rather than the beta cells of the islets of Langerhans, which are destroyed. (Interestingly, a natural picornavirus of wild mice, EMC virus, does destroy beta cells and induces hyperglycemia; it may be a useful model for the study of diabetes.)

Hepatitis. The recently described 27 nm hepatitis A virus may turn out to be either an enterovirus or, more likely, a parvovirus; it is described at length in Chapter 23. Echovirus type 19 has been isolated from overwhelming infections of newborn infants in which the cause of death has been assessed as hepatic necrosis.

Gastroenteritis. The role of enteroviruses in epidemic diarrhea is still open to question. Certainly, echovirus types 6, 11, 14, 18, 22, and others, as well as several coxsackie A viruses, have been recovered from the feces of patients during epidemics of gastroenteritis more frequently than from controls. However, a causal correlation has not been established for any except the "Norwalk agent," which is probably a parvovirus, and the "rotavirus" which certainly belongs to the family Reoviridae. These two agents which appear to be the major agents of enteritis in man are discussed in Chapter 23.

Pathogenesis and Immunity

Enteroviruses multiply primarily in the pharynx and small intestine; they are shed in the feces for up to a month and in respiratory secretions for a few days. Dissemination via the blood stream is doubtless the route of spread to the wide range of target organs susceptible to attack by coxsackieviruses. The incubation period is about a week in the case of such generalized infections but may be as short as 2 days for respiratory or conjunctival disease. The only cases that come to autopsy, namely those of fatal encephalitis or myocarditis (Plate 18-4C) often reveal evidence of hepatitis and pancreatitis as well as carditis and encephalitis. The histopathology is reminiscent of that seen in the baby mouse. Acquired immunity to enterovirus infections is type-specific and long-lasting.

Laboratory Diagnosis

As would be expected, feces are the best source of enteroviruses, though virus can often be recovered for shorter periods from the throat also. In meningitis, virus may be isolated from the cerebrospinal fluid as well as from feces, and virus can usually be recovered from the heart or brain of fatal cases of carditis or encephalitis respectively. Enterovirus 70 is readily grown at 33° from conjunctival swabs or scrapings taken from patients with acute hemorrhagic conjunctivitis.

Specimens should be inoculated into primary cultures of monkey kidney or human embryonic kidney, and into a diploid strain of human fibroblasts, which constitutes the most sensitive system for many enteroviruses. Certain coxsackie A viruses will grow *in vitro* only in human cells. Cytopathic effects resemble those described for poliovirus, but develop more slowly, commencing with foci of rounded refractile cells which then lyse and fall off the glass. Whereas some serotypes destroy the monolayer completely within a few days, with others CPE becomes apparent only after a week or so and does not progress to total cell destruction.

Identifying the isolate is a laborious procedure which can be short-cut by a number of laboratory tricks. Initially the field can be narrowed by testing the ability of the virus to agglutinate human erythrocytes. The actual typing is carried out by hemagglutination-inhibition if practicable, but usually by neutralization. Here the task can be lightened by employing an "intersecting" series of "polyvalent" serum pools, each of which contains antisera (raised in horses) against about 10 serotypes; since antiserum to any given serotype is present in several pools and absent from several others, the isolate can be positively identified in the light of the "pattern" of neutralization, i.e., which pools neutralize the virus and which do not.

Use of the infant mouse is diminishing, but some coxsackie A viruses can still be grown only in this way. Newborn mice (less than 24 hours of age) are inoculated intraperitoneally and intracerebrally. Coxsackie A produces a fulminating generalized myositis, resulting in flaccid paralysis (Plate 18-4,A) and death within a week. By contrast, coxsackie B more slowly produces scattered focal myositis (Plate 18-4,B), brown fat necrosis, and often pancreatitis, hepatitis, and encephalitis, the latter resulting in spastic paralysis.

Theoretically, immunofluorescence provides a rapid method of diagnosing enteroviral meningitis; leukocytes spun out of CSF may be treated with combination antiserum pools than stained with fluorescein-labeled anti-horse globulin. In practice, difficulties occur because of the large numbers of non-cross-reacting enteroviral serotypes involved.

For the same reason retrospective diagnosis of enteroviral infections by screening for antibodies in the patient's serum is rarely employed.

Epidemiology and Control

Enteroviruses are transmitted mainly via the fecal–oral route, spreading rapidly and efficiently within families. Echoviruses appear in the alimentary tract of most infants shortly after birth. In underdeveloped countries with low standards of hygiene enteroviruses may be recovered from the feces of up to 80% of children at any time; by contrast the "Virus Watch" program exposed an enterovirus carriage rate of only 2.4% in New York, while a comparable study of six other U.S.

cities gave a similar figure. The fact that the proportion is closer to 10–20% in southern U.S. cities has been taken to indicate that tropical climatic conditions may be relevant in addition to poor sanitation.

Droplet spread also occurs, more so with the coxsackieviruses, and may be more relevant to the acquisition of the upper respiratory infections for which enteroviruses are quite often responsible. Acute hemorrhagic conjunctivitis is highly contagious in the crowded unhygienic conditions of Asia and Africa spreading directly to the eye with an incubation period of about 24 hours. Over half a million cases occurred in Bombay alone during the 1971 pandemic.

In general, coxsackieviruses are more highly transmissible than echoviruses, probably because they are shed for longer in respiratory secretions as well as feces. The probability of transmission to a nonimmune sibling was demonstrated in one study to be 76% for coxsackieviruses and 43% for echoviruses.

Enteroviruses, particularly coxsackieviruses, are most prevalent in late summer and autumn. Outbreaks of coxsackie A9, 16, 23, and B1–5 and of echovirus types 4, 6, 9, and 16 often reach epidemic proportions. Children usually contract an inapparent infection, or one of the exceedingly common "undifferentiated summer febrile illnesses," or perhaps a rash.

RHINOVIRUSES

In this era of heart transplantation and other such dramatic demonstrations of the powers of medical science there is an irony that appeals to the man-in-the-street in the inability of modern medicine to make the slightest impact on that most common and trivial of all man's ailments, the common cold. The initiated see an even greater irony in the fact that years of searching for *the* elusive common cold virus have led us to, not one, but over 100 separate rhinoviruses. Moreover, it is now clear that the rhinoviruses are by no means the only viruses capable of producing this syndrome; they appear to be responsible for only about one quarter of all the mild upper respiratory infections of man.

The rhinovirus virion differs from that of the enteroviruses in a number of respects, the most significant of which is that it is destroyed by acid (pH 3–5), whereas enteroviruses are not.

Over 100 serotypes are already known, although only 89 of these were officially recognized at the time of writing. They are distinguished from one another by neutralization tests. Several of them agglutinate sheep red blood cells at 4°.

Human rhinoviruses are transmissible experimentally to chimpanzees, and related rhinoviruses have been isolated from natural infections of a number of domestic animals but it is not yet clear whether the latter play any role in the epidemiology of human colds.

Otherwise, the rhinoviruses can only be grown in cell culture, and even there, with a certain amount of difficulty. Tyrrell and his colleagues were eventually able to grow these viruses consistently by mimicking the conditions prevailing in the nasal mucosa, namely, (a) 33° (c.f., the usual 37°), (b) pH around 7.0 (c.f., 7.4), achieved by lowering the concentration of $NaHCO_3$ in the medium, and (c) good oxygenation (achieved by rotating the cultures on a roller drum at low speed). These precautions are still required for growing rhinoviruses in human embryonic kidney, but the pH and temperature restrictions are not essential in diploid strains of human embryonic lung fibroblasts (HEL) or in Tyrrell's line of HeLa cells, both of which support growth of the virus just as well.

Clinical Features

We are all too familiar with the symptomatology of the common cold. A prodromal irritation and fullness in the nose leads to a profuse, watery discharge, often accompanied by a headache, cough or sore throat. There is little or no fever. After 4 or 5 days the cold either resolves or becomes complicated by a mucopurulent discharge commonly assumed (often wrongly), to be due to secondary bacterial infection. Sinusitis or otitis media may supervene. Children occasionally develop a febrile lower respiratory tract infection, including croup, bronchitis, and even bronchopneumonia. There is evidence also that rhinoviruses may be associated with exacerbations of chronic bronchitis and asthma in predisposed adults.

Pathogenesis and Immunity

Rhinoviruses usually remain localized to their point of entry, viz., the nasal mucosa, where they provoke inflammation with edema and copious secretion within 48 hours (occasionally up to 4 days). Perhaps they are so restricted because of their adaptation to growth at 33°. They multiply more satisfactorily in organ cultures of ciliated nasal epithelium than in corresponding cultures from the trachea.

Acquired immunity is type-specific, and lasts for perhaps a couple of years. It correlates more closely with the titer of IgA in the nasal mucus than of IgG in the serum. Immunity to heterologous serotypes is solid for a month postinfection, decreasing to zero over the ensuing few months; this short-lived resistance may possibly be attributable to interferon.

Laboratory Diagnosis

One would hardly ask the laboratory to assist in the diagnosis of the common cold, except perhaps in the course of epidemiological research. However, rhinoviruses are sometimes responsible for significant lower respiratory disease and are often isolated from nasopharyngeal aspirates or throat swabs following inoculation into nonconfluent cultures of human embryonic kidney (HEK) and diploid strains of human embryonic lung (HEL), rolled slowly at 33° in a low bicarbonate, low serum medium. A few serotypes ("M" strains) will also grow in monkey kidney cells. Cytopathic effects develop more slowly and are less complete than seen with other picornaviruses. Foci of rounded or irregular, refractile, granular cells become evident within 2–10 days; "microplaques" can often be counted even in nonoverlaid monolayers. Organ cultures of human embryonic nasal or tracheal epithelium may be necessary for the isolation of particularly fastidious strains; viral growth is detected by cessation of ciliary action.

Epidemiology and Control

The multiplicity of serotypes suggests that the rhinoviruses undergo antigenic drift. Unlike influenza, however, it has been clearly demonstrated that even in relatively small communities, no single strain replaces all the others; rather, several rhinovirus serotypes circulate simultaneously. Hence, the average person contracts 2–6 colds per annum, each caused by a separate serotype. The fact that many of these occur in the colder half of the year has led to the belief that "colds" are etiologically associated with cold and wet. However, this notion was not supported by experiments conducted at the Common Cold Research Establishment in England (Plate 18-5). Andrewes vividly relates the story of the honeymoon couples who were offered free accommodation in return for serving as guinea pigs. Those exposed to inclement weather conditions in fact contracted no more colds than the "controls," who were not so exposed. While it is possible that cold weather or abrupt changes of temperature and humidity may alter the susceptibility of the nasal mucosa, it seems equally likely that the increased incidence of respiratory virus infections in winter may be a direct result of people's reaction against the weather, namely to coop themselves up together in ill-ventilated houses where they can

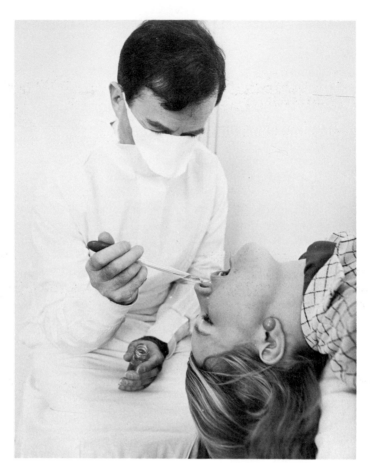

PLATE 18-5. *Intranasal inoculation of human volunteer with rhinovirus. (Courtesy Dr. D. A. J. Tyrrell.)*

more effectively swap parasites. Actually, common colds attributable to rhinoviruses seem to be more prevalent in late summer and autumn than in the colder months of the year, and principally in adults; many, and perhaps most winter colds may turn out to be due to coronaviruses, paramyxoviruses, etc. The incubation period is 1–3 days. Virus is shed for about a week but maximum transmissibility occurs during the first 2 days of coryza. The minimum infectious dose for man is very small indeed (a few particles only). Other features of the epidemiology of the common cold were discussed in Chapter 10.

Clearly, the diversity of antigenic types virtually rules out any possibility of a satisfactory vaccine. Needless to say, the commercially available "Christmas stocking" vaccines directed against miscellaneous secondary bacterial invaders are worthless. Antiviral chemotherapy seems to be the only real hope for the future, but no effective agent is yet in sight. Interferon and various interferon inducers have shown evidence of marginal prophylactic efficacy in clinical trials (see Chapter 13), but have no therapeutic value.

FURTHER READING

Baltimore, D. (1971). Polio is not dead. *Perspect. Virol.* **7**, 1.

Baltimore, D., Huang, A., Manly, K. F., Rekosh, D., and Stampfer, M. (1971). The synthesis of protein by mammalian RNA viruses. *Strategy Viral Genome, Ciba Found. Symp.* p. 101. Churchill Livingstone, Edinburgh.

Fenner, F., McAuslan, B. R., Mims, C. A., Sambrook, J. F., and White, D. O. (1974). "The Biology of Animal Viruses," 2nd ed. Academic Press, New York.

Gard, S. (1967). Inactivated poliomyelitis vaccine—present and future. *In* "First International Conference on Vaccines against Viral Rickettsial Diseases of Man," Sci. Publ. No. 147, p. 161. Pan Amer. Health Organ., Washington, D.C.

Gear, J. H. S., and Measroch, V. (1973). Coxsackievirus infections of the newborn. *Progr. Med. Virol.* **15**, 42.

Hamre, D. (1968). Rhinoviruses. *Monog. Virol.* **1**, 1.

Jackson, G. G., and Muldoon, R. L. (1973). Viruses causing common respiratory infections in man. II. Enteroviruses and paramyxoviruses. *J. Infec. Dis.* **128**, 387.

Kibrick, S. (1976). Current status of coxsackie and echoviruses in human disease. *Progr. Med. Virol.* (in press).

Koprowski, H. (1957). Discussion of properties of attenuated poliovirus and their behavior in human beings. *In* "Cellular Biology, Nucleic Acids and Viruses" (T. M. Rivers, ed.)., Spec. Publ., Vol. 5, p. 128. N.Y. Acad. Sci., New York.

Levintow, L. (1974). Reproduction of picornaviruses. *In* "Comprehensive Virology" (H. Fraenkel-Conrat and R. R. Wagner, eds.), Vol. 2, p. 109. Plenum Press, New York.

Melnick, J. L. (1971). Poliomyelitis vaccine: Present status, suggested use, desirable developments. *In* "International Conference on the Application of Vaccines against Viral, Rickettsial and Bacterial Diseases of Man," Sci. Publ. No. 226, p. 171. Pan Amer. Health Organ., Washington, D.C.

Melnick, J. L. (1975). Enteroviruses. *In* "Viral Infections of Man: Epidemiology and Control" (A. S. Evans, ed.), Plenum, New York.

Melnick, J. L., and Phillips, C. A. (1970). Enteroviruses: Vaccines, epidemiology, diagnosis, classification. *Crit. Rev. Clin. Lab. Sci.* **1**, 87.

Mirkovic, R. R., Kono, R., Yin-Murphy, M., Sohier, R., Schmidt, N. J., and Melnick, J. L. (1973). Enterovirus type 70: The etiologic agent of pandemic acute haemorrhagic conjunctivitis. *Bull. W. H. O.* **49**, 341.

Ogra, P. L., and Karzon, D. T. (1971). Formation and function of poliovirus antibody in different tissues. *Progr. Med. Virol.* **13**, 156.

Rueckert, R. R. (1971). Picornaviral architecture. *In* "Comparative Virology" (K. Maramorosch and E. Kurstak, eds.), p. 256. Academic Press, New York.

Sabin, A. B. (1957). Properties of attenuated polioviruses and their behavior in human beings. *In* "Cellular Biology, Nucleic Acids and Viruses" (T. M. Rivers, ed.), Spec. Publ., Vol. 5, p. 113. N.Y. Acad. Sci., New York.

Sabin, A. B. (1967). Poliomyelitis: accomplishments of live virus vaccine. *In* "First International Conference on Vaccines against Viral and Rickettsial Diseases of Man," Sci. Publ. No. 147, p. 171. Pan Amer. Health Organ., Washington, D.C.

Salk, J. E. (1958). Basic principles underlying immunization against poliomyelitis with a noninfectious vaccine. *In* "Poliomyelitis: Papers and Discussions presented at the Fourth International Poliomyelitis Conference," p. 66. Lippincott, Philadelphia, Pennsylvania.

Stott, E. J., and Killington, R. A. (1972). Rhinoviruses. *Annu. Rev. Microbiol.* **26,** 503.

Tyrrell, D. A. J. (1968). Rhinoviruses. *Virol. Monogr.* **2,** 67.

Wenner, H. A. (1973). Virus diseases associated with cutaneous eruptions. *Progr. Med. Virol.* **16,** 269.

Wenner, H. A., and Behbehani, A. M. (1968). Echoviruses. *Virol. Monogr.* **1,** 1.

World Health Organization. (1972). WHO Expert Committee on Biological Standardization, 24th Report. *World Health Organ., Tech. Rep. Ser.* **486.**

CHAPTER 19
Arboviruses: Togaviridae, Bunyaviridae, Reoviridae

INTRODUCTION

An arbovirus, by definition, is one that multiplies in a blood-sucking arthropod and is transmitted by bite to a vertebrate. Such "arthropod-borne viruses" comprise a convenient epidemiological grouping, but embrace viruses from families as taxonomically diverse as the Togaviridae, Bunyaviridae, Reoviridae, and Rhabdoviridae. The great majority of those that infect man are either togaviruses or bunyaviruses, although a few orbiviruses (family Reoviridae) also cause human disease. Members of these three families will be discussed in a single chapter to emphasize the unitary approach to their ecology that has contributed so much to our understanding of their behavior in nature.

There are over 300 known arboviruses, most of them recovered from mosquitoes or ticks during the 1950's and 1960's as a result of a concerted world-wide drive sponsored by the Rockefeller Foundation. Only about 100 of these are capable of infecting man and only about 40, mainly togaviruses, produce significant disease, some of which, notably the encephalitides and the hemorrhagic fevers, rank among the most lethal of all viral diseases.

In their natural environment, arboviruses alternate between an invertebrate *vector* (mosquito, tick, sandfly, or gnat) and a vertebrate *host* (mammal or bird). The mosquito-borne arboviruses, which comprise the majority, are endemic in the tropics, but are responsible for sporadic summer epidemics in temperate zones, where arthropod transmission is seasonal; several of the tick-borne arboviruses are endemic in temperate regions. Man does not play a role in the natural cycle of the arboviruses; he becomes involved quite by accident, when he intrudes into the natural ecosystem (e.g., by penetrating a swamp or a jungle) or interferes with it (e.g., by clearing a forest or irrigating a desert).

PROPERTIES OF THE VIRUSES

Physicochemical Characteristics

Togaviridae. Togaviruses consist of a small icosahedral core, 20–40 nm in diameter, enclosed within a tightly fitting envelope (Plate 19-1). Alphaviruses (once known as "group A arboviruses") have an overall diameter of about 50–70 nm, while flaviviruses ("group B arboviruses") are somewhat smaller. Both contain a single molecule of RNA, molecular weight 4 million daltons (see Table 19-1).

By virtue of the glycoprotein peplomers projecting from the envelope, togaviruses hemagglutinate, but only under very restricted conditions. Nevertheless, the reaction is widely used for laboratory and epidemiological studies. The virus is usually concentrated and freed from nonspecifically agglutinating lipids and lipoproteins by acetone extraction, and may need to be pretreated by sonic vibration, trypsin, or fluorocarbon to achieve satisfactory hemagglutinin titers. Erythrocytes of geese, pigeons, and 1-day-old chicks are the most sensitive. The pH is also critical, the optimum differing for different togaviruses

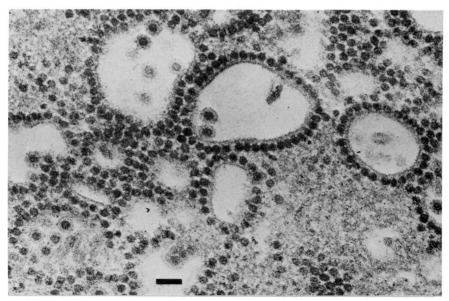

PLATE 19-1. *Togaviridae. Alphavirus nucleocapsids acquiring an envelope as they bud through membranes into cytoplasmic vacuoles (bar = 100 nm). (Courtesy Dr. I. H. Holmes.)*

TABLE 19-1
Properties of Togaviruses

Spherical virion, diameter 40–70 nm
Icosahedral symmetry
Envelope
RNA, single-stranded, mol. wt. 4 million

TABLE 19-2
Properties of Bunyaviruses

Spherical virion, diameter 90–100 nm
Helical symmetry
Envelope
RNA, single-stranded, segmented, mol. wt. 6 million
Transcriptase in virion

---

within the general range of 6.0–7.0. Some agglutinate at a temperature of 4°, some at 37°. No spontaneous elution occurs.

Togaviruses are rather susceptible to inactivation by both heat and lipid solvents. Indeed, sensitivity to ether, chloroform, or bile salts, which indicate that the virion is enveloped, has long been used as a first step in identification of the togaviruses and bunyaviruses, primarily to differentiate them from enteroviruses occurring in the mice in which isolation was usually attempted.

Bunyaviridae. Bunyaviruses are larger (90–100 nm), fragile enveloped RNA viruses of helical symmetry, more closely resembling an orthomyxovirus than a togavirus (Table 19-2 and Plate 19-2).

Orbivirus. Orbiviruses comprise a genus of the family Reoviridae, which is discussed at greater length in Chapter 22. Briefly, they may be defined as icosohedral viruses with a single-shelled capsid 55 nm in diameter containing a double-stranded segmented RNA of molecular weight 15 million (see Table 19-3).

Antigenic Composition

The arboviruses have traditionally been classified into a large number of separate groups (now nearly 40) on the basis of shared antigens demonstrable by hemagglutination-inhibition or complement fixation. There is no serological cross-reactivity between groups, some

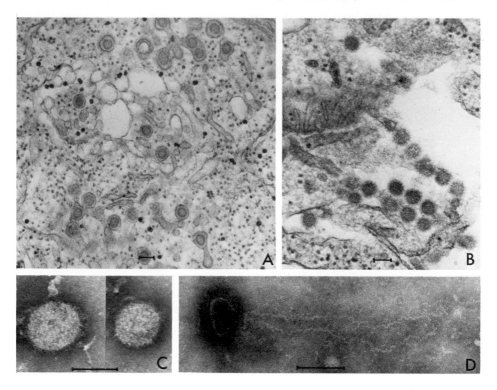

PLATE 19-2. *Bunyaviridae (bars = 100 nm). (A) Virions in Golgi vesicles. (B) Extracellular virions (sectioned). (C) Virions (negatively stained). (D) Virions, spontaneously disrupted by storage at 4°, showing released helical ribonucleoprotein. (A, B, C, courtesy Dr. F. A. Murphy; D, courtesy Prof. N. Oker-Blom.)*

of which have now been designated as genera, e.g., *Alphavirus* and *Flavivirus,* or even families, e.g., Bunyaviridae. Paradoxically, within arbovirus groups, the hemagglutinin-inhibition test shows broader reactivity than complement fixation. The neutralization test is usually the most specific, and is employed to differentiate the viruses within each group. By convention, the 300 or more individual viruses are regarded as separate "species," rather than "types." This practice is out of line with that prevailing for other groups of viruses; for example, the numbered serotypes of picornaviruses show less cross-reaction than arboviruses that have been accorded the status of named species, such as the closely related flaviviruses responsible for Japanese, Australian, West Nile, and St. Louis encephalitis, found on four separate continents. It must be admitted, however, that virology

TABLE 19-3
Properties of Orbiviruses

Spherical virion, diameter 55–65 nm
Icosahedral symmetry, 32 capsomers in inner shell
No true envelope but outer protein shell
RNA, double-stranded, mol. wt. 15 million, in 10 segments
Transcriptase in virion

---

would be the poorer without such exotic names as Bunyamwera, Bush Bush, chikungunya, or o'nyong-nyong. Note, incidentally, that arboviruses named after a geographic locality in which they occur are given a capital letter, but those named after the disease they cause are not.

Host Range

By definition, all arboviruses are capable of multiplication in a hematophagous (blood-feeding) arthropod vector. This is usually a mosquito (especially the genera *Culex* and *Aedes*), or less commonly a tick, sandfly, or midge. Many other types of hematophagous arthropod, well-known for their ability to transmit rickettsiae, bacteria, protozoa, or even helminths, have not been incriminated as vectors of arboviruses.

The natural vertebrate host of most arboviruses is a wild mammal or bird; usually several species are involved. Most frequently, the infection is inapparent, reflecting the prolonged evolution of a perfect host-parasite relationship. A quite remarkable adaptation, the evolution of which probably dates back even further, is the capacity of arboviruses to multiply alternately in cold-blooded arthropods and warm-blooded vertebrates with equal facility and equal speed, despite the pronounced temperature differential.

In the laboratory, subclinical infections can often be set up in monkeys. Much more generally useful however, is the infant mouse, which is the most sensitive laboratory animal. Following intracerebral inoculation, the mice die of encephalitis. The 1-day-old chick is also useful. Indeed, "sentinel" mice, chicks, hamsters, or monkeys may be suspended in cages in the jungle canopy, to be bitten by mosquitoes. Embryonated hen's eggs are also susceptible to some arboviruses, but cultured cells from mammals, or sometimes birds are more commonly used for diagnostic and experimental work. Insect cell cultures are being increasingly exploited, as is the direct injection of virus into mosquitoes.

Viral Multiplication

Alphaviruses have been extensively employed as models for the study of the multiplication of RNA viruses. They offer the following advantages: (a) the genome is a single strand of RNA which codes for a relatively small number of proteins; (b) infectious RNA is easily isolated and rapidly assayed by sensitive plaque assay; and (c) many temperature-sensitive mutants, induced by nitrous acid or nitrosoguanidine, have been demonstrated to fall into a small number of complementation groups reflecting distinguishable viral functions.

In general, the togaviruses seem to replicate very much after the fashion of picornaviruses. Rapid shutdown of cellular macromolecular synthesis is often a feature in infections of vertebrate cells, but not in insect cells. Virus-coded RNA polymerase(s) catalyze(s) the synthesis of viral RNA via a replicative intermediate. However, since they are enveloped, the virions mature differently from the picornaviruses. Electron-dense "cores" corresponding to the naked nucleocapsid acquire an envelope by budding through the plasma membrane (alphaviruses) or into cytoplasmic vacuoles (flaviviruses) (Plate 19-1).

Little is yet known about the multiplication of bunyaviruses. They develop in the cytoplasm and bud into Golgi vesicles (Plate 19-2). Orbiviruses also mature in the cytoplasm; in most important respects their replication probably resembles that outlined for other members of the family Reoviridae (see Chapter 22).

HUMAN INFECTIONS

Clinical Features

Fortunately the great majority of arbovirus infections of man are subclinical, e.g., for every reported case of encephalitis there are hundreds of asymptomatic "seroconversions" (acquisition of antibody) indicative of inapparent infection. Furthermore, clinical disease, when it does occur, most commonly takes the form of a relatively harmless "undifferentiated fever" which, incidentally, is often misdiagnosed as influenza, malaria, hepatitis, leptospirosis, or some other illness commonly encountered in tropical areas. Nevertheless, some 40 of the arboviruses are capable of producing more serious illness, and about half of these are potentially lethal. The clinical picture can be quite varied but is conveniently considered under four main headings.

Fever, Arthralgia ± Rash (Prototype: Dengue Fever). This is the most common manifestation of arboviral disease. Indeed, it is seen not only with the viruses listed under this leading in Table 19-4, but to some

TABLE 19-4
Arboviral Fever-Arthralgia-Rash

VIRUS/DISEASE	GENUS	DISTRIBUTION	VECTOR	RESERVOIR
Chikungunya	*Alphavirus*	Asia, Africa	Mosquito	Monkeys?
O'nyong-nyong	*Alphavirus*	Africa	Mosquito	?
Ross River	*Alphavirus*	Australia	Mosquito	Mammals?
Sindbis	*Alphavirus*	Widespread	Mosquito	Birds, mammals
Dengue (types 1–4)	*Flavivirus*	Widespread, esp. Asia, Pacific, Caribbean	Mosquito	Monkeys?
West Nile	*Flavivirus*	Africa, Asia	Mosquito	Birds
Sandfly fever	*Bunyavirus*	Mediterranean	Sandfly	?
Oropouche	*Bunyavirus*	South America	Mosquito	?
Rift Valley fever	*Bunyavirus*	Africa	Mosquito	Sheep, cattle
Colorado tick fever	*Orbivirus*	USA	Tick	Rodents

degree also with most of the viruses causing more serious illness (Tables 19-5 and 19-6). Dengue fever, the most widespread of the arboviral diseases, will be taken as the prototype. The characteristic features are sudden onset of fever, chills, headache (especially behind the eyes), conjunctivitis, and lymphadenitis, accompanied by excruciating pains in the back, muscles, and joints, which give the disease its popular name of "break-bone fever." The fever often falls and rises again a few days later producing the characteristic "saddle-back" temperature chart. Leukopenia is usually demonstrable. A scarlatiniform or maculopapular rash appears on the third or fourth day, and fades 2–3 days later. Convalescence is slow but certain.

The viruses listed in Table 19-4 all produce somewhat similar syndromes, although some, like chikungunya, o'nyong-nyong, and Ross River viruses cause more marked joint involvement than others. Bunyamwera, Oropouche, Rift Valley fever, Colorado tick fever, and sandfly (phlebotomus) fever usually give no rash.

Hemorrhagic Fever. As the name implies, the striking feature of this lethal syndrome is hemorrhage, which makes itself apparent as petechiae and ecchymoses on the skin and mucous membranes (Plate 19-3) with bleeding from all the body orifices. Examination of the blood reveals thrombocytopenia (and usually leukopenia). In fatal cases the patient collapses abruptly from hypotensive shock; indeed in the highly lethal "dengue shock syndrome" the child may die before hemorrhagic signs have become manifest.

The hemorrhagic fevers are a diverse group which should probably not be considered as a single entity. On the one hand the Russian hemorrhagic fevers and Kyasanur Forest disease of India are primary

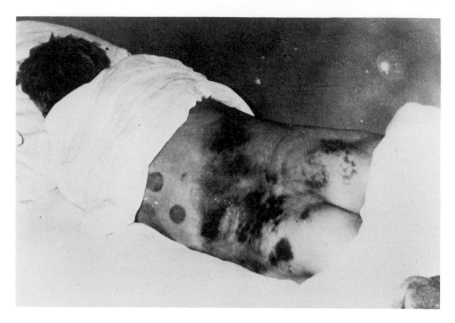

PLATE 19-3. *An acute case of Crimean hemorrhagic fever, showing sites of capillary fragility tests.* [*From J. Casals, B. E. Henderson, H. Hoogstraal, K. M. Johnson, and A. Shelekov, J. Infect. Dis.* **122,** *437 (1970); courtesy Dr. A. Shelekov.*]

infections with known tick-borne flaviviruses listed in Table 19-5. By contrast dengue shock syndrome may represent a hypersensitivity reaction to reinfection with a second serotype of dengue virus (see Immunity). Moreover, there are several other hemorrhagic fevers (Argentinian, Bolivian, Lassa) which, though clinically similar, are epidemiologically quite distinct, being caused by nonarthropod-borne arenaviruses (see Chapters 22 and 24).

Hemorrhagic Fever with Hepatitis and Nephritis (Prototype: Yellow Fever). Fortunately only one virus, that of yellow fever, seems capable of producing this potentially lethal syndrome. The disease has been feared for centuries, and immortalized in literature and music as the curse of the Ancient Mariner and Flying Dutchman, respectively. Devastating outbreaks decimated the crews of English sailing ships anchored off the coast of West Africa and the Frenchmen engaged on the construction of the Panama Canal. The name of Walter Reed, a United States Army doctor, is known to every medical student for his success in demonstrating first, that the disease was mosquito-transmitted, and second, that it was caused by an agent that passed through filters

TABLE 19-5
Arboviral Hemorrhagic Fever

VIRUS/DISEASE	GENUS	DISTRIBUTION	VECTOR	RESERVOIR
Chikungunya	*Alphavirus*	Africa, Asia	Mosquito	Monkeys?
Dengue 1–4	*Flavivirus*	Widespread	Mosquito	Monkeys?
Yellow fever	*Flavivirus*	Africa, South America	Mosquito	Monkeys, man
Kyasanur Forest disease	*Flavivirus*	India	Tick	Monkeys, rodents
Omsk hemorrhagic fever	*Flavivirus*	U.S.S.R.	Tick	Mammals
Crimean HF-Congo	*Bunyavirus*	Central Asia, Africa	Tick	Mammals?

holding back bacteria. As recently as 1960–1962 an epidemic of yellow fever killed 15,000–30,000 people in Ethiopia.

Yellow fever begins with fever, chills, headache, backache, and vomiting—not so very different from the milder arboviral syndromes. However, there soon follows extensive hemorrhages, especially gastrointestinal, hence the rather revolting nickname, "black vomit." Blood pressure naturally drops, but strangely, the pulse rate does too. Jaundice and proteinuria are ominous signs. Mortality averages about 10%, and survivors recover completely. Fortunately, many infections are mild or even inapparent. Yellow fever is, of course, a notifiable and quarantinable disease.

Encephalitis. A large number of togaviruses are capable of causing this serious disease (Table 19-6). The disease begins with the usual arboviral symptoms of fever, chills, headache, widespread aches, and vomiting. However, within a day or two the patient develops a telltale drowsiness, often accompanied by neck rigidity, indicative of meningeal involvement. The worst cases progress to confusion, paralysis, convulsions, coma, and death.

Japanese and Eastern equine encephalitis (EEE) are frequently fatal, St. Louis, WEE, VEE, and Australian encephalitis somewhat less so, while California encephalitis is not commonly lethal. Survivors are often left with neurological sequelae such as mental retardation, epilepsy, paralysis, deafness, or blindness.

Actually, many of the "nonencephalitogenic" togaviruses have encephalitogenic potential. This is apparent, not only from the neurological damage produced by arboviruses in suckling mice, but also from the cases of encephalitis that resulted accidentally when "nonencephalitogenic" togaviruses were tested for their oncolytic ability in terminal cancer patients.

TABLE 19-6
Arboviral Encephalitis

VIRUS/DISEASE	GENUS	DISTRIBUTION	VECTOR	RESERVOIR
Eastern equine encephalitis (EEE)	*Alphavirus*	Americas	Mosquito	Birds
Western equine encephalitis (WEE)	*Alphavirus*	Americas	Mosquito	Birds, reptiles?
Venezuelan equine encephalitis (VEE)	*Alphavirus*	Americas	Mosquito	Rodents
St. Louis encephalitis (SLE)	*Flavivirus*	Americas	Mosquito	Birds
Japanese encephalitis (JE)	*Flavivirus*	East Asia	Mosquito	Birds
Australian encephalitis	*Flavivirus*	Australia	Mosquito	Birds
West Nile encephalitis (WN)	*Flavivirus*	Africa, Europe	Mosquito	Birds
Tick-borne encephalitis (TBE)	*Flavivirus*	Eastern Europe	Tick	Mammals
Russian spring–summer encephalitis (RSSE)	*Flavivirus*	U.S.S.R., Europe	Tick	Rodents
Louping ill	*Flavivirus*	Britain	Tick	Sheep
Powassan	*Flavivirus*	North America	Tick	Rodents
California encephalitis (CE)	*Bunyavirus*	North America	Mosquito	Rodents, rabbits

Pathogenesis

The infected mosquito introduces its proboscis directly into a capillary beneath the skin, and injects saliva that contains the virus. Viral multiplication ensues in vascular endothelium and reticuloendothelial cells in lymph nodes, liver, spleen, and elsewhere. Virus liberated from these organs sets up a viremia, which precipitates the "systemic phase" of the resulting illness (fever, chills, aches) after an incubation period of 4–7 days. The disease process may not progress further. Alternatively, it may go on over the next few days to involve any of a number of systems: the joints and muscles, to produce arthritis and myositis; the skin, giving a rash, sometimes hemorrhagic; the liver and kidneys, causing a hepatitis and nephritis which may be fatal; or the brain, resulting in a potentially lethal encephalitis.

The histopathology of the common syndromes will be considered briefly. The skin rash of dengue fever and related infections is characterized by swelling of the capillary endothelium, perivascular edema, and infiltration of mononuclear cells. By contrast, the petechiae and ecchymoses of the hemorrhagic fevers are not accompanied by inflamation, but probably result from a thrombocytopenia that leads to bleeding from mucous membranes all over the body. In yellow fever the lethal change is an acute liver atrophy, characterized by hyaline necrosis of parenchyma, particularly in the mid-zonal region of the lobules. Two types of inclusion may be seen: eosinophilic intranuclear inclusions, known as Torres bodies, and larger eosinophilic masses of hyaline material resulting from fusion of necrotic cells, known as Councilman bodies. Fatty degeneration is also prominent in the tubules of the kidney, and to a lesser extent in the spleen, heart, lymph nodes, and brain. In the encephalitides, inflammatory foci distributed throughout the brain show necrosis of neurons, perivascular cuffing, capillary thrombosis, infiltration of lymphocytes, and glia, as well as a variable degree of meningitis.

Racial origin seems to influence susceptibility to some arboviral diseases. Yellow fever, for example, is milder in West Africans than in Caucasians, probably due to natural selection for genetic resistance. A parallel situation can be demonstrated in the laboratory; the PRI strain of mice is resistant to the 17D yellow fever virus by virtue of a single gene, inheritable in cross-breeding experiments with Swiss mice as a simple Mendelian recessive.

Immunity

Type-specific acquired immunity to the arboviruses probably lasts for life, which is attributable to two facts: (a) during the original infection the virus multiplies extensively in the reticuloendothelial system itself and generates a solid immune response, and (b) at the time of reinfection, the challenge virus encounters these high concentrations of antibodies during its viremic phase. Furthermore, immunity can be boosted periodically by clinical or subclinical infection with different viruses from the same serological group. For example, repeated infection with one or more flaviviruses (e.g., dengue) broadens the patient's spectrum of neutralizing antibodies to embrace other flaviviruses (e.g., yellow fever); this may explain in part why yellow fever is not common in endemic areas of Africa amongst the indigenous people in whom antibodies against various other flaviviruses are almost universal.

This is an appropriate point to discuss Halstead's theory about the etiology of the dengue shock syndrome. The four serotypes of dengue virus have cross-reacting antigenic determinants, but the cross-reactivity does not provide lasting protection against heterologous types and sequential infections with different types do occur. Although occasional cases of hemorrhagic fever had been described in earlier outbreaks of dengue, a remarkable change in their incidence was noticed in the Philippines in 1954, and since then in many other countries of South and Southeast Asia. Severe and often fatal febrile disease characterized by hemorrhage and shock has become a relatively common form of dengue especially among the local inhabitants of highly endemic areas. Epidemiological and serological observations on dengue shock syndrome suggest that this "new" disease may usually be associated with a second dengue infection in young children occurring about 6 months to 2 years after an earlier infection with a different serotype. Halstead has postulated that the shock syndrome represents some sort of hypersensitivity response, e.g., the accelerated antibody response during the second infection may lead to the formation of virus-antibody complexes with consequent fixation of complement leading to the formation of the potent permeability factor C3a. The relatively recent and sudden appearance of the dengue shock syndrome in SE Asia correlates well with the increased opportunity for infection of children with heterologous serotypes of dengue virus, which is a natural consequence of the increasing movement of people between villages and towns due to better transport facilities, hence increasing dissemination of new serotypes. If this is all true it may be anticipated that hemorrhagic dengue will shortly appear in other areas where it is at present unknown, such as coastal New Guinea. However we must stress that Halstead's theory is still unproven, and the possibility remains that the shock syndrome in children, as well as some cases of hemorrhagic fever in children and adults, are simply an expression of the appearance of unusually virulent strains of dengue virus.

Laboratory Diagnosis

Arboviruses are so difficult to isolate from man that the task can only be attempted under optimal conditions. Virtually the only satisfactory specimen is blood taken during the first 2–4 days of the illness. More commonly, virus is only recovered from brain or other appropriate organs removed at autopsy. The infant mouse (inoculated intracerebrally at 1–3 days of age) is the most sensitive host, but cell cultures should also be employed, one blind passage being desirable if the results are negative.

Primary cultures of green monkey or hamster kidney, and of chick embryo fibroblasts are susceptible to most arboviruses, and human amnion to many. Most convenient, however, are continuous cell lines derived from monkey kidney (e.g., Vero or LLC-MK2), pig kidney (PK), or hamster kidney (BHK-21). Cytopathic effects are frequently incomplete and rather unsatisfactory. Granular, cytoplasmic inclusions may be seen; hemadsorption is sometimes used.

The isolated virus may first be allocated to a group by HI or CF, then typed definitively by neutralization tests against known antisera. Minor differences between strains isolated in different geographical localities can be discerned by means of "kinetic HI," a test which measures the rate at which avid virus-antibody interaction occurs.

Usually, attempts at viral isolation are unsuccessful and the patient's illness is diagnosed serologically. A rising titer of specific antibodies is detectable by HI, CF, or neutralization tests.

Serum surveys have been widely used to determine the frequency with which particular arboviruses infect man in various regions of the world. While undoubtedly informative, such surveys must be interpreted with caution for two reasons: firstly, evidence of infection years ago can only be reliably obtained using the sensitive neutralization test—CF tests are worthless for this purpose, and even HI titers tend to fall off with time. Secondly, the dominance of antibodies to a particular togavirus may mask the presence of antibodies to several closely related agents.

Epidemiology

Arboviruses can only survive in a complex ecosystem of interdependent factors, including a suitable blood-sucking *vector* (invertebrate host) and an effective *reservoir* (vertebrate host). Infected vertebrates usually recover rapidly, eliminate the virus, and develop a lasting immunity to reinfection. Arthropods, on the other hand, may carry the virus for life, but they rarely live for more than a few months. Accordingly, the virus survives by virtue of alternating between vertebrate and invertebrate hosts in a "cycle" of the type depicted in Fig. 10-1.

Man can be infected "tangentially," when he intrudes on the natural ecological cycle and is bitten by a blood-sucking arthropod. The consequences of infection may be considerably more serious for him than for the natural vertebrate host that has presumably benefited from thousands of years of natural selection for genetic resistance. Infection of man usually represents a blind alley, terminating the chain of transmission.

It is not difficult to understand how arboviruses can persist indefinitely in tropical countries, where mosquitoes and ticks coexist with forest mammals and birds and can breed throughout the year. Under these conditions the virus is endemic and requires only a continuing supply of juvenile nonimmune vertebrates and a sufficient population density of arthropods for its continued survival.

The survival of arboviruses in temperate climates presents more of a problem. Transmission of the mosquito-borne arboviruses during the summer months follows the same pattern as in the tropics, but the mechanism of "overwintering" is not completely understood. The question is discussed in Chapter 10; here it suffices to recall that, although the virus may survive for long periods in occasional long-lived mosquitoes, this is very rare. Transovarial transmission from one generation of mosquito to the next has been demonstrated for several of the bunyaviruses, but not consistently for togaviruses. An alternative overwintering mechanism in the cool temperate zones is probably the infection of hibernating mammals (e.g., bats) and reptiles, including snakes. It has been shown that viremia, which disappears during hibernation, may recur when hibernation ceases several months later. Moreover, some togaviruses are transmitted congenitally in snakes.

Tick-borne arboviruses have fewer problems. Ticks are present throughout the year even in temperate climates and often live through more than a single breeding cycle of their host. Immature ticks usually parasitize birds or small mammals, e.g., rodents, hence the epidemics of tick-borne encephalitis that accompany plagues of voles or mice in Eastern Europe. Mature ticks often prefer larger animals, such as domestic cattle or goats, or the deer and antelopes abounding in nature reserves, hence the concentration of Kyasanur Forest disease along cattle tracks through the Mysore State forests. Domestic animals represent extremely important *amplifier hosts* in the maintenance, and particularly the spread to man, of many arboviruses. It is not essential for tick-borne viruses to multiply in such amplifier hosts because they can be passed from one developmental stage (instar) to another (transstadial transmission) and also from one generation of tick to the next (transovarial transmission). Arboviruses are also transmitted transovarially in sandflies.

The pattern of arboviral disease in man will tend to reflect the basic cycle into which he intrudes. For instance, many come to our notice as "occupational diseases" of armies or work forces encountering an ecological situation they have not met before. Lacking the natural or acquired immunity of the indigenous population, the invaders are completely susceptible. On the other hand, man may so change his natural environment that the arboviruses come to him. Perhaps the best examples involve the introduction of irrigation to arid regions like southern California; the new conditions attract both waterbirds and mosquitoes, which comprise the vertebrate and invertebrate hosts of a number of arboviruses.

To illustrate some of the principles of arbovirus ecology, let us look briefly at the life history of viruses responsible for each of the clinical syndromes discussed above.

Dengue and Urban Yellow Fever. These two most important arboviral diseases have the simplest cycles. As far as is known, man is the main vertebrate reservoir in both cases; in this respect these two togaviruses, together perhaps with o'nyong-nyong (which affected two million East Africans during an epidemic in 1959–60) are most atypical because they are not true zoonoses, at least in an urban setting. The urban species of mosquito, *Aedes aegypti,* comprises the invertebrate vector (Fig. 10-1B). Dengue is endemic in cities of Southeast Asia large enough to ensure a continuing supply of new susceptible humans. Urban yellow fever used to be endemic in many cities of Africa and South America, but has now been eradicated from all but the smaller settlements. Arboviruses cannot survive for long in an urban environment if man is the only reservoir and if modern methods are employed to control mosquitoes.

Jungle Yellow Fever. Jungle yellow fever, on the other hand (and probably rural dengue also), is maintained in cycles involving mosquitoes and monkeys (Fig. 10-1A; Plate 19-4). Close and frequent contact between the two is established in the forest canopy. Since mosquitoes rarely travel far, man is bitten only when he trespasses onto

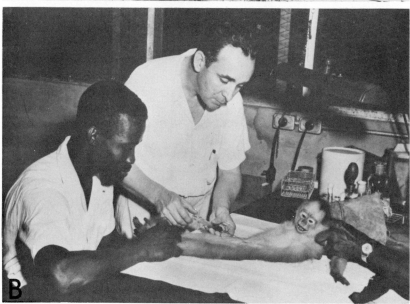

PLATE 19-4. *Yellow fever research. (A) Serological survey being conducted among Indians of the Amazon forest. (B) Cebus monkey being bled in Trinidad. (Courtesy of WHO and Cambridge University Press. From F. M. Burnet and D. O. White, "Natural History of Infectious Disease." Cambridge Univ. Press, London and New York, 1972.)*

monkey territory in the course of building a canal, road, or airstrip, clearing a forest, or pursuing animal or human adversaries.

Mosquito-Borne Encephalitides. Extensive epidemics of encephalitis broke out without warning in the valley of the Murray River system of Australia during the summers of 1918, 1951, and 1974. All three years were preceded by exceptionally heavy flooding which encouraged the build-up of large populations of mosquitoes and waterbirds in the area. The virus of Australian encephalitis is enzootic in the relatively unpopulated tropical north of Australia and New Guinea where the principal vector, *Culex annulirostris,* is plentiful throughout the year, and it makes its presence felt in the temperate southern region only following unusually wet seasons.

As can be seen from Table 19-6 many of the other encephalitis flaviviruses have a similar ecology. Japanese encephalitis virus, for example, which is responsible for regular late summer epidemics in Japan, China, and Korea, is also enzootic in a mosquito–heron cycle, with the domestic pig as an important amplifier host. Japanese, Australian, St. Louis and West Nile encephalitis viruses are so similar antigenically as well as ecologically that they may well represent variants of the same ancestral virus which has, over the course of millenia, become adapted to different host-vector associations on different continents.

Tick-Borne Hemorrhagic Fevers. Kyasanur Forest disease may be taken as the best understood example of this group. The natural vertebrate hosts include certain species of Indian monkey, but forest rodents may also be implicated, and cattle ensure a build-up of infected ticks along the forest paths. In one sense the tick itself can be regarded as a true reservoir, because ticks readily pass on arboviruses to their offspring via the transovarial route. The ecology of the Russian tick-borne hemorrhagic fevers and encephalitides is basically similar (Fig. 10-1C). Rodents are thought to be the principal natural hosts; agricultural and forest workers are the usual victims.

One must appreciate that the majority of human arboviruses have been discovered by a most circuitous route. Being unusually difficult to isolate directly from man, they are normally recovered from quite a different source. Arthropods are trapped, then macerated, and inoculated into infant mice or cell cultures. Alternatively, "sentinel" mice or chickens are left overnight in insect-infested areas as a short-cut to the same end. By throwing the net so widely, arbovirologists harvest a diverse collection of agents with but two characteristics in common—their ability to grow in arthropods and in mice (or chicks or cell culture). Only a proportion of these viruses turn out to have relevance to humans. Hence, the next step is to use the newly isolated virus (ideally having been passaged only by direct inoculation into mosquitoes) as antigen in serological tests, in an attempt to identify the corresponding antibody in sera from man, birds, or mammals collected from the same area. This is the sequence of logical steps whereby the ecology of most arboviruses has been deduced. However, still further investigations are required before the new arbovirus can be allocated any role in human disease. Characteristically, well over 90% of all human infections with a given arbovirus are subclinical. Correspondingly few of those people randomly found to have antibody will recall any meaningful episode of illness. Therefore, one must await an epidemic, sizeable enough to attract the attention of virologists, before being able to pin the disease on the virus by demonstrating rising titers of antibody in a number of patients with the same illness. With any luck, viral isolations will also be made during such outbreaks from the vector, reservoir or even some of the human cases.

It is interesting to speculate about the evolution of arboviruses. Some virologists believe that most viruses may have originated in insects, which have the unique ability to transmit viruses to and from mammals, birds, and reptiles, on the one hand, and plants on the other. The capacity of viruses to cross kingdom barriers and multiply in such very different cells was the first evidence for the universality of the genetic code. Eventually, it is conceivable that some arboviruses may lose their dependence upon the arthropod. For example some of the flaviviruses that are naturally tick-borne may sometimes be transmitted via milk (Fig. 10-1,C). Perhaps the bat salivary gland virus and rubella, both of which are togaviruses by all criteria but are not arthropod-borne, may represent more extreme instances of adaptation to respiratory spread.

Prevention and Control

Control of arbovirus infection can be considered under four headings.

Avoidance of Exposure. Nonimmune individuals should take pains to avoid mosquito- or tick-infested areas. Protective clothing and repellants should be used while awake; mosquito nets while asleep.

Eradication of Vector (or reduction below the "critical population density"). The favorite breeding places of mosquitoes should be destroyed. Stagnant water tends to collect in domestic water storage jars, old tires and tin cans which accumulate in the vicinity of human settlement. Pools and surroundings should be sprayed with insecticides such as malathion or fenitrothion. Ultra-low volume (short-acting) spraying of organophosphorous insecticides (e.g., "Abate") from low-flying aircraft is used to stem outbreaks of mosquito-borne disease in larger towns and cities (e.g., during the 1964 epidemic of St. Louis encephalitis in Houston). Unfortunately however most urban mosquitoes are now resistant to the commonly used insecticides, such as D.D.T.

Careful prospective studies should be undertaken of proposed irrigation projects or new settlements in tropical forests to ensure that the development is not opening up a Pandora's box of medical problems by thrusting nonimmune people into a dangerous new environment.

Control of Natural or Amplifier Hosts. By and large it is not practicable to attempt to eliminate the natural hosts of arboviruses but some steps can be taken to reduce their population in rural townships. For example, rodent control could have a significant impact on the tick-borne fevers of Eastern Europe.

Immunization. Years before the sophisticated technology of modern virology had been developed, Theiler produced, on a purely empirical basis, a first-rate vaccine against yellow fever. This triumph of practical common sense was rewarded with the Nobel Prize. The vaccine is a living, avirulent variant of the Asibi strain of yellow fever virus, attenuated by successive passage through the mouse, mouse embryo tissue culture, chick embryo culture, tissue culture of chicks from which the brain and spinal cord had been removed, and, finally, embryonated eggs. The end product was the 17D strain of yellow fever virus that comprises the present-day vaccine.

Like all live vaccines it must be stored cold. It is administered subcutaneously, after reconstitution from the lyophilized state. Side effects are minimal; encephalitis in children is a very rare complication with modern egg-grown vaccines; allergy to eggs is a contraindication. Protection is long-lasting, but boosters every 10 years are required by law for those entering endemic areas.

Experimental vaccines have since been developed against several other arboviruses but, in general, the pharmaceutical companies see insufficient financial incentive to pursue the matter to its logical conclusion. The market for most such vaccines would be confined to a restricted population of indigenous people and visitors to a particular locality in which the virus in question is endemic. Some degree of cross-immunity might be expected however between closely related viruses enzootic in different parts of the world, e.g., the flavivirus encephalitides. A formalin-inactivated, mouse brain-derived Japanese encephalitis vaccine is routinely administered to Japanese children and has dramatically reduced the incidence of that disease in Japan. There is a good case also for routine immunization of the principal amplifier host, the domestic pig, and a live attenuated vaccine is currently being used for this purpose. Live vaccines available against WEE and VEE have not been licensed for administration to humans in the U.S. but are widely employed for the immunization of horses.

The desirability or otherwise of a vaccine against dengue depends critically on acquiring an unequivocal answer on the pathogenesis of the dengue shock syndrome. If indeed the syndrome is a hypersensitivity response to second infection, then widespread active immunization against dengue could turn out to be a disaster. If on the other hand the recent appearance of the shock syndrome and hemorrhagic dengue simply reflects the emergence of a more virulent strain of the virus, then immunization becomes a top priority.

FURTHER READING

Chamberlain, R. W. (1968). Arboviruses, the arthropod-borne animal viruses. *Curr. Top. Microbiol. Immunol.* **42**, 38.

Doherty, R. L. (1974). Arthropod-borne viruses in Australia and their relation to infection and disease. *Progr. Med. Virol.* **17**, 136.

Fenner, F., McAuslan, B. R., Mims, C. A., Sambrook, J. F., and White, D. O. (1974). "The Biology of Animal Viruses," 2nd ed. Academic Press, New York.

Gibbs, A. J., ed. (1973). "Viruses and Invertebrates." North-Holland Publ., Amsterdam.

Halstead, S. B. (1970). Observations related to pathogenesis of dengue hemorrhagic fever. VI. Hypotheses and discussion. *Yale J. Biol. Med.* **42**, 350.

Horzinek, M. C. (1973). The structure of togaviruses. *Progr. Med. Virol.* **16**, 109.

Johnson, K. M., Halstead, S. B., and Cohen, S. N. (1967). Hemorrhagic fevers of Southeast Asia and South America: A comparative appraisal. *Progr. Med. Virol.* **9**, 105.

McLean, D. M. (1968). Arboviruses. *In* "Textbook of Virology" (A. J. Rhodes and C. E. van Rooyen, eds.), 5th ed., p. 633. Williams & Wilkins, Baltimore, Maryland.

Pfefferkorn, E. R., and Shapiro, D. (1974). Reproduction of togaviruses. *In* "Comprehensive Virology" (H. Fraenkel-Conrat and R. R. Wagner, eds.), Vol. 2, p. 171. Plenum, New York.

Reeves, W. C. (1974). Overwintering of arboviruses. *Progr. Med. Virol.* **17**, 193.

Simpson, D. I. H. (1972). Arbovirus diseases. *Brit. Med. Bull.* **28**, 10.

Supplement to Catalogue of Arthropod-Borne Viruses of the World. (1970). *Amer. J. Trop. Med. Hyg.* **19**, 1082.

Taylor, R. M., ed. (1967). "Catalogue of Arthropod-Borne Viruses of the World." U.S. Dept. of Health, Education and Welfare, Washington, D.C.

Theiler, M., and Downs, W. G. (1973). "The Arthropod-Borne Viruses of Vertebrates." Yale Univ. Press, New Haven, Connecticut.

World Health Organization. (1967). Arboviruses and human disease. *World Health Organ., Tech. Rep. Ser.* **369**.

World Health Organization. (1971). WHO Expert Committee on Yellow Fever, Third Report. *World Health Organ., Tech. Rep. Ser.* **479**.

CHAPTER 20

Orthomyxoviridae

INTRODUCTION

Few viruses have played a more central role in the historical development of virology than that of influenza. The pandemic that swept the world in 1918, just as the Great War ended, killed more people than the war itself. The eventual isolation of the virus in ferrets by Smith, Andrewes, and Laidlaw in 1933 was a milestone in the development of virology as a laboratory science. During the ensuing two decades Burnet pioneered technological and conceptual approaches to the study of the virus in embryonated eggs. His system became the accepted laboratory model for the investigation of viral multiplication and genetic interactions until the early 1950's, when newly discovered cell culture techniques transferred the advantage to poliovirus, but embryonated eggs are still used for the production of influenza vaccines. Hemagglutination, discovered accidentally by Hirst when he tore a blood vessel while harvesting influenza-infected chick allantoic fluid, provided a simple assay method, subsequently extended to many other viruses. Indeed, it was the affinity of the influenza viral hemagglutinin and neuraminidase for mucins that gave birth to the original group name "myxovirus." The group has now been accorded the status of a family, the Orthomyxoviridae, embracing all the influenza viruses (Table 20-1).

PROPERTIES OF THE VIRUS

The "typical" virion is spherical and about 100 nm in diameter, but larger, more pleomorphic forms are commonly seen (Plate 20-1). The latter are often totally or partially deficient in RNA, hence are noninfectious ("incomplete virus"). Filaments up to several microns in length, clearly visible by darkfield microscopy, are characteristic of

TABLE 20-1
Properties of Orthomyxoviruses

Filamentous or spherical virion, diameter 100 nm
Nucleocapsid with helical symmetry
Envelope, containing hemagglutinin and neuraminidase
RNA, single-stranded, 7 molecules, total mol. wt. 5 million
Transcriptase in nucleocapsid

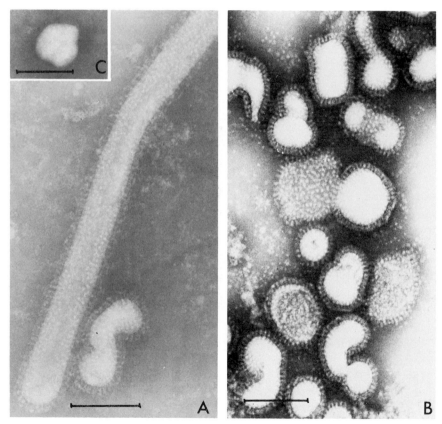

PLATE 20-1. *Orthomyxoviridae. Negatively stained preparations of type A influenza virus (bars = 100 nm). (A) Filamentous form, characteristic of strains recently isolated from man. (B) Pleomorphic "spherical" particles with well defined peplomers. (C) Particle lacking peplomers, obtained by bromelain treatment. [A from P. W. Choppin, J. S. Murphy, and W. Stoeckenius, Virology* **13,** *548 (1961); B by courtesy of Dr. W. G. Laver; C from R. W. Compans, N. J. Dimmock, and H. Meier-Ewert, in "The Biology of Large RNA Viruses" (R. D. Barry and B. W. J. Mahy, eds.), p. 87. Academic Press, New York, 1970).]*

some strains upon first isolation. The viral envelope, derived by budding from plasma membrane, contains lipids characteristic of the cell in which the virus was grown. Host proteins, on the other hand, are lacking, having been displaced by virus-coded peplomers. These are of two distinct types, closely packed into a regular mosaic. One is the hemagglutinin, a rod-shaped glycoprotein polymer; the other, the neuraminidase, is a mushroom-shaped glycoprotein polymer quite distinct from the corresponding enzyme found in some normal cells (Fig. 20-1).

The hemagglutinin enables the virus to adsorb to glycoprotein receptors present on the surface of erythrocytes of many mammalian and avian species. The association is relatively stable at 4°, but at 37° the virus rapidly elutes as a result of destruction of the receptors by the viral neuraminidase.

The seven influenza genes occur as separate RNA molecules in association with nucleoprotein and transcriptase (Figure 20-1).

Like all enveloped viruses of helical symmetry influenza is rather susceptible to inactivation by drying, freezing-thawing and, of course, lipid solvents.

The influenza viruses are subdivided into three "types": A, B, and C. The types are further subdivided into a number of "subtypes" (A0, A1, A2), and the subtypes into "strains." The three types have no shared antigens. All strains of influenza within each type share common internal proteins (the nucleoprotein, NP, and the "membrane" protein, M; Fig. 20-1) but differ in their surface proteins, the hemagglutinin (HA or H) and the neuraminidase (NA or N). Thus, the major pandemic subtype of influenza A which arose in Hong Kong in 1968 is designated A/Hong Kong/1/68 (H3N2) to indicate the type, A; the location, Hong Kong; the isolate (first) and the year, 1968. The bracketed letters and numbers indicate that its hemagglutinin, but not its neuraminidase antigen, was distinctly different from that of the earlier subtype isolated in Singapore at the beginning of the 1957 pandemic, A/Singapore/1/57 (H2N2). Such major "discontinuous" changes in H or N antigens (H2 → H3) are known as "antigenic shift," whereas minor changes detectable from year to year, all within either H2 or H3, for example, are referred to as "antigenic drift." The reader is advised to return to Chapter 11 for a detailed discussion of these important phenomena. Suffice to say here that a mutation affecting a single amino acid in the key antigenic determinant of the hemagglutinin glycoprotein is sufficient to bring about significant antigenic drift, whereas antigenic shift results from the replacement of the gene for HA with one coding for a totally different amino acid sequence. By and large, changes in the HA are of greater epidemiological signifi-

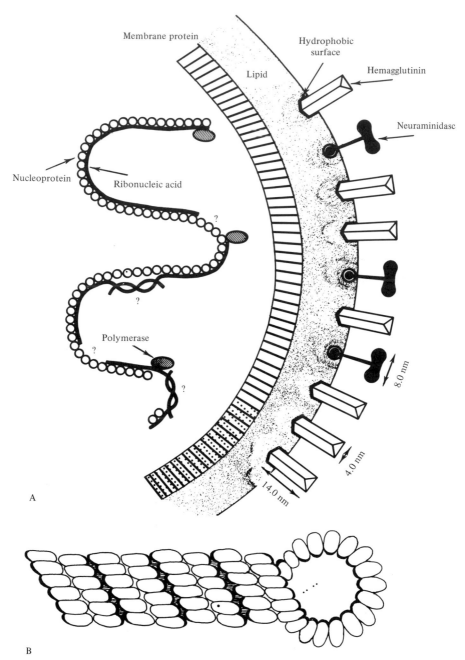

A

B

FIG. 20-1. *Orthomyxoviridae. Schematic diagrams illustrating the structure of (A)
the influenza virion [from W. G. Laver, Advan. Virus Res.* **18,** *57 (1973)], and (B)
one of the pieces of ribonucleoprotein [from R. W. Compans, J. Content, and P. H.
Duesberg, J. Virol.* **10,** *795 (1972)].*

cance than changes in the NA, because antibodies directed against the HA are considerably more effective in neutralizing the infectivity of the virion.

All three types of influenza virus infect man; type *A*, but not B and C, has a wide host range and commonly infects dozens of species of birds, including fowls (in which the disease is highly lethal and is known as "fowl plague"), ducks, and turkeys, as well as horses, pigs, and other animals. Animal strains are capable of multiplying in heterologous species, and genetic reassortment (see Chapter 4) between various pairs of animal and bird influenza strains has been demonstrated to occur under simulated "natural conditions." Antibodies to the H and N antigens characteristic of all of the recent human strains have been found in wild birds (Plate 20-2). All this evidence serves to reinforce the credibility of the theory that pandemic subtypes of influenza A arise from viruses indigenous to birds or animals, perhaps directly, or perhaps by genetic reassortment (see Chapter 11).

Viral Multiplication

The multiplication cycle of influenza virus presents a number of unique features that merit consideration. Following up his observa-

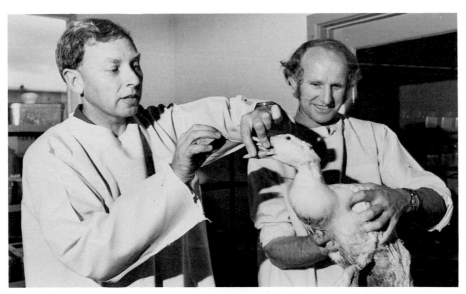

PLATE 20-2. *Drs. Laver and Webster from the Australian National University swabbing the throat of an avian influenza patient.*

tions on hemagglutination over 30 years ago, Hirst was able to demonstrate that the glycoprotein receptors for the viral hemagglutinin were present not only on erythrocytes, but also on epithelial cells of the mouse, chicken, or human respiratory tract. Clearly, this is the mechanism of attachment of virion to host cell. Penetration is thought to be accomplished by fusion of the lipid membrane of the virus with that of the cell, resulting in release of nucleocapsid directly into the cytoplasm; phagocytosis doubtless also occurs (see Plate 3-1).

The transcriptase associated with the ribonucleoprotein then transcribes mRNA from each of the 7 separate RNA molecules comprising the influenza genome. An unusual feature of the multiplication of orthomyxoviruses is their apparent requirement for the synthesis of some species of cellular RNA of unknown function early in the cycle. The intracellular migration of newly synthesized influenza proteins, their incorporation into membranes, glycosylation, and budding have been the subject of close study. The viral neuraminidase is not involved in the budding process itself, but may expedite the elution of released virions from the outer surface of the cell as well as digesting glycoprotein inhibitors present in the mucus lining the respiratory tract.

As a consequence of its segmented genome, influenza virus readily undergoes genetic reassortment (see Chapter 4) if it finds itself inside a cell simultaneously infected by another strain. For similar reasons, synchronous infection of a cell with several particles of a single strain of influenza virus leads to the formation of "incomplete virus" (defective virions) which are noninfectious because they lack the largest of the 7 RNA molecules.

HUMAN INFECTIONS

Clinical Features

"Flu" is a term that is cast around rather loosely by the layman, and even, one is ashamed to relate, by some of his medical brethren. One hears it applied more or less at random to any or all of that indefinable miscellany of upper respiratory tract infections for which we already have perfectly satisfactory siglas like URTI and ARD! In fact, influenza is a distinct clinical entity, with quite marked systemic manifestations including fever, chills, headache, generalized muscular aches, and sometimes prostration. A sore throat is sometimes, but not always, followed by hoarseness, cough, nasal obstruction, and discharge. Although the disease is rarely serious in a healthy individual, infants quite commonly develop croup, and death may rarely occur at any age from viral pneumonia. The greater danger, however, is from bacterial superinfection in the very young, the very old, or in those debilitated by chronic pulmonary or cardiac disease. Winter epidemics are often associated with a substantial number of "excess deaths"—a statistical concept referring to the connection between the epidemic and an increase in "nonspecific" mortality in the community. Though not always confirmed by viral isolation, influenza is circumstantially incriminated as "the straw that breaks the camel's back." During the great 1957 pandemic the mortality was largely attributable to staphylo-coccal and pneumococcal pneumonia, particularly among the elderly. There is no certain explanation of the much higher mortality recorded in the 1918 pandemic, in which millions of healthy young adults suc-cumbed. Superinfection by *Hemophilus influenzae* was so common that this was erroneously thought at the time to be the causal agent of influenza itself.

Pathogenesis

Upon entering the respiratory tract influenza virus may encounter immediate hazards in the form of specific IgA produced in response to earlier infections, interferon and a glycoprotein inhibitor present in the mucus coating the ciliated epithelium. If these are surmounted, widespread inflammation of the upper respiratory tract is produced within 1–2 days. Involvement of the lower respiratory tract is less common, but death from influenzal pneumonia occasionally occurs; at autopsy one sees hemorrhagic necrosis and desquamation of the epithelium of the alveolae, bronchioles, and trachea, together with widespread interstitial edema and inflammatory cell infiltration. Influ-enza also lowers the resistance of the respiratory tract to secondary bacterial infection by pneumococci, staphylococci, or group A strep-tococci. The detailed pathogenesis of influenza is considered as a model for the respiratory viruses in Chapter 6.

Immunity

Acquired immunity to a given strain is adequate for a year or two, depending on the titer of neutralizing IgA in the respiratory mucus. By definition this specific immunity is insufficient to prevent reinfec-tion with new mutants emerging as a result of antigenic drift. The nature of the antibody response to this second infection is dictated by the individual's previous experience. The mutant usually shows enough antigenic overlap with earlier strains to elicit an anamnestic antibody response to them. In consequence, repeated infections with successive strains of influenza virus continually reinforce the antibodies to the antigenic determinants of earlier related strains. Inevitably, therefore, the dominant antibodies in any individual's serum are those directed against the strain he first experienced. The phenomenon has been called, rather irreverently, the "doctrine of original antigenic sin" (see Chapter 7). Of course, this does not apply to antigenic shift, when the new strain is totally unrelated to the old. However, it does have two important repercussions. First, immunization with a contemporary vaccine may, if the dosage is inadequate, elicit antibodies to an older strain no longer relevant. Second, serological surveys can yield deceptive data about individuals' presumed experience of various influenza strains.

Such serological surveys have unearthed one intriguing finding which has been taken by some to indicate that there is a limited number of possible HA configurations and that they reappear in a cyclical fashion every few human generations. Examination of stored sera taken from people *before* the advent of "Asian" influenza in 1957 or "Hong Kong" influenza in 1968 revealed that antibodies to these strains were prevalent in individuals who were alive more than 65 years before the pandemic in question. The simplest, but not the only, explanation of the data is that the great pandemic of 1889 was caused by a strain of influenza similar to or identical with that responsible for the 1957 pandemic some 68 years later and that the less devastating pandemic that occurred around the turn of the century was caused by the Hong Kong (1968) strain.

Laboratory Diagnosis

A nasopharyngeal aspirate or throat swab is the most convenient specimen. Material should be inoculated into the amniotic cavity of 10-day-old chick embryos and also into primary cultures of monkey (or human) kidney cells; continuous cell lines do not support the growth of wild influenza strains. After 3 days at 33°–35° amniotic and allantoic fluid is tested for hemagglutination with guinea pig or human erythrocytes. After 3–7 days at 33°–35° the monkey kidney cultures should show hemadsorption at 4°, even though CPE is minimal. The isolate is then identified by hemagglutination inhibition (HI) (Plate 14-4). Definitive description of a novel strain demands additional testing by neuraminidase-inhibition (NI) and subtype-specific complement fixation (CF) tests using standard antisera. Radial immunodiffusion is being used increasingly (see Chapter 14).

Epidemiology

Epidemics of influenza A occur about every second or third winter (Fig. 10-3), those of influenza B somewhat more sporadically. Antigenically novel strains of influenza A are disseminated through the community with such speed that an epidemic may run its course in a major city within two or three months. Up to 70% of the population may become infected before the epidemic burns itself out. Such speed and efficiency of spread may be attributed to the susceptibility of the population, the extremely short incubation period (1–2 days), and the large doses of virus shed in droplets discharged by sneezing and coughing. The situation is not helped by the fact that most of the victims are neither sick enough nor thoughtful enough to remove themselves from circulation—and up to half the infections may be subclinical anyway. As always, the school is the marketplace for "trading parasites," and the virus is taken home to the family. Minor outbreaks occur almost every winter, when new young susceptibles, as well as some of the people who escaped infection the previous year, encounter the same strain of virus, which has been lying dormant in the community during the summer months.

Every generation or so, a major antigenic shift in the virus of influenza A precipitates an epidemic of world-wide proportions. Such "pandemics" were recorded in 1889, 1918, 1957, and 1968. The 1918 holocaust killed more than 20 million people. That of 1957 was almost as widespread but far less lethal. The mechanisms involved in antigenic drift, and in the more major discontinuous changes known as antigenic shift, were discussed at length in Chapter 11 in the context of the evolution and survival of viruses in nature.

Prevention

Needless to say, the critical requirement for any influenza vaccine must be that it incorporates the current strains of influenza A and B—at the time of writing A/England/42/72 (H3N2) and B/Hong Kong/5/72. World Health Organization Influenza Surveillance Centers in over 50 countries maintain an "early warning system," which enables new mutants to be detected almost as soon as they arise, so that once a radically new strain has been identified, the virus may be distributed without delay to vaccine manufacturers around the world. Only by means of such "tailor-made" vaccines can there be any hope of nipping a pandemic in the bud, since, *ipso facto*, the novel strain is unaffected by antibodies to all previous strains.

The inactivated vaccines in current use are manufactured as follows. The virus is grown in the chick embryo allantois, and purified from allantoic fluid by continuous flow zonal ultracentrifugation. Its infectivity is then destroyed by formaldehyde. The resulting inactivated vaccine is inoculated subcutaneously or intramuscularly. A booster is required 1–2 months later, then annually each autumn, or in the face of an oncoming epidemic. Particularly at risk are people over the age of 65 or those with chronic respiratory or cardiac ailments. There is also a case for immunizing infants, but blanket inoculation of the community at large is generally not advocated because the protection afforded by most recent vaccines has been unimpressive, varying from an efficacy of 50–75% down to zero in some studies. Results depend very much on the recipient's experience of the homologous and heterologous strains of influenza virus and on whether the prevalent agent has undergone some degree of antigenic drift since the vaccine was made.

Pyrogenic reactions to the vaccine are quite common, especially in young children—local tenderness, regional lymphadenopathy, sometimes accompanied by fever, chills, malaise, and headache. They can be minimized by (a) intramuscular inoculation, (b) zonally purified vaccines from which have been removed bacterial endotoxins and other nonvirion proteins (including egg proteins to which some people are allergic), (c) "split" vaccines in which the virions have been dissociated with a lipid solvent such as sodium deoxycholate.

Theoretically it might be supposed that the ideal vaccine would consist of purified HA ($\pm$NA) antigen. In fact such "subunit vaccines" can be made quite readily by chromatographic separation from detergent-disrupted virions but they are adequately immunogenic only when emulsified with an adjuvant. As yet no adjuvant has been licensed for use in man on a world-wide basis, but it is clear that this is a top priority, because antibody responses to inactivated viral vaccines in general are enhanced at least tenfold in their presence.

Recently a number of research groups have been reexamining the value of live intransal influenza vaccines, originally introduced experimentally by Burnet during World War II. Such vaccines have been in general use in U.S.S.R. for several years. Chanock and his colleagues at the U.S. National Institutes of Health selected a 5-fluorouracil-induced temperature-sensitive (*ts*) mutant of the H2N2 subtype which was unable to replicate above 37° in cultured calf kidney cells, and which when administered intranasally to hamsters grew poorly in their lungs but quite satisfactorily in the turbinates of the upper respiratory tract, and protected the animals against subsequent challenge with wild-type virus. The mutant was then recombined with a calf kidney-grown

A2/Hong Kong/68 (H3N2) strain to yield reassortants in which the RNA molecule carrying the *ts* lesion had been transferred to the H3N2 virus. One of the reassortants, *ts*-1 (E), proved sufficiently attenuated to make a suitable vaccine. It grows (to rather low levels) in the upper respiratory tract of human volunteers, producing no symptoms in most of the men and mild coryza in the rest, and induces moderate titers of anti-HA neutralizing antibody in both the nose and the serum. The resulting protection against challenge with wild-type virus is not all that could be desired, but an important principle has been established and it may prove possible to derive other mutants which strike the right balance between adequate attenuation and immunogenicity.

FURTHER READING

Compans, R. W., and Choppin, P. W. (1975). Reproduction of myxoviruses. *In* "Comprehensive Virology" (H. Fraenkel-Conrat and R. R. Wagner, eds.), Vol. 4, p. 179. Plenum, New York.

Davenport, F. M. (1971). Killed influenza virus vaccines: Present status, suggested use, desirable developments. *In* "International Conference on the Application of Vaccines against Viral, Rickettsial and Bacterial Diseases of Man," Sci. Publ. No. 226, p. 89. Pan Amer. Health Organ., Washington, D.C.

Dowdle, W. R., Coleman, M. T., and Gregg, M. B. (1974). Natural history of influenza type A in the United States. *Progr. Med. Virol.* **17,** 91.

Fazekas de St. Groth, S. (1969). New criteria for selection of influenza vaccine strains. *Bull. W.H.O.* **41,** 651.

Fazekas de St. Groth, S. (1970). Evolution and hierarchy of influenza viruses. *Arch. Environ. Health* **21,** 293.

Fenner, F., McAuslan, B. R., Mims, C. A., Sambrook, J. F., and White, D. O. (1974). "The Biology of Animal Viruses," 2nd ed., pp. 25, 111–117, and 243–251. Academic Press, New York.

Fox, J. P., and Kilbourne, E. D. (1973). Epidemiology of influenza — summary of National Institutes of Health Influenza Workshop IV. *J. Infec. Dis.* **128,** 361.

Jackson, G. G., and Muldoon, R. L. (1975). Viruses causing common respiratory infections in man. V. Influenza A (Asian). *J. Infec. Dis.* **131,** 308.

Kilbourne, E. D. (1973). The molecular epidemiology of influenza. *J. Infect. Dis.* **127,** 478.

Kilbourne, E. D., Chanock, R. M., Choppin, P. W., Davenport, F. M., Fox, J. P., Gregg, M. B., Jackson, G. G., and Parkman, P. D. (1974). Influenza vaccines — summary of National Institutes of Health Influenza Workshop V. *J. Infec. Dis.* **129,** 750.

Laver, W. G. (1973). The polypeptides of influenza viruses. *Advan. Virus Res.* **18,** 57.

Skehel, J. J. (1974). The origin of pandemic influenza viruses. *Symp. Soc. Gen. Microbiol.* **24,** 321.

Webster, R. G. (1972). On the origin of pandemic influenza viruses. *Curr. Top. Microbiol. Immunol.* **59,** 75.

Webster, R. G., and Laver, W. G. (1971). Antigenic variation in influenza virus. Biology and chemistry. *Progr. Med. Virol.* **13,** 271.

White, D. O. (1974). Influenza viral proteins: Identification and synthesis. *Curr. Top. Microbiol. Immunol.* **63,** 1.

World Health Organ. (1972). Influenza in animals. *Bull. W.H.O.* **47,** 439.

CHAPTER 21

Paramyxoviridae

INTRODUCTION

The paramyxoviruses derive their name from a morphological resemblance to the orthomyxoviruses, with which they were originally classified. However, it is now recognized that the two groups differ in several important characteristics, having to do particularly with the nature of the viral nucleic acid and its replication (see Table 21-1). Though few in number, all the human paramyxoviruses are important causes of respiratory infection in children. The parainfluenza and respiratory syncytial viruses are responsible for most of the croup, bronchiolitis, and pneumonitis in infants today. Measles and mumps are familiar to every mother, though the first of these is gradually receding in the face of widespread immunization.

PROPERTIES OF THE VIRUSES

Larger than the orthomyxoviruses and more pleomorphic, the paramyxoviruses range from 150–300 nm in diameter. They are enclosed by a loose lipoprotein envelope, which is extremely fragile, rendering the virion vulnerable to destruction by storage, freezing and thawing, or even preparation for electron microscopy. Accordingly, particles often appear distorted in electron micrographs and may rupture to reveal their internal nucleocapsid (Plate 21-1). Paramyxovirus RNA, unlike that of the orthomyxoviruses, occurs as a single molecule. Like the orthomyxoviruses however, the RNA is "negative" in polarity and is accompanied by a transcriptase. A hemagglutinin is to be found in the envelope of all paramyxoviruses, a neuraminidase in most, and a hemolysin in some (Table 21-2).

There is no paramyxovirus group antigen, although mumps and the five parainfluenza types are each related serologically to at least one

TABLE 21-1
Properties of Paramyxoviruses

Pleomorphic virion, diameter 150–300 nm
Nucleocapsid with helical symmetry
Envelope, containing hemagglutinin and often neuraminidase
RNA, single-stranded, one molecule, mol. wt. 7 million
Transcriptase in virion

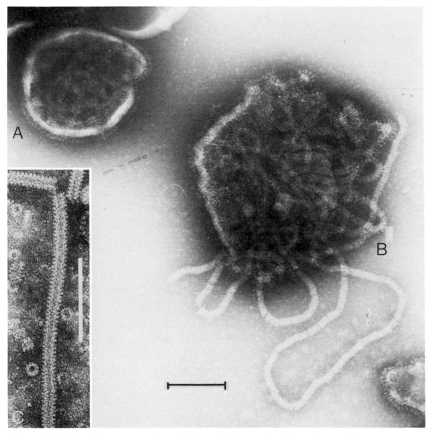

PLATE 21-1. *Paramyxovirus. Negatively stained virions of mumps virus (bars = 100 nm). (A) Intact virion; peplomers visible at lower edge. (B) Partially disrupted virion, showing nucleocapsid. (C) Enlargement of portion of nucleocapsid, in longitudinal and cross section. (Courtesy Dr. A. J. Gibbs.)*

TABLE 21-2
Biological Properties of Human Paramyxoviruses

	MEASLES	MUMPS	PARAINFLUENZA	RESPIRATORY SYNCYTIAL VIRUS
Number of serotypes	1	1	5	1[a]
Serological relationships	Distemper Rinderpest	Parainfluenza	Mumps	—
Hemagglutinin	+[b]	+	+	—
Neuraminidase	—	+	+	—
Hemolysin	+	+	+	—
Cell fusion	+	+	+	+
Hemadsorption	+	+	+	—
Inclusions[c]	N and C	C	C	C
Growth in eggs	+	+	+	—
Disease in animals	Monkey	Monkey	Monkey, mouse, cattle	Chimpanzee

[a] Several "strains" distinguishable by sensitive neutralization tests.
[b] Monkey erythrocytes only, at 37°.
[c] N = nuclear; C = cytoplasmic.

other member of the group. Measles virus shares antigenic determinants with the viruses of distemper of dogs and rinderpest of cattle. Respiratory syncytial virus (RSV) is sufficiently distinct from all the other members of the Paramyxoviridae family to suggest it should be placed in a separate genus.

All the paramyxoviruses except RSV can be grown in embryonated eggs, sometimes only after adaptation. Nevertheless, they are all more easily handled in cultured cells. Primary cultures of human or simian kidney, diploid strains of human fetal fibroblasts, and, for some serotypes, heteroploid human cell lines are all suitable. The paramyxoviruses do not cause extensive cell destruction; indeed carrier cultures readily arise (see Chapter 8). Syncytium formation is a regular feature, with acidophilic inclusions in the cytoplasm, and sometimes the nuclei, of these giant cells (Plate 21-2). Indeed ultraviolet-inactivated paramyxoviruses have been widely employed to fuse different species of cells together to form artificial hybrids (heterokaryons).

Since the genome of the paramyxoviruses is in the form of a single "negative" strand of RNA, monocistronic RNA molecules must first be transcribed by the virion's transcriptase. In most other respects their replication resembles that of other RNA viruses of negative polarity. The whole process takes place in the cytoplasm; indeed, virus will grow in enucleated cells or in the presence of actinomycin D.

The paramyxoviruses have served as valuable models for the study of the budding process, described in Chapter 3 and illustrated in Plate 3-3, as well as the study of cell fusion (see Chapter 5) and carrier cultures (Chapter 8).

MEASLES

Introduction

Measles is perhaps the best known of all the common childhood diseases. The characteristic maculopapular rash, conjunctivitis, and

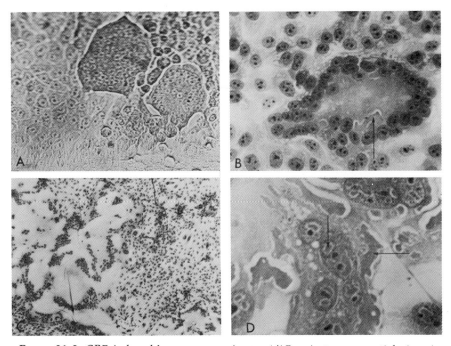

PLATE 21-2. *CPE induced by paramyxoviruses.* (A)*Respiratory syncytial virus in HEp-2 cells (unstained;* ×200). *Note two syncytia, resulting from cell fusion.* (B) *Respiratory syncytial virus in HEp-2 cells (hematoxylin and eosin stain;* ×400). *Syncytium containing many nuclei and acidophilic cytoplasmic inclusions (arrow).* (C) *Measles virus in human kidney cells (hematoxylin and eosin stain;* ×30). *Huge syncytium containing hundreds of nuclei. This monolayer was embedded in nitrocellulose before being stripped from the culture tube and stained.* (D) *Measles virus in human kidney cells (hematoxylin and eosin stain;* ×400). *Note multinucleated giant cell containing acidophilic nuclear (vertical arrow) and cytoplasmic (horizontal arrow) inclusions.* (*Courtesy I. Jack*)

TABLE 21-3
Diseases Caused by Paramyxoviruses

| | | DISEASE | |
VIRUS	AGE	COMMON	LESS COMMON
Measles	Young children	Measles (rash, coryza, conjunctivitis)	Otitis media Pneumonia Encephalitis SSPE[b]
Mumps	Children and young adults	Parotitis Orchitis Meningitis	Pancreatitis Oophoritis Encephalitis Thyroiditis
Parainfluenza 1–4	Any age Infants Infants	URTI,[a] pharyngitis Croup Bronchiolitis, pneumonitis	
Respiratory syncytial	Any age Infants	URTI Bronchiolitis, pneumonitis	
Newcastle disease[c]	Adults		Conjunctivitis

[a] URTI = upper respiratory tract infection.
[b] SSPE = subacute sclerosing panencephalitis.
[c] Rare laboratory infection contracted from fowls.

coryza (Table 21-3) are familiar to all. It is less widely appreciated that this can be a dangerous disease. Encephalomyelitis occurs often enough to make measles a greater killer than poliomyelitis was in the days before vaccination against either disease. The rare and fatal subacute sclerosing panencephalitis is a late manifestation of an earlier measles infection. The widespread use of an effective live vaccine, pioneered by Enders, has greatly reduced the incidence of measles in many parts of the western world.

Clinical Features

Inapparent infection being quite rare, measles regularly presents as the classical florid disease (Plate 21-3). After an incubation period of 9–12 days one sees a prodromal syndrome of fever together with an upper respiratory tract infection marked by coryza, cough, and conjunctivitis, all of which combine to make a thoroughly miserable looking child. At this early stage the diagnosis can be established by the detection of Koplik's spots, which are red macules or ulcers with a

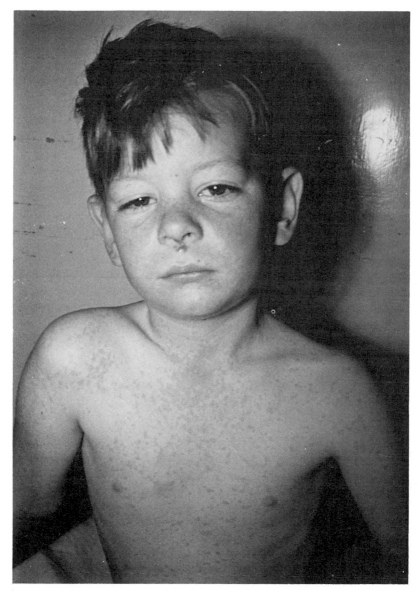

PLATE 21-3. *Measles. Note maculopapular rash and conjunctivitis in young child with upper respiratory infection. (Courtesy Dr. J. Forbes, Fairfield Hospital for Infectious Diseases, Melbourne.)*

bluish-white center, seen on the mucous membrane of the inside of the cheek. Three days later the exanthem appears on the head, then spreads progressively during the ensuing 1 or 2 days, over the chest, trunk, and then limbs. The rash consists of flat macules that fuse to form blotches rather larger than those of other viral exanthems. They are slow to fade, and often leave the skin temporarily stained. Recovery is rapid and complete.

Complications are common (10–20%) and sometimes severe. Most frequent are otitis media and pneumonia, attributable to secondary invasion by bacteria. Purely viral complications include croup and bronchitis. More rarely a fatal giant-cell pneumonia, occurring often without a rash in immunologically deficient children, can present a difficult diagnostic problem. The most feared complication of measles is encephalomyelitis, which develops in rather less than one out of every 1000 cases, inflicting a mortality of about 15%. Permanent sequelae, including epilepsy and personality changes, often follow.

Subacute sclerosing panencephalitis (SSPE) is a rare disease, developing in children or adolescents, usually males from rural areas, a few years after an earlier measles infection. The patient suffers a progressive loss of cerebral function ending in spasticity, coma, and inevitable death after 6–12 months.

Pathogenesis and Immunity

Following inhalation, the virus multiplies in the epithelial cells of the respiratory tract. The earliest spread is probably to the draining lymph nodes, then to the rest of the reticuloendothelial system, where typical Warthin–Finkeldey giant cells develop. These giant cells can be found in the liver, spleen, tonsils, adenoids, appendix, Peyer's patches, lymph nodes, and lungs (see Plate 24-1). The virus multiplies extensively in macrophages and lymphocytes. This may account for (a) the wide dissemination of virus throughout the body, (b) the leukopenia, so characteristic of the prodromal stage of the disease, (c) the chromosomal aberrations detectable in about half of all the leukocytes examined at this stage, and (d) the temporary loss of tuberculin hypersensitivity seen in patients with measles.

Some 2 weeks after the virus first enters the body, antibody becomes detectable in the serum, and a macular rash appears on the skin. The rash is probably a hypersensitivity reaction, resulting from the interaction of sensitized lymphocytes or complement-fixing antibody with viral antigen in the plasma membrane of infected cells in the skin or capillary endothelium. When measles occurs in children with T cell immunological deficiencies they frequently die of giant cell pneumonia without any skin rash at all.

Delayed hypersensitivity may be responsible for the encephalomyelitis that follows 1–2 weeks after the rash in a small proportion of measles cases. Virus cannot be isolated from the brain, yet one sees lymphocytic infiltration followed by demyelination very reminiscent of the histopathology of experimental allergic encephalomyelitis (see Plate 24-2B).

Subacute sclerosing panencephalitis (SSPE) is a much rarer and quite different condition, developing some years after the original attack of measles. The brain contains a few particles of a virus morphologically indistinguishable from measles, together with plentiful nucleocapsids, measles antigen demonstrable by immunofluorescence, and inclusion bodies inside typical giant cells (see Plate 24-2C). Plasma cells are abundant in the brain, and the CSF contains high levels of measles antibody, which is diagnostic. Infectious measles virus can be demonstrated only after biopsy specimens of SSPE brain are "cocultivated" with a measles-sensitive "indicator" cell line. Heterokaryons form, due to the effect of measles "cell-fusion factor," and measles virions then appear in considerable numbers. These SSPE strains are temperature-sensitive mutants.

Exogenous reinfection with measles is virtually unknown; acquired immunity seems to be lifelong, as it is for most of the systemic viral diseases (see Chapter 7). High levels of IgG persist in the serum for many years. There is experimental data that, when an individual's antibody titer begins to wane, it can be boosted by an inapparent infection. However, striking evidence that this is not necessary to ensure solid immunity comes from a number of illuminating epidemiological studies on isolated communities, in which some individuals were found to retain their immunity for up to 65 years in the total absence of subsequent exposure to measles (see Chapter 10).

Laboratory Diagnosis

The clinical diagnosis of measles is so straightforward that it is rarely necessary to call on the laboratory for help. However, an early diagnosis can be established before the rash appears by demonstrating multinucleated giant cells in stained smears of nasal mucus; their specificity can be established by fluorescent antibody staining.

Virus can be isolated with difficulty from the nose, throat, conjunctiva, or blood during the prodromal stage and up to 2 days after the appearance of the rash. Thereafter, it is futile to attempt virus isolation, except perhaps from the urine. Primary human fetal kidney or amnion, or monkey kidney, are the most useful cells. Cytopathic changes may take up to a week to develop. The multinucleated giant cells containing numerous acidophilic inclusions in both the cytoplasm and the nuclei are characteristic (Plate 21-2C and D). If difficulty is experienced distinguishing the CPE from that of cytomegalovirus or respiratory syncytial virus, the antigen may be identified by immunofluorescence or other serological procedures.

Epidemiology

In Western urban society measles has been a common disease of childhood, occurring endemically with fairly regular winter-spring epidemics about every second year. Spread occurs via droplets shed from the respiratory tract during the few days before and after the onset of the rash. It is so highly infectious that most children contract the disease during the first decade of life. Infection acquired at kindergarten or primary school is passed on to younger siblings at home with very high efficiency (90%).

In isolated communities measles presents a completely different picture. Classical studies have been carried out on islands known to have been out of contact with the rest of the world for many years (see Chapter 10). When accidentally introduced into such a community, measles produces 100% morbidity and high mortality in the nonimmune indigenes born since the last epidemic.

In the developing nations of West Africa and South America, measles is a highly lethal disease, sometimes killing up to 5% of children. This devastating mortality is probably due to the synergistic effects of infection and severe malnutrition, the diets being grossly deficient in proteins (see Chapter 7).

Prevention and Control

The rationale for advocating universal immunization against this superficially benign disease is that measles encephalitis occurs sufficiently frequently to make measles an important killing disease, even in advanced societies, while respiratory complications cause substantial morbidity. In the undernourished children of Africa, Asia, and South America measles is still one of the chief causes of death.

Active immunization is now widely employed. The first attenuated vaccine was developed by Enders as a result of serial passage of measles virus successively through human kidney and human amnion cell cultures, then the amnion of the developing chick embryo, and finally chick embryo tissue culture. The resulting "Edmonston" vaccine strain was insufficiently attenuated; in about a third of all recipients it gave rise to a mild attack of measles which could only be diminished by administering human immunoglobulin simultaneously into the other arm. Further attenuation of the Edmonston strain by Schwarz led to the production of a more innocuous vaccine, which even so, produces mild fever in 10–20% of recipients and a transient rash (5–12 days after immunization) in a smaller percentage. Minor electroencephalographic changes are sometimes detectable following immunization, but permanent neurological sequelae are extremely rare (<1 in 10^6). Nevertheless, there is some case for a further attenuated vaccine, such as the CAM-EX strain being tested in Japan.

The Schwarz and comparable attenuated vaccines are administered subcutaneously at about 12–18 months of age, after maternal antibody has completely disappeared. Seroconversion follows a single injection in over 95% of cases. The resulting antibody titers are somewhat lower than following natural infection, but immunity is long-lived and protection levels of over 90% are maintained for several years at least, despite declining antibody. Vaccination is contraindicated in children with immunological deficiencies, including malignancy, as well as those undergoing immunosuppressive therapy, or suffering from acute fevers or tuberculosis, or those allergic to eggs, streptomycin or neomycin.

Mass immunization has been practised in many parts of the world since the mid 1960's. The resulting decline in the occurrence of measles in the major cities of the U.S.A. during the late 1960's was quite dramatic, but there have been problems since that time in administering the *coup de grace* to the disease. Theoretically, eradication of measles on a national scale is a realizable objective, but apathy and ignorance allow the virus to persist in unimmunized children living in sizable pockets of urban poverty. Conscientious surveillance for early indications of outbreaks, followed by prompt "ring vaccination" would go a long way toward reducing the residual problem. Measles vaccine is often administered in combination with other live viral vaccines, e.g., with rubella and mumps in the U.S.A., or with smallpox and yellow fever (by jetgun) in Africa.

The introduction of active immunization has led to a marked decline in the use of immunoglobulin. Nevertheless, passive immunization is still indicated for the prophylaxis of measles following exposure of an unimmunized very young or sickly child, particularly one with a history of repeated respiratory infections. If given immediately after exposure, normal human immunoglobulin may abort the disease completely; if given a few days later, the disease may occur in a modified form.

Bacterial superinfections such as pneumonia or otitis media require vigorous chemotherapy with the antibiotics indicated by drug-sensitivity tests following isolation of the causative organism. A case can be made for the prophylactic use of antibiotics in measles in children with a history of repeated otitis media or serious respiratory infections.

Children with measles should be excluded from school for a week after the onset of the rash.

MUMPS

Introduction

Mumps is a common contagious disease of children and young adults characterized by inflammation of the salivary glands, especially the parotid. Frequent complications include orchitis in young men and meningoencephalitis. A satisfactory live vaccine is now available.

Clinical Features

The comical spectacle of the unhappy young man with face distorted by painful, edematous enlargement of both parotid glands, unable to eat or talk without discomfort, is familiar music-hall fare, often spiced with the innuendo that the case is complicated by a well-deserved orchitis. Less frequently only one parotid is inflamed. The submaxillary glands are quite commonly involved, the sublingual glands rather less frequently (Plate 21-4). Epididymoorchitis is a painful development in 20–25% of all cases in post-pubertal males. It may lead to atrophy of the affected testicle, though rarely to total sterility, as it is usually unilateral. Aseptic meningitis, or more rarely meningoencephalitis occurs often enough in all age groups to make mumps the most common single cause of meningitis. Not infrequently, the case presents as a primary meningitis without parotid involvement. Fortunately, the prognosis is very much better than with bacterial meningitis, and sequelae are not common, unilateral deafness being the most frequent. A wide variety of other glands including the pancreas (quite commonly), ovary, thyroid, and breast are more rarely involved.

Pathogenesis and Immunity

It is not known whether the infecting virus multiplies first in the respiratory tract. However, there is good evidence for a viremia, which carries the agent to the various salivary glands, as well as the testes or ovaries, pancreas, and brain. An incubation period of 16–18 days precedes the development of symptoms referrable to these organs. Virus can be recovered from the saliva and blood for the first few days of the illness and from the urine for somewhat longer. Acquired immunity is virtually lifelong.

Laboratory Diagnosis

The classic case of mumps can be identified without help from the laboratory, but atypical cases often present a diagnostic problem. Appropriate specimens will then consist of saliva or throat swab, cerebrospinal fluid (in meningitis), or urine, mumps being one of the few viruses regularly isolated from the latter source. The material may be inoculated without any preliminary treatment directly into primary cultures of human or monkey kidney epithelium, or into the amniotic cavity of 7 to 9-day-old chick embryos. After 3–5 days the cell cultures

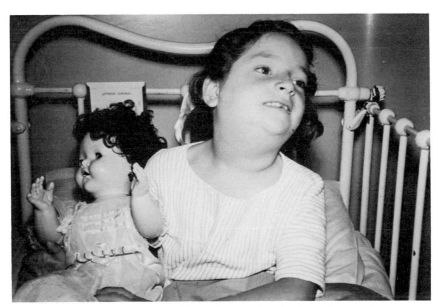

PLATE 21-4. *Mumps. Note swelling of parotid and other salivary glands. (Courtesy Dr. J. Forbes.)*

are monitored by hemadsorption, and the amniotic fluid by hemagglutination.

Epidemiology

Mumps is transmissible by direct contact with saliva, or by droplet
spread, from a few days before the onset of symptoms until about a
week after. It is not as infectious as the other paramyxoviruses and the
incubation period is longer, with the result that the disease tends to
occur rather later in childhood. About a quarter of all cases, therefore,
are seen in adolescents and young adults. For the same reasons,
mumps does not show the dramatic periodicity of the other paramyxoviruses, but tends instead to cause sporadic cases throughout all
seasons with only occasional winter-spring epidemics every few years.
About one infection in every three is inapparent.

Prevention and Control

A live attenuated mumps vaccine, derived by passage of the Jeryl–
Lynn strain in chick embryos and then in cultured chick fibroblasts,
has been in use in the United States since 1968. Over 95% of recipients
develop antibodies (to rather low titer), and over 95% acquire resistance to subsequent challenge. The vaccine is free of side-effects. Some
physicians recommend routine administration to 1-year-old children,
but many doubt whether it is worthwhile to immunize against such a
trivial disease. A case can be made for the use of mumps vaccine in
military recruits and other groups of vulnerable male adolescents or
adults particularly prone to epidemics. Children with mumps should
be excluded from school for 2 weeks altogether.

PARAINFLUENZA VIRUSES

Introduction

First isolated in the mid-1950's, the parainfluenza viruses were originally known variously as "influenza D," "hemadsorption" viruses,
and "croup-associated" viruses. These names offer us some insight
into the properties of the agents, for the parainfluenza viruses are not
altogether unlike influenza, they are recognized in cultured cells by
hemadsorption and they constitute the main cause of croup (laryngo-
tracheo-bronchitis) in young children. In older children and adults
they more commonly produce milder upper respiratory tract infections
(URTI).

Clinical Features

Primary infection in a child usually takes the form of low fever, coryza, pharyngitis, and often some degree of bronchitis. Acute laryngotracheobronchitis (croup), though less common, is the most serious manifestation of the infection. A young child presents with cough, stridor, and severe respiratory distress that may go on to respiratory obstruction requiring tracheotomy to save its life. Parainfluenza types 1, 2, and 3 are between them responsible for almost half of all cases of croup, type 1 being the most important. Equally dangerous is the bronchiolitis or bronchopneumonia thay may occur especially when type 3 infects infants. Overall the parainfluenza viruses are responsible for nearly 20% of the severe lower respiratory infections of children.

Pathogenesis and Immunity

Parainfluenza virus infections are confined to the epithelium of the respiratory tract, hence the incubation period is invariably short (2–5 days). In adults, only the mucosa of the nose and throat may be affected; in young children encountering the virus for the first time, infection often spreads to the larynx, trachea, bronchi, bronchioles, or alveoli.

Acquired immunity to the parainfluenza viruses is poor, being dependent upon continuing production of IgA in the respiratory tract, and reinfection with the same serotype is quite common. Such reinfection, occurring typically in older children and adults, is usually mild and confined to the upper respiratory tract.

Laboratory Diagnosis

Because of the lability of parainfluenza viruses, throat swabs should be kept cold but not frozen, and inoculated into cultured cells without delay. Human kidney cultures are preferable to monkey kidney, as the latter often carry a simian parainfluenza virus. If, however, only monkey kidney is available, SV5 antiserum can be incorporated in the medium used to grow the original monolayer, or else uninoculated control cultures can be monitored for hemadsorption at the end of the test. Human parainfluenza viruses multiply slowly, producing minimal CPE in the case of types 1, 3, and 4, and conspicuous syncytia in the case of type 2. Guinea pig or fowl red cells are added to the cultures after 4, 7, and 10 days (types 1 and 3 being somewhat faster growers than types 2 and 4). Unequivocal differentiation from influenza or mumps virus requires hemagglutination-inhibition tests against known antisera.

Epidemiology

Unlike influenza and respiratory syncytial viruses, the parainfluenza viruses do not cause large epidemics, but are responsible for a percentage of respiratory disease throughout the year, with a winter peak. Occasionally outbreaks may occur, notably in institutions, where type 3 in particular spreads very effectively.

Parainfluenza type 3 is somewhat more common that types 1 and 2, with types 4 and 5 being rather rare. Accordingly, most children contract a type 3 infection in the first 2 years of life, then have their immunity to this agent reinforced by subsequent infection with types 1 and 2 during the next few years. Nevertheless, the resultant immunity is not adequate to prevent the occurrence of subsequent attacks of coryza and pharyngitis during adult life.

RESPIRATORY SYNCYTIAL VIRUS

Introduction

Unknown before 1956, when first described as the "chimpanzee coryza agent," the respiratory syncytial virus (RSV) is now recognized to be the most important cause of serious lower respiratory tract infection in infants. Regular winter epidemics produce bronchiolitis and pneumonitis in babies, and harmless upper respiratory tract infection in adults. The virus derives its name from its capacity to fuse cells into a multinucleated syncytium (Plate 21-2A and B).

Clinical Features

RSV is responsible for about half of all cases of bronchiolitis and a quarter of all pneumonia occurring during the first few months of life. Characteristically, a 1- to 2-month-old baby develops signs of dyspnea and cyanosis, with or without wheezing or emphysema due to bronchiolar obstruction (Plates 21-5 and 21-6). More rarely the infant may die without warning ("cot death") presenting the coroner with a mystifying diagnostic problem. In general, the prognosis is worst in children with a familial history of allergy, and infants who contact RSV bronchiolitis have a threefold higher probability of becoming asthmatics than do other children. Bacterial superinfection is the exception, rather than the rule in RSV bronchiolitis. In older children RSV infection tends to be confined to the upper respiratory tract, as a febrile rhinitis and pharyngitis with limited involvement of the bronchi; otitis media occasionally supervenes. In adults, RSV infection is quite innocuous, consisting of an afebrile rhinitis indistinguishable from "the common cold."

Pathogenesis and Immunity

The virus multiplies in the mucous membrane of the nose, throat, and sometimes the larynx, but in infants may spread to the trachea, bronchi, bronchioles, and alveoli. Fatal cases usually show extensive bronchiolitis and pneumonitis with scattered areas of atelectasis and emphysema resulting from bronchiolar obstruction (see Plate 24-1).

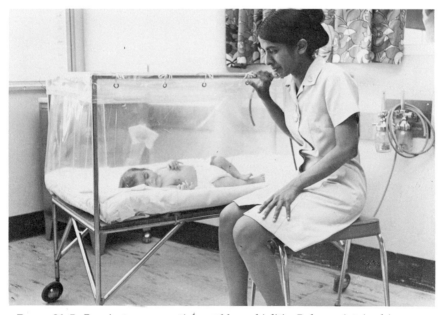

PLATE 21-5. *Respiratory syncytial viral bronchiolitis. Baby maintained in oxygen tent. (Courtesy Drs. H. Williams and P. Phelan, Royal Children's Hospital, Melbourne.)*

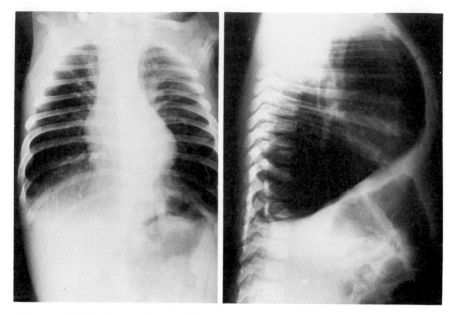

PLATE 21-6. *Radiographs of a baby with respiratory syncytial viral bronchiolitis and pneumonitis. Note grossly overinflated lungfields with depression of the diaphragm and bulging of the anterior chest wall in the lateral view. (Courtesy Drs. H. Williams and P. Phelan.)*

The short incubation period of RSV (4 days) may be attributed to the fact that infection remains localized to the respiratory mucosa.

Maternal antibody is inadequate to protect infants against RSV infection, because IgA does not cross the placenta, and this is probably the only class of immunoglobulin of relevance in protection against localized infections of the respiratory tract. Following infection, the baby develops some local IgA, but the response is generally poor. This may be attributable in part to the confinement of virus to syncytia, and in part to the relative underdevelopment of the mechanisms of immunity at that age, as well as to the localized nature of the infection.

The most challenging unanswered question is why severe lower respiratory disease occurs only in babies, virtually all of whom have anti-RSV IgG derived transplacentally from their mothers. Clearly such serum IgG is nonprotective, but it could be it is actually responsible for the bronchiolitis. Chanock has postulated that RSV bronchiolitis is a hypersensitivity reaction of the Arthus type, akin to "farmer's lung," in which virus-antibody complexes are precipitated in the lung. The vaccination accident described below gives further support to this hypothesis. The association between RSV bronchiolitis and recurrent wheezing (asthma) suggests that the particular infants who react to RSV infection may be those with a hereditary predisposition to allergy.

Laboratory Diagnosis

The best way of collecting mucus from the nasopharynx of small infants is by aspiration using a device such as that depicted in Plate 14-3. The extreme lability of RSV dictates that the aspirate be added directly to cultured cells without preliminary freezing. Some virologists even recommend that specimens be transferred straight from the patient into the culture tube at the bedside. Human cells are the most sensitive, Hep-2, HEL, and HEK being the most commonly employed. A quick answer will not be forthcoming, because up to 2 weeks may elapse before the characteristic syncytia becomes obvious; nevertheless a trained eye can usually detect early changes by about the fifth day. Fixation and staining reveals extensive syncytia containing acidophilic cytoplasmic inclusions. Absence of hemadsorption distinguishes RSV from all the other paramyxoviruses. Definitive identification can be established by immunofluorescence, using either the infected cell culture when early CPE first becomes apparent, or exfoliated cells aspirated directly from the child's nasopharynx.

Epidemiology

Newborn infants are so susceptible, and the virus so highly transmissible, that RSV causes a sharply-defined epidemic every winter (see Fig. 10-3). Accordingly, most children become infected at home during the first year or two of life. The majority of cases of serious lower respiratory infection occur in infants less than 12 months old, with the peak incidence at 1–2 months. Outbreaks occasionally occur in the neonatal wards of maternity hospitals, sometimes inflicting a high mortality.

Prevention and Control

No vaccine is available, though the need is clear. The difficulty will be to achieve sufficiently high antibody titers in the mucus bathing the respiratory tract of young children, for such a vaccine would be required for use in the first 6 months of life when the infant is not highly responsive to antigenic stimulation. In a clinical trial of a highly concentrated, inactivated vaccine, given by parenteral inoculation, immunized children actually suffered more serious lower respiratory tract disease during a subsequent RSV epidemic than did controls. This paradox led to the hasty and permanent withdrawal of the vaccine and alerted virologists to the potential dangers of parenteral inoculation of inactivated vaccines against respiratory viruses in general.

Turning toward the alternative of a live vaccine Chanock and his colleagues reasoned that *ts* mutants might be safe because of the temperature differential between the upper and lower respiratory tract. A *ts* mutant was isolated which was unable to grow at 37° in cultured cells but multiplied asymptomatically in the upper respiratory tract of adults and protected them against subsequent challenge with virulent RSV. Though the immunogenicity of this particular mutant was minimal the study established the feasibility of *ts* mutants as human respiratory vaccines.

FURTHER READING

Chanock, R. M., Parrott, R. H., Kapikian, A. Z., Kim, H. W., and Brandt, C. D. (1968). Possible role of immunological factors in pathogenesis of RS virus lower respiratory tract disease. *Perspect. Virol.* **6,** 125.

Choppin, P. W., and Compans, R. W. (1975). Reproduction of paramyxoviruses. *In* "Comprehensive Virology" (H. Fraenkel-Conrat and R. R. Wagner, eds.), Vol. 4, p. 95. Plenum, New York.

Choppin, P. W., and Klenk, H.-D., Compans, R. W., and Caliguiri, L. A. (1971). The parainfluenza virus SV5 and its relationship to the cell membrane. *Perspect. Virol.* **7,** 127.

Compans, R. W., and Choppin, P. W. (1971). The structure and assembly of influenza and parainfluenza viruses. *In* "Comparative Virology" (K. Maramorosch and E. Kurstak, eds.), p. 407. Academic Press, New York.

Enders, J. F., and Katz, S. L. (1967). Present status of live rubeola vaccines in the United States. *In* "First International Conference on Vaccines against Viral and Rickettsial Diseases of Man," Sci. Publ. No. 147, p. 295. Pan. Amer. Health Organ., Washington, D.C.

Fenner, F., McAuslan, B. R., Mims, C. A., Sambrook, J. F., and White, D. O. (1974). "The Biology of Animal Viruses," 2nd ed. Academic Press, New York.

Hilleman, M. R. (1970). Mumps vaccination. *Mod. Trends Med. Virol.* **2,** 241.

Jackson, G. G., and Muldoon, R. L. (1973). Viruses causing common respiratory infections in man. II. Enteroviruses and paramyxoviruses. III. Respiratory syncytial viruses and coronaviruses. *J. Infec. Dis.* **128,** 387 and 674.

Krugman, S. (1971). Present status of measles and rubella immunization in the U.S.: A medical progress report. *J. Pediat.* **78,** 1.

Morley, D. C. (1967). Measles and measles vaccine in industrial countries. *Mod. Trends Med. Virol.* **1,** 141.

Stokes, J., Weibel, R. E., Villarejos, V. M., Argeudas, J. A., Buynak, E. B., and Hilleman,

M. R. (1971). Trivalent combined measles-mumps-rubella vaccine. *J. Amer. Med. Ass.* **218,** 57.

Ter Meulen, V., Katz, M., and Muller, D. (1972). Subacute sclerosing panencephalitis: A review. *Curr. Top. Microbiol. Immunol.* **57,** 1.

Wright, P. F., Mills, J., and Chanock, R. M. (1971). Evaluation of a temperature-sensitive mutant of respiratory syncytial virus in adults. *J. Infec. Dis.* **124,** 505.

CHAPTER 22

Other Families of Viruses

INTRODUCTION

There remain for consideration several other families of viruses that do not merit separate chapters in a textbook of this size. For the most part they are families which have been relatively recently defined taxonomically and which contain relatively few important human pathogens. Having said that, we hasten to add that the rhabdovirus, rabies, is the most lethal of all viruses and the arenaviruses, Lassa and Junin, are not much less so; a reovirus is now recognized to be the most common cause of enteritis in infants, and the coronaviruses are the second most common cause of the common cold, while a recently described papovavirus, ubiquitous in man, has been associated with a degenerative disease of the CNS complicating advanced malignancies.

REOVIRIDAE

Introduction

The name reovirus is a sigla, short for "*r*espiratory *e*nteric *o*rphan," because these viruses are found in both the respiratory and enteric tracts of man but in general are "orphans" in the sense that they have not been associated with human disease. However, in 1973 a new reovirus was discovered which is in fact the major etiological agent of infantile enteritis.

Reoviruses have attracted much attention from molecular biologists because of the unique nature of their genome—double-stranded RNA, segmented into ten separate genes; experiments leading to the *in vitro* synthesis and identification of the protein product of each of these genes were described in Chapter 3. Taxonomists are also intrigued by the ubiquity of this family of viruses, with representatives known to infect plants and arthropods as well as mammals.

TABLE 22-1

Properties of Reoviruses

Icosahedron, diameter 55–75 nm
92 hollow "capsomers" (or 180 subunits)
Inner icosahedral protein shell
RNA, 10–12 double-stranded molecules, total mol. wt. 15 million
Transcriptase in virion

Properties of the Virus

The family Reoviridae is defined by the properties listed in Table 22-1. The double-stranded RNA genome is fragmented into 10–12 pieces, each corresponding to a single gene. Virions of the genus *Reovirus* have two icosahedral shells (total diameter 75 nm), while those of the genus *Orbivirus* (*orbis* = ring) are smaller (55–65 nm) because they have only an indistinct outer layer covering the inner icosahedral capsid.

The genus *Reovirus* contains 3 mammalian serotypes which share a common CF antigen but are distinguishable by neutralization or HI tests (using human erythrocytes). There are numerous members of the genus *Orbivirus* (38 at the time of writing), all of them arboviruses which are usually isolated from arthropods or animals, e.g., bluetongue and African horse sickness viruses. Colorado tick fever virus is the only orbivirus so far recognized as a human pathogen. The newly described virus of human infantile enteritis is morphologically and serologically related to the agents of epizootic diarrhea of infant mice (EDIM), Nebraska calf scours, pig diarrhea, and the simian virus SA11. It has been proposed that all these viruses should be grouped together in a new genus to be called either *"Rotavirus"* (because the circular outline of the outer capsid suggests the appearance of a wheel), or *"Duovirus"* (viruses with double-stranded RNA enclosed within a double-shelled capsid).

Clinical Features

Rotaviruses. The rotaviruses are responsible for about half of the cases of enteritis in infants throughout the world; the virus is less commonly found in older children or adults. Diarrhea is the cardinal feature, with or without fever and/or vomiting. The condition is generally mild and self-limiting but death occasionally follows untreated dehydration in severe cases.

Colorado Tick Fever Virus. This is caused by an orbivirus and follows the bite of an infected tick. The onset is sudden with a characteristic "saddle-back" fever, chills, severe generalized aches, especially in the back, head, and eyes, and leukopenia.

Reoviruses 1, 2, and 3. These viruses have not been proved to have any causal relationship to human disease, although they have been isolated frequently enough from patients with respiratory infections, gastroenteritis, or rashes. The recovery of reoviruses from the rare pneumonitis–hepatitis–encephalitis syndrome as well as from occasional patients with steatorrhea is intriguing in the light of the fact that these are precisely the pathological effects of reoviruses on infant mice (see below).

Laboratory Diagnosis

Rotaviruses. These viruses are readily identifiable in ultracentrifuged deposits of clarified fecal suspensions by direct electron microscopy following negative staining (Plate 22-1). Immunoelectron microscopy is feasible but not necessary because the virions are so characteristic.

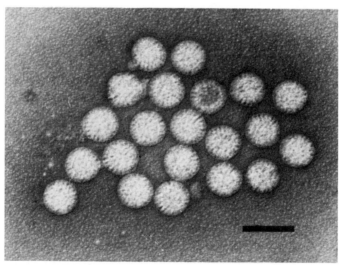

PLATE 22-1. *Reoviridae (bar = 100 nm). Characteristic virions of the "rotavirus" ("duovirus"), which is the most common etiological agent in infantile enteritis. Diagnosis is established by direct electron microscopic examination of an ultracentrifuged deposit of clarified feces. (Courtesy Dr. I. H. Holmes.)*

The rotavirus has been grown in organ cultures of human fetal intestine but sophisticated techniques (immunofluorescence or electron microscopy) are required to identify its presence. Rotavirus-infected organ cultures can also be used to identify antibody in human convalescent sera by immunofluorescence. Immunoelectron microscopy (aggregation of virions by convalescent sera) is a rather simpler method of screening for antibodies in man. Complement fixation tests can also be used, employing as antigen any rotavirus, e.g., calf scours virus, since the human and animal diarrhea viruses share at least one CF antigen, but CF tests are less sensitive than immunofluorescence for detecting antibodies in the sera of young infants.

Colorado Tick Fever Virus. This is isolated from the blood by intracerebral inoculation of infant mice or in cell culture lines such as BHK-21.

Reoviruses 1, 2, and 3. These are quite frequently isolated from feces and occasionally from the throat of normal people following inoculation into conventional monolayer cultures. Cytopathic effects are rather slow to develop, and not particularly striking; infected cells acquire a granular appearance, rather like the nonspecific degeneration seen in normal aging cultures, and one "blind passage" may be necessary to reveal the presence of the virus. When the monolayer is stained however, pathognomonic inclusions are seen. Beginning as a crescentic structure in the perinuclear region of the cytoplasm, these huge acidophilic inclusions spread until they may come to encircle the nucleus completely (see Plate 5-1, B). Typing of the isolate is achieved by HI using human erythrocytes.

Reoviruses also grow in infant mice, indeed were originally discovered by isolation in this host. Stanley named them the "hepatoencephalomyelitis" viruses in recognition of the pathology they produce in suckling mice. Pancreatitis, also a feature, leads to steatorrhea, which causes the well-known "oily-hair" appearance.

Pathogenesis

Rotaviruses were originally discovered (by electron microscopy and immunofluorescence) in biopsied duodenal mucosa. The virus multiplies in the cytoplasm of the cells lining the villi of the small intestine from the duodenum to the jejunum. Enormous numbers of virions (up to 10^{10} per gram of feces) are excreted during the few days of illness which follow the 2 day incubation period.

Epidemiology and Control

Rotaviruses. The rotaviruses are ubiquitous, being responsible for at least 50% of cases of enteritis among infants under the age of 5 the world over. Virtually all children have antibodies by the second year of life. In temperate climates infection occurs throughout the year but is most prevalent during the winter months. Spread via the fecal-oral route has been demonstrated in human volunteers.

There is a clear need for a vaccine against the human rotavirus, which affects countless millions of children annually and probably kills large numbers of undernourished infants in developing countries. The first requirement is, of course, to grow the agent in conventional monolayer cultures. Thereafter the prospects are probably favorable, because we already have an encouraging precedent in a live attenuated oral calf diarrhea vaccine which is quite effective in preventing that disease.

Colorado Tick Fever Virus. As the name implies, this is a tick-borne arbovirus, transmitted from rodents (especially squirrels) to man during the spring and early summer in western U.S.A.

Reoviruses 1, 2, and 3. These are among the most widespread of all viruses in nature. Most animals, wild and domestic, have antibodies indicative of past infection. Two-thirds of all humans acquire antibodies before reaching adulthood. The primary mode of spread is assumed to be fecal as the virus is often excreted by this route for several weeks and is very stable. Respiratory spread doubtless also occurs.

RHABDOVIRIDAE

Introduction

The family Rhabdoviridae embraces a fascinating variety of agents parasitizing hosts from right across the animal and plant kingdoms. These uniquely bullet-shaped viruses are to be found in plants, insects, and cold-blooded as well as warm-blooded vertebrates. Two rhabdoviruses of wild animals, rabies and the Marburg agent, are lethal for man.

Throughout history man has lived in terror of the "mad dog," which can transmit the most lethal of all infectious diseases, rabies (*rabidus* = mad). The agonizing muscular spasms accompanying any attempt to swallow water provided the disease with its notorious pseudonym "hydrophobia." However, rabies has an even more important place in history. In 1884, long before the recognition of the nature of viruses,

Pasteur developed the world's first man-made viral vaccine. His dramatic demonstration of its efficacy in saving the life of a peasant boy bitten by a rabid dog heralded a new era in medicine.

Properties of the Virus

The rhabdoviruses are shaped like a bullet, 76 nm wide and about 175 nm long; with some, such as the Marburg agent (see Chapter 12), bizarre branched filaments up to several microns in length are found. Truncated bullets ("T particles") encountered in many rhabdovirus preparations are noninfectious. Hemagglutinating peplomers project from the lipoprotein envelope. Single-stranded RNA is enclosed in a typical internal nucleocapsid showing helical symmetry (Table 22-2 and Plate 22-2).

Rabies virus appears to be capable of infecting every warm-blooded animal, though in nature it is found mainly in dogs, cats, wolves, foxes, skunks, mongooses, and squirrels, for all of which it is lethal. Bats are often symptomless carriers.

Clinical Features

Prodromal symptoms of rabies include malaise, anxiety, and hyperesthesia around the wound. Muscles then become hypertonic; painful spasms of the muscles of deglutition give the patient his "hydrophobia." Paralysis may or may not be a feature. Delirium, coma, and death follow within a week. Death is usual, but not inevitable as previously thought.

Pathogenesis

Man becomes infected when bitten (or even licked on an abraded area) by a rabid animal which secretes the virus in its saliva. Disease follows in somewhat less than half of such exposures. It usually becomes manifest in about 1–2 months, although incubation periods as

TABLE 22-2
Properties of Rhabdoviruses

Bullet-shaped virion, 175 × 75 nm
Nucleocapsid with helical symmetry
Envelope, containing hemagglutinin
RNA, single-stranded, mol. wt. 4 million
Transcriptase in virion

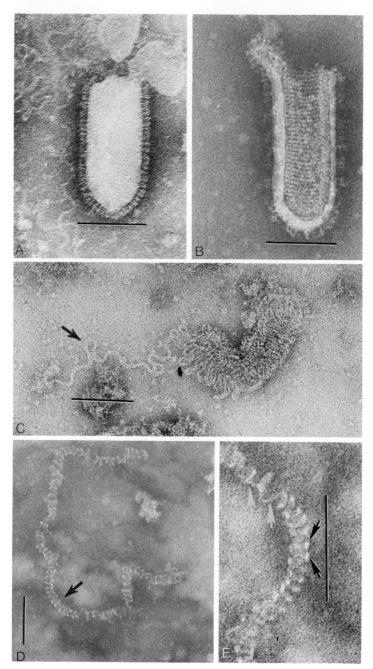

PLATE 22-2. *Rhabdovirus* (bars = 100 nm). (A) *Flattened particle showing peplomers at periphery.* (B) *Particle penetrated by stain and showing characteristic cross-striations.* (C) *Unwinding nucleocapsid; a continuous 5 nm wide strand with repeating subunits (arrow).* (D) *Portion of nucleocapsid showing helical arrangement (arrow).* (E) *Higher magnification of (D).* [*From R. W. Simpson and R. E. Hauser, Virology* **29,** *654 (1966). Courtesy Dr. R. W. Simpson.*]

short as a week or as long as a year have been reported. The shorter incubation periods tend to occur in children, often following bites to the face. This finding has been variously interpreted to reflect the relatively high doses of virus received in this sort of encounter, or the proximity of the wound to the brain. There is strong evidence from experiments in animals that the virus reaches the brain by neural spread. Virus inoculated peripherally cannot be found in the blood stream, but first becomes demonstrable by immunofluorescence in the dorsal root ganglion of the peripheral nerve serving the inoculated area. If this nerve is severed, disease is prevented. The corresponding posterior horn and eventually the rest of the spinal cord and brain become infected and the patient dies from an overwhelming encephalitis affecting some areas, such as the hippocampus, more than others. Only the neurons themselves are infected, as evidenced by immunofluorescence, or by hematoxylin and eosin staining for the highly characteristic "Negri bodies," which are pathognomonic of rabies. These are cytoplasmic inclusions, 2–30 μm in diameter, round or oval, eosinophilic (red) with occasional basophilic (blue) granules inside them.

Laboratory Diagnosis

Most commonly the laboratory will be called upon to diagnose rabies in the animal known to have inflicted the bite. For instance, dogs or cats are routinely held for observation and released if they are still symptomless after 10 days. Should signs of illness develop however, they are killed and brain smears examined by immunofluorescent staining.

Postmortem diagnosis can be confirmed in man by demonstrating rabies antigen in the neurons of the hippocampus by immunofluorescence. In addition, classical Negri bodies can be displayed with Sellers stain. Virus can be grown from brain (or sometimes from saliva before death) by intracerebral inoculation into suckling mice, which become paralyzed and die within 1–3 weeks with the typical histopathology; their brains show positive immunofluorescence within a week. The virus can also be grown in cultured cells, setting up a persistent non-cytocidal infection.

Epidemiology

Man is irrelevant to the ecology of rabies. To the virus he constitutes a blind alley, for humans cannot transmit the infection. In nature, rabies is endemic in a wide variety of animals. The domesticated dog and cat are by far the most important sources of human infection, but "sylvan rabies" is even more common in certain wild animals, notably foxes, skunks, raccoons, and related species in America, jackals and mongooses in India and South Africa, and wolves in Eastern Europe. In recent years there has been a disturbing spread of rabies into Western Europe, where it has become endemic in foxes. Only certain islands like Australia and, until recently, Britain, are free of rabies. The fact that rabies is lethal for all of these animals raises interesting questions. How have such vulnerable species survived a highly lethal virus that perpetuates itself by inducing the animal to run around in a mad frenzy biting everything in sight? Does the induction of this temperamental change in the victim advantage or disadvantage the virus in *its* battle for survival. Is there a more important reservoir for which the virus is relatively harmless? There is some evidence that the bat may be such a reservoir, although most infected bats do probably die from rabies. Vampire bats kill thousands of cattle each year in Latin America, while speleologists occasionally die following inhalation of aerosols created by the secretions of insectivorous bats roosting in caves.

Prevention and Control

Protection of man from rabies depends on a number of precautionary measures. Quarantine, rigidly enforced, continues to exclude rabies from islands like Australia, aptly described by an author (with other things in mind) as "the lucky country." Dogs, for example, are isolated for 6 months before entry. Most of the rest of the world must content itself with trying to confine the virus to its native fauna and minimizing its spread to domestic animals by insistence on registration and immunization of all pet dogs and cats and destruction of stray animals. Active immunization of humans is confined to those exposed to occupational risk, such as veterinarians, dog-handlers, trappers, and park rangers.

Rabies can usually be prevented even after an unimmunized person has been bitten by a rabid animal. Following thorough cleansing and disinfection of the wound, a large dose of human antirabies immunoglobulin is injected into and around the bite. Active immunization with inactivated rabies vaccine is then instituted. Rabies vaccination practice is quite unique in that active immunization may be commenced *after* an individual has been bitten by a potentially rabid animal, a procedure that is successful only because of the unusually long incubation period of the disease. The remarkable number of different rabies vaccines is partly attributable to the very long history of research into this disease, going back to the days of Pasteur himself; indeed, that is the era to which some of the current vaccines belong, for one cannot conceive of them being licensed were they to be submitted for initial approval today.

The Semple vaccine is a 10% emulsion of phenol-inactivated rabies virus-infected rabbit brain which carries a risk of allergic encephalomyelitis variously assessed at 0.01%–3%, of whom 20% die and another 30% sustain permanent sequelae. Such a vaccine is totally unacceptable even though it is said that most of the encephalitogenic material can be removed with fluorocarbon. Vaccines made from suckling mouse or rat brain, on the basis of the fact that brains from young animals contain less myelin, also cause far too many CNS reactions. The β-propiolactone-inactivated duck embryo vaccine is much less hazardous, but produces a high incidence of Arthus-type local reactions and a moderate number of generalized allergic reactions, especially in those presensitized to chick embryo tissue, e.g., by prior yellow fever immunization. Moreover, the antibody response to the duck embryo vaccine is too low during the crucial interval between 1 and 2 months after exposure to be certain of its value. The Flury-HEP living attenuated strain, derived by prolonged passage in chick embryos, is widely and successfully used in domestic animals but gives completely inadequate antibody responses in man.

Koprowski and his colleagues at the Wistar Institute have devoted many years to the development of a safer rabies vaccine. In an extensive trial they compared almost a dozen separate vaccines grown in various nonneural tissues for their efficacy in monkeys. All gave superior immunity to that provided by the rabbit brain, mouse brain, and duck embryo vaccines which were used as reference standards. The best was a concentrated β-propiolactone-inactivated vaccine grown in the human diploid fibroblast strain WI-38. A single dose injected 4–6 hours after rabies virus had been inoculated into the neck muscles of monkeys saved the lives of the animals. The regime now recommended for therapeutic use in man is 3 injections given at 3–4 day intervals. This constitutes a major advance because the practice in the past has been to continue daily injections of vaccine for at least 2 weeks, a procedure that promotes the development of hypersensitivity. Meanwhile work is proceeding on alternative types of experimental vaccines consisting of partially purified immunogenic subunits of the rabies virion, freed of all nucleic acid.

ARENAVIRIDAE

Introduction

This recently defined family consists mainly of viruses associated with chronic inapparent infections of rodents. Their relevance to human medicine is minimal but a number of recent episodes have reminded us that viruses harmless to the host to which they have adapted over the course of thousands or millions of years may nevertheless be lethal if they should accidentally spread to man.

Properties of the Virus

The arenaviruses derive their name from the fact that they contain cellular ribosomes (presumably picked up fortuitously during their maturation by budding) which, in thin sections examined by electron microscopy, look somewhat like grains of sand (*arenosus* = sandy). Other characteristics of the virions are listed in Table 22-3.

The prototype is lymphocytic choriomeningitis (LCM) virus, which causes the classical persistent infection of mice discussed in detail in Chapter 8, and occasionally causes meningitis in man (hence its name). Other arenaviruses, such as the "Tacaribe complex," indigenous to South America, and Lassa virus, found in West Africa, are probably analogous tolerated infections of their natural rodent hosts.

Clinical Features

Lymphocytic choriomeningitis virus infection of man, which is thought to be uncommon, may take the form of either a "grippe-like" illness (fever, chills, and myalgia) or meningoencephalitis. Junin, Machupo, and Lassa viruses on the other hand produce highly lethal hemorrhagic fevers (Table 22-4), although milder and subclinical infections quite commonly occur also. Argentinian–Bolivian hemorrhagic fever is characterized by fever, a hemorrhagic rash, myalgia, leukopenia, renal involvement, and shock. Lassa fever, on the other hand,

TABLE 22-3
Properties of Arenaviruses

Spherical virion, diameter 85–120 nm, sometimes up to 300 nm
Contains ribosomes
Envelope
RNA, single-stranded, segmented, total mol. wt. 5 million
Transcriptase in virion

TABLE 22-4
Arenaviruses

VIRUS	DISTRIBUTION	RESERVOIR	CLINICAL FEATURES IN MAN
LCM	Widespread	House mouse	Meningitis
Lassa	West Africa	Multimammate rat	Hemorrhagic fever
Junin	Argentina	Field mouse	Hemorrhagic fever
Machupo	Bolivia	Field mouse	Hemorrhagic fever
?	U.S.S.R., Korea	Vole	Hemorrhagic fever with renal syndrome (hemorrhagic nephrosonephritis)

shows fewer hemorrhagic manifestations but severe pharyngitis, enteritis, myocarditis, and pneumonitis, and in fatal cases, circulatory collapse.

Laboratory Diagnosis

Only highly experienced laboratories with adequate security safeguards should attempt to isolate such dangerous arenaviruses as that of Lassa fever. All work on this lethal agent was halted following the death at Yale of laboratory workers involved in the identification of Lassa virus from the first-described outbreak in Nigeria.

Arenaviruses grow satisfactorily in a number of cultured cell lines, such as Vero, in which they produce CPE of variable extent, with cytoplasmic inclusions. They are also readily isolated in infant mice but disease may only be apparent following inoculation of adult mice.

Epidemiology and Control

Lymphocytic choriomeningitis is indigenous to the domestic house mouse, *Mus musculus,* and probably spreads to man via urine and feces contaminating food or creating an aerosol. Junin and Machupo viruses cause regular autumn epidemics of hemorrhagic fever among rural workers, particularly maize harvesters in Argentina and farm workers in Bolivia, respectively. The natural hosts are assumed to be the common field mice of the genus *Calomys.* Such rodents are permanently viremic and shed virus persistently in their urine, which contaminates grain and water. Recent serological surveys of West African animals indicate that the natural host of Lassa virus is also a rodent, *Mastomys natalensis.* There is clear evidence that Lassa virus also spreads readily from human to human, being excreted in saliva, urine, and possibly feces.

Because of the extreme danger of cross-infection of hospital personnel, patients with Lassa fever must be barrier-nursed in strict isolation. Convalescent serum has been successfully employed in treatment.

CORONAVIRIDAE

Introduction

In 1965 a new type of virus, morphologically resembling infectious bronchitis virus of chickens, was isolated by inoculating organ cultures of human embryonic trachea with nasopharyngeal washings from patients with common colds. The "coronaviruses" are now recognized to be a significant cause of minor respiratory infections in man. They are also regularly visualized by electron microscopy in human feces, but have not been positively incriminated as causal agents of human gastroenteritis as they have in piglets and calves.

Properties of the Virus

At first glance these viruses appear to resemble orthomyxoviruses, being enveloped RNA viruses with nucleocapsids of helical symmetry, 80–120 nm in diameter. However, the projections from the envelope are seen to be more widely spaced, and rounded with a terminal bulb, often being described as petal- (or club-) shaped (Plate 22-3, and Table 22-5). The name "coronavirus" refers to the fringe of petal-shaped projections surrounding the virion, reminiscent of the solar corona.

At least 3 human serotypes have been isolated. Human coronaviruses show some serological cross-reaction with mouse hepatitis virus, but not with avian infectious bronchitis virus.

Clinical Features

Coronavirus infections occur chiefly during the winter months, affecting adults in the main. A common cold-like illness is the usual

TABLE 22-5
Properties of Coronaviruses

Spherical virion, diameter 80–120 nm
Nucleocapsid with helical symmetry
Envelope, containing "petal-shaped" peplomers
RNA, single-stranded, one molecule, mol. wt. 9 million
Transcriptase in virion

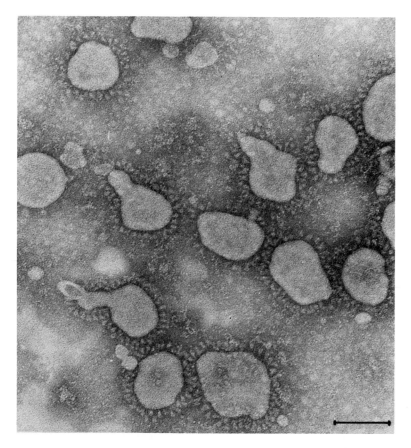

PLATE 22-3. *Human coronavirus (bar = 100 nm). Negatively stained. (Courtesy Dr. A. Z. Kapikian.)*

result. Regular symptoms are coryza, sore throat, cough, cervical adenitis, malaise, chills, headache, and low fever. The lower respiratory tract is rarely involved, although there is some evidence that these viruses may be incriminated in exacerbations of recurrent wheezing in asthmatic children and perhaps chronic bronchitis in adults.

Pathogenesis and Immunity

In volunteer studies the incubation period of coronavirus colds has been found to range from 2–5 days, with an average of 3 days. Acquired immunity is poor, hence reinfection is common, even with the same serotype.

Laboratory Diagnosis

Coronaviruses are extremely fastidious agents. Only one major sero-type (229E) has so far been isolated in conventional cell cultures, e.g., human embryonic fibroblast strains such as HEL, HEK, WI-38, or Bradburne's strain, L132, or HEI (human embryonic intestine). Other types can still be isolated only in organ cultures of human embryonic trachea. Detection of viral growth in such organ cultures is tedious; cilia cease to beat and coronavirions can be demonstrated in the supernatant fluid by immunoelectron microscopy.

Many strains can be adapted to grow in infant mice or hamsters, as well as in human diploid cell strains. CPE is usually minimal in cultured cells but syncytia are sometimes observed. "Pseudohemadsorption" is demonstrable using erythrocytes from man or vervet monkey at 4°, or from chicken, rat, or mouse at any temperature between 4° and 37°. Since coronavirions bud only internally (into cytoplasmic vesicles) the outer plasma membrane does not contain viral hemagglutinin but some recently liberated virions adhere to the cell surface and provide a bridge for red cells.

The average diagnostic laboratory is not well set up for isolating coronaviruses. Serology (HI or CF) is a simpler alternative.

Epidemiology

Coronaviruses are most prevalent in winter and early spring and, hence, are responsible for a much higher proportion of "winter colds" than "summer colds." Overall, they are the second most common cause of the common cold, being recoverable from about 10–15% of cases, particularly in adults. Most children become infected before they reach school age and suffer repeated winter infections thereafter. Intrafamilial spread occurs readily.

PAPOVAVIRIDAE

Introduction

During the 1950's and 1960's the papovaviruses, polyoma (of mice) and SV40 (of monkeys), were the subject of intensive investigation by tumor virologists seeking to understand the way in which these agents induce cancer in baby rodents and transform cultured cells *in vitro* (see Chapter 9). The only human papovavirus then known was that responsible for the comparatively mundane hyperplastic condition, warts. More recently, however, interest in the possible role of papova-

TABLE 22-6

Properties of Papovaviruses

Icosahedron, diameter 45[a] or 55[b] nm, 72 capsomers
DNA, double-stranded, circular, mol. wt. 3[a] or 5[b] million

[a] *Polyomavirus.*
[b] *Papillomavirus.*

viruses in human cancer has been revived by the isolation of an agent resembling SV40 from the brains of patients with progressive multifocal leukoencephalopathy (PML).

Properties of the Viruses

The family Papovaviridae (Table 22-6) contains two genera, *Papillomavirus* and *Polyomavirus,* the latter having a smaller virion (45 nm as compared to 55 nm diameter) and a smaller genome (3×10^6 as compared to 5×10^6 daltons). Plate 22-4 depicts the characteristic icosahedral virions of *Papillomavirus* and *Polyomavirus* respectively (see Plate 1-2 for the unique circular DNA molecule).

Clinical Features

A single *Papillomavirus* is responsible for all the clinically distinct varieties of human warts. The common wart (*verruca vulgaris*) usually

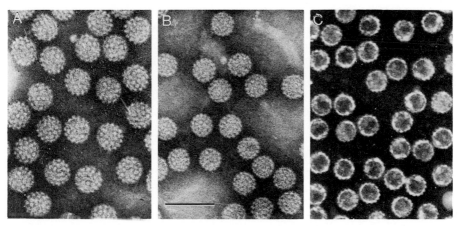

PLATE 22-4. *Papovaviridae (bar = 100 nm). (A) Papillomavirus (human wart virus). (B) Polyomavirus. (C) Polyomavirus, empty virions. (Courtesy Dr. E. A. Follett.)*

occurs in localized crops, especially on exposed areas subject to abrasion, such as hands and knees. Other clinical forms are the plane (juvenile), the filiform, the plantar, and the genital wart. The latter, also known as *condyloma accuminatum,* is a more moist, soft pedunculated excrescence found on the external genitalia. There is some evidence to suggest that laryngeal papillomas of children may be a viral disease acquired at the time of birth from mothers with genital warts.

There appear to be at least two serotypes of the human *Polyomavirus* associated with PML; one is closely related to SV40; the more ubiquitous strain, known as JC, is only distantly related. *Progressive multifocal leukoencephalopathy (PML)* is a degenerative disease of the brain occurring as a rare complication of advanced disseminated malignant conditions such as Hodgkin's disease, lymphosarcoma or chronic lymphatic leukemia, or of immunosuppressive therapy.

Pathogenesis and Immunity

Warts. These are generally regarded as a simple hyperplasia rather than as a true benign neoplasm, but the distinction cannot be absolute; *condyloma accuminata* sometimes become malignant. Virus is found mainly in the keratinized layers of young warts and gradually disappears from older ones before they eventually regress. Warts virus can be transmitted to volunteers by intradermal inoculation of filtrates from any type of human wart, but transmission occurs most readily from young plantar warts with plentiful inclusions. The incubation period varies from 6 weeks to a year.

Antibody against the human warts virus is detectable by immunodiffusion or immunoelectron microscopy. Infected individuals synthesize IgM for as long as virus continues to be produced in growing warts. The mechanism of the legendary sudden and simultaneous regression of all lesions after months or even years is unknown, though some recent evidence suggests it may be predicted when the anti-wart virus IgG:IgM ratio exceeds a certain level.

Progressive Multifocal Leukoencephalopathy. The characteristic histopathology of PML is demyelination of neurons accompanied by proliferation of giant bizarre astrocytes. The nuclei of the oligodendrocytes which surround the foci of neuronal demyelination contain very large numbers of papovavirus particles, clearly visible by electron microscopy and identifiable by immunofluorescence. It seems clear that the disease is precipitated by immunodeficiency (Chapter 8), but it is uncertain whether it usually represents a primary exogenous infection or an activation of an endogenous persistent infection.

Laboratory Diagnosis

The virus laboratory is never called upon to diagnose a human wart. Gross appearance and histology are usually pathognomonic. There is marked proliferation of prickle cells in the Malpighian layer of the epidermis, resulting in a protruding papilloma, topped by extensive hyperkeratosis. Acidophilic inclusions may be observed in nuclei, and more rarely in the cytoplasm. Warts virus can be cultured only with difficulty, in organ cultures of human skin epithelium.

PML viruses have been grown in cultured human fetal spongiblasts. On inoculation into newborn hamsters these viruses have been reported to produce malignant gliomas of the brain, but not neuronal degeneration resembling PML in man.

Epidemiology and Control

Warts are transmitted by direct contact (*condylomata accuminata* by venereal spread). The vulnerability to common warts of active young boys is directly proportional to their enthusiasm for wrestling and football. Secondary spread to other areas of skin can result from autoinoculation by scratching. Individual warts may be removed by chemical or thermal cautery, X-irradiation, or surgery. Multiple lesions are more of a problem, but in view of the fact that they almost always regress eventually it is not unreasonable to let nature take its course.

The human *Polyomavirus*, J.C., is ubiquitous in the community, producing subclinical infection in the majority of children, as evidenced by the fact that most adolescents already have antibody. Presumably the great majority of infections are subclinical.

PARVOVIRIDAE

The smallest of all viruses, the Parvoviridae (*parvo* = small), contain such a small amount of genetic information that several of them are defective. The human adeno-associated viruses (AAV) multiply only in the presence of a helper adenovirus (Plate 15-3). There are 4 serotypes and they are quite commonly found in the human throat but have not yet been related to any disease. Isolation is accomplished in cell cultures concomitantly infected with an adenovirus; in the absence of such a helper, infection is "cryptic" (nonproductive) and the AAV genome may be rescued by subsequent superinfection. Molecular biologists have taken an interest in the unique finding that the single molecule of single-stranded DNA encapsidated by an AAV virion may be of positive *or* negative polarity (Table 22-7).

TABLE 22-7
Properties of Parvoviruses

Icosahedron, diameter 20 nm, 32 capsomers
DNA, single-stranded, polarity + or −, mol. wt. 1.5 million

There is current controversy about whether two important human pathogens, hepatitis A virus and the Norwalk agent of gastroenteritis, are parvoviruses or picornaviruses, but the former seems more likely. The definitive answer must await proper analysis of the nucleic acids of the respective viruses. The medical significance of these two viruses is discussed in Chapters 23 and 24.

FURTHER READING

REOVIRIDAE

Editorial. (1975). Rotaviruses of man and animals. *Lancet* **1**, 257.
Jackson, G. G., and Muldoon, R. L. (1973). Viruses causing common respiratory infections in man. IV. Reoviruses and adenoviruses. *J. Infec. Dis.* **128**, 811.
Joklik, W. K. (1974). Reproduction of Reoviridae. *In* "Comprehensive Virology" (H. Fraenkel-Conrat and R. R. Wagner, eds.), Vol. 2, p. 231. Plenum, New York.
Kapikian, A. Z. (1974). Acute viral gastroenteritis. *Prev. Med.* **3**, 535.
Rosen, L. (1968). Reoviruses. *Virol. Monogr.* **1**, 73.
Shatkin, A. J. (1971). Viruses with segmented ribonucleic acid genomes: Multiplication of influenza versus reovirus. *Bacteriol. Rev.* **35**, 250.
Stanley, N. F. (1974). The reovirus murine models. *Progr. Med. Virol.* **18**, 257.
Verwoerd, D. W. (1970). Diplornaviruses: A newly recognized group of double-stranded RNA viruses. *Progr. Med. Virol.* **12**, 192.

RHABDOVIRIDAE

Baer, G. M., ed. (1975). "The Natural History of Rabies." Academic Press, New York.
Debbie, J. G. (1974). Rabies. *Progr. Med. Virol.* **18**, 241.
Howatson, A. F. (1970). Vesicular stomatitis and related viruses. *Advan. Virus Res.* **16**, 196.
Hummeler, K. (1971). Bullet-shaped viruses. *In* "Comparative Virology" (K. Maramorosch and E. Kurstak, eds.), p. 361. Academic Press, New York.
Kaplan, M., and Koprowski, H., eds. (1973). "Laboratory Techniques in Rabies," 3rd ed., World Health Organ., Monogr. Ser. No. 23. World Health Organ., Geneva.
Plotkin, S. A., and Clark, H. F. (1971). Prevention of rabies in man. *J. Infec. Dis.* **123**, 227.
World Health Organization. (1973). Expert Committee on Rabies, Sixth Report. *World Health Organ., Tech. Rep. Ser.* **523**.

ARENAVIRIDAE

Hotchin, J. (1974). Slow virus diseases. *Progr. Med. Virol.* **18**, 64.

CORONAVIRIDAE

Bradburne, A. F., and Tyrrell, D. A. J. (1971). Coronaviruses of man. *Progr. Med. Virol.* **13**, 373.

Jackson, G. G., and Muldoon, R. L. (1973). Viruses causing common respiratory infections in man. III. Respiratory syncytial viruses and coronaviruses. *J. Infec. Dis.* **128,** 674.

Kapikian, A. Z. (1975). The coronaviruses. *In* "Immunity to Infections of the Respiratory System in Man and Animals."

McIntosh, K. (1973). Coronaviruses: A comparative review. *Curr. Top. Microbiol. Immunol.* **63,** 85.

PAPOVAVIRIDAE

Salzman, N. P., and Khoury, G. (1974). Reproduction of papovaviruses. *In* "Comprehensive Virology" (H. Fraenkel-Conrat and R. R. Wagner, eds.), Vol. 3, p. 63. Plenum, New York.

Weiner, L. P., and Narayan, O. (1974). Virologic studies of progressive multifocal leukoencephalopathy. *Progr. Med. Virol.* **18,** 229.

Zu Rhein, G. M. (1969). Association of papovavirions with a human demyelinating disease (progressive multifocal leukoencephalopathy). *Progr. Med. Virol.* **11,** 185.

PARVOVIRIDAE

Berns, K. I. (1974). Molecular biology of the adeno-associated viruses. *Curr. Top. Microbiol. Immunol.* **65,** 1.

Hoggan, M. D. (1970). Adenovirus associated viruses. *Progr. Med. Virol.* **12,** 211.

Rose, J. A. (1974). Parvovirus reproduction. *In* "Comprehensive Virology" (H. Fraenkel-Conrat and R R. Wagner, eds.), Vol. 3, p. 1. Plenum, New York.

CHAPTER 23

Hepatitis, Rubella, "Slow" Viruses

INTRODUCTION

Some important human viruses remain to be considered: those of hepatitis, rubella, and the spongiform viral encephalopathies. The viruses of hepatitis A ("infectious" hepatitis) and hepatitis B ("serum" hepatitis), both recently discovered, are quite dissimilar in physicochemical properties. Hepatitis A virus has been tentatively classified as a parvovirus; hepatitis B virus is unlike any other known animal virus. Rubella is classified as a togavirus, but we preferred not to deal with it in Chapter 19 (Arboviruses) because it differs from most other togaviruses in not being arthropod-transmitted. We conclude this chapter with a description of a miscellaneous group of recently discovered viruslike agents that appear to be responsible for certain "slow" degenerative diseases of the human brain, known collectively as the spongiform viral encephalopathies.

HEPATITIS

Introduction

Hepatitis is the only severe viral disease of man that is actually increasing in incidence, as a result of changing social conditions and our lack of effective vaccines. Progress toward the latter objective has accelerated greatly in the last decade, following the discovery of the causative viruses of hepatitis A and B. Neither virus has yet been grown in conventional cell culture, which is the usual prelude to the development of a vaccine, but if the current rate of progress continues the production of effective vaccines against one or both kinds of viral hepatitis may be anticipated within a decade.

TABLE 23-1
Comparison of Hepatitis A and B

	HEPATITIS A	HEPATITIS B
Virion		
Diameter	27 nm	43 nm
Symmetry	Icosahedral	Icosahedral
Nucleic acid	?	DNA, DS, circular, 1.6 million
Animal models		
Chimpanzee	Subclinical	Hepatitis
Other monkeys	Marmoset (subclinical)	Rhesus (subclinical)
Transmission		
Primary route	Fecal–oral	Parenteral
Other routes	—	? Transplacental, feces, saliva
Clinical		
Incubation period	2–6 weeks (average 4)	6–26 weeks (average 10)
Mortality	0.1%	1–10%
Transaminases	+, 1–3 weeks	+, 4–30 weeks
Chronic active hepatitis	Rare	More common

Properties of the Viruses

There are two quite distinct groups of hepatitis viruses. Hepatitis viruses A and B, now well characterized, are so different from one another that they clearly belong to different viral families (Table 23-1).

Hepatitis A Virus. This virus was discovered in 1973 in the feces of patients with infectious hepatitis by immunoelectron microscopy. The 27 nm icosahedral virion (Plate 23-1) has been tentatively classified as a parvovirus, but confirmation of this suggestion must await characterization of its nucleic acid. As has been known for many years the infectivity of the virion is remarkably stable to inactivation by heat.

Hepatitis A virus has still not been grown in cell culture, despite many optimistic but subsequently unconfirmed reports of success. Apart from man, it will grow in marmoset monkeys and chimpanzees, both of which develop biochemical, histological, and serological evidence of infection.

Hepatitis B Virus. Sometimes known as the Dane particle, this virus is a 43 nm double-shelled icosahedron, consisting of an inner 27 nm core surrounded by an outer coat 7–8 nm thick, which contains some lipid. Most specimens of the virus found in human serum also contain a much greater number of smaller particles, some spherical and some filamentous, of diameter 22 nm, which consist of outer coat capsomers

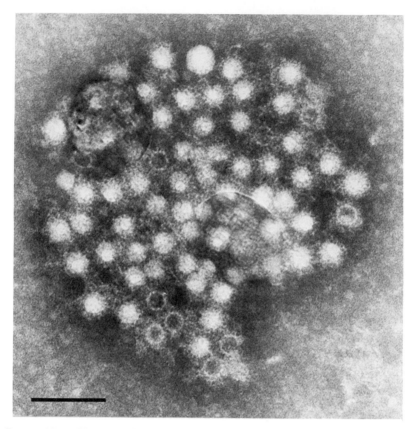

PLATE 23-1. *Hepatitis A virions aggregated by immune serum. Bar = 100 nm.* (*Courtesy Dr. A. Z. Kapikian.*)

that have been synthesized in excess and have spontaneously aggregated (Plate 23-2). The Dane particle contains, in its core, a circular molecule of double-stranded DNA with a molecular weight of 1.6 million daltons and a DNA polymerase. Clearly, the hepatitis B viruses represent the prototype of a new, as yet unnamed family of viruses.

There are several serotypes of hepatitis B virus defined by different antigenic determinants identifiable in the outer coat ("surface antigen," or HB_sAg) of the Dane particle. The subtype-specific determinants d and y are mutually exclusive allelic alternatives, as are the other pair, w and r. Hence, there are four major serotypes: HB_sAg/adr, *ayr, adw,* and *ayw,* (*a* being a group-specific determinant present in the coat of all serotypes). A number of minor variants also occur. The "core antigen," HB_cAg, is, as might be expected, common to all serotypes.

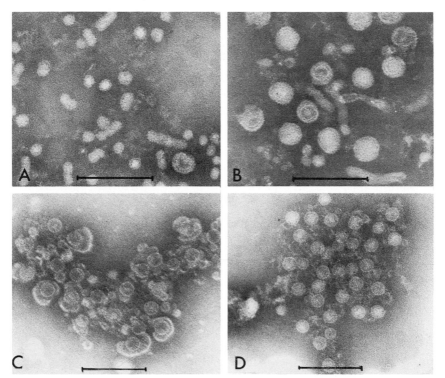

PLATE 23-2. *Hepatitis B virus.* (A) *Hepatitis surface antigen B* ($HB_sAg =$ *Australia antigen); spherical and tubular particles; single Dane particle at bottom right.* (B) HB_sAg *specimen with many Dane particles.* (C) *Specimen B treated with Tween 80.* (D) *Particles from liver homogenate of a case of hepatitis; same diameter as internal core of* (C) (*Courtesy Dr. J. D. Almeida.*)

Hepatitis B virus has still not been grown with any degree of reproducibility *in vitro*, though there is some evidence that it may be capable of undergoing a few cycles of multiplication in organ cultures of human fetal liver. However, the virus readily infects primates, which are used for laboratory investigations of pathogenesis and immunity. Infection of rhesus monkeys is subclinical but chimpanzees often develop frank hepatitis. Indeed, chimpanzees are so susceptible that they readily contract hepatitis from their human keepers; it is difficult to maintain "clean" (anti-HB_sAg-free) colonies of the animals.

Clinical Features

The clinical pictures of "infectious" hepatitis (A) and "serum" hepatitis (B) are almost indistinguishable, and will be described together.

Prodromal symptoms are chiefly referable to the gastrointestinal tract. Anorexia and malaise are early dominant features, with intermittent fever, then nausea, vomiting, diarrhea, and liver tenderness. About a week later the patient often feels temporarily better for a day or so, until jaundice becomes evident, the urine darkens, and the feces may turn pale and offensive. Recovery takes at least 4–6 weeks, and lassitude and depression often linger on for even longer.

The disease is milder in children than in adults. Indeed, most infections in children are anicteric and many more are completely inapparent. Mortality varies from 0.1% with hepatitis A up to 10% or higher in post-transfusion hepatitis B. Most of the deaths from hepatitis A occur in adults, particularly in pregnant or older women. Rarely the patient may die rapidly as a result of fulminant hepatitis, once known as "acute yellow atrophy." More commonly, infection progresses to a subacute or chronic active hepatitis, which may become complicated by posthepatic cirrhosis.

In hepatitis B the clinical picture is indistinguishable but in general more serious. Extrahepatic manifestations, attributable to the formation of immune complexes (see Chapters 7 and 8) between HB$_s$Ag and the corresponding antibody (mainly IgM), include serum sickness, polyarteritis nodosa and glomerulonephritis. Prodromal serum sickness, consisting of transient polyarthralgia and an urticarial rash, is seen in 10–20% of cases some days or weeks before "hepatic" symptoms develop. Somewhat less than 10% of hepatitis B cases progress to subacute or chronic active or persistent hepatitis, which may go on to cirrhosis and, some believe, to malignant hepatoma. The overall mortality from hepatitis B is about 1%, largely due to the very high death rate (10–20%) in post-transfusion hepatitis, which is attributable partly to the large dose of virus received and partly to the underlying disease from which such patients are already suffering.

Pathogenesis

Hepatitis A Virus. Usually entering the body by *ingestion,* it is generally assumed that hepatitis A virus multiplies in the intestinal epithelium before spreading through the bloodstream to the parenchymal cells of the liver. Virus is detectable in the blood and feces during the week prior to the appearance of the cardinal sign, dark urine, and disappears after the serum transaminase levels reach their peak. The incubation period is about 4 weeks (within the limits of 2–6 weeks).

Hepatitis B Virus. This virus enters the body by *artificial inoculation* of human serum, as well as by other routes not yet clearly understood.

The incubation period is longer than with hepatitis A, namely 6 weeks to 6 months (average 10 weeks). Although hepatitis B virus can be detected in the feces, the more striking finding is that in a small proportion of cases it persists in the serum for months or years after the patient has fully recovered. Up to 10^{10} or more particles of HB_SAg (but very much less infective virus) may be demonstrable by electron microscopy in every milliliter of serum of such carriers.

Liver biopsy at the height of hepatitis A or B reveals necrosis of liver parenchyma (Plate 23-3A). Rarely the patient develops a fulminating

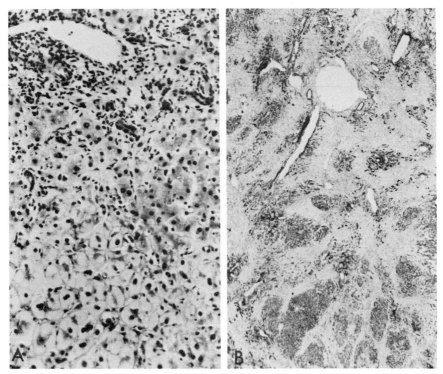

PLATE 23-3. *Viral hepatitis. (A) Liver biopsy 2 weeks after onset of acute infectious hepatitis (hematoxylin and eosin stain; ×125). Note focal necrosis within the liver lobule with characteristic ballooning of hepatocytes (bottom left), biliary stasis (center right), and inflammatory cell infiltration around a widened portal tract (top left). (B) Liver at autopsy of patient who died 3½ weeks after onset of infectious hepatitis (hematoxylin and eosin stain; ×12). Note massive necrosis of liver parenchyma and subsequent collapse leaving portal tracts with proliferating ductules and nodules of regenerating liver cells. (Courtesy Dr. P. S. Bhathal and I. Jack.)*

hepatitis and dies within days, from acute yellow atrophy (Plate 23-3B). Alternatively, subacute or chronic active hepatitis may lead to death or to cirrhosis and then sometimes to carcinoma of the liver. More usually, regeneration of liver parenchyma is complete within 2–3 months.

In cases of hepatitis B, cores (HB$_c$Ag) are demonstrable by immunofluorescence and electron microscopy in the nuclei of affected liver cells, while HB$_s$Ag is seen, together with Dane particles, in the cytoplasm. It seems clear that the cores are assembled in the nucleus of parenchymal cells and acquire their outer coat of HB$_s$Ag in the cytoplasm.

Immunity

Antibodies develop regularly after infection and confer lasting protection against reinfection with the same serotype but only partial protection against heterologous serotypes so that an individual may have more than one attack of acute viral hepatitis.

The most detailed studies of the immune response have so far been done with hepatitis B. Anti-core antibodies (anti-HB$_c$) rise rapidly and persist only as long as HB$_c$Ag continues to be synthesized. They are therefore a sensitive indicator of active infection, being the first antibodies to appear during clinical or subclinical acute infection, and persisting throughout the HBAg carrier state. Anti-HB$_s$ antibodies, on the other hand, rise about 3 weeks to 6 months later than anti-HB$_c$ and persist almost indefinitely in about half of all HBV infections; they are the only neutralizing antibodies and are therefore the best indicator of both past infection and immunity.

As in other chronic infections (see Chapter 8), antibodies may contribute to the disease state, certain aspects of which may have an "autoimmune" basis. During the preicteric stage large numbers of intact hepatocytes contain HBV cores, surface antigen and virions demonstrable by immunofluorescence and electron microscopy. Liver damage may result from immune cytolysis of noncytocidally infected hepatocytes by complement-fixing antibodies or sensitized T lymphocytes. Moreover, deposition of antigen–antibody complexes in glomeruli, arterioles and joints produces various manifestations of immune complex disease.

Laboratory Diagnosis

Jaundice can result from such a variety of causes — infective, toxic, or obstructive — that laboratory investigations are mandatory. Until techniques are developed for the isolation of the causative viruses in cultured cells, we must rely on serology and liver function tests to distinguish viral hepatitis from other causes of jaundice. The serum glutamic oxaloacetic transaminase (SGOT) test provides a sensitive early index of liver damage, rising to a peak titer of 500–2000, corresponding in time with the appearance of jaundice. Serum glutamic pyruvic transaminase (SGPT) values may be more specific. Transaminase levels tend to rise earlier and fall earlier in hepatitis A than in hepatitis B. The cephalin flocculation and thymol turbidity tests become positive, the serum albumin:globulin ratio inverts, and bilirubin appears in the urine and blood. Though these liver function tests, particularly transaminase levels, may be diagnostic of viral hepatitis, today one has available a battery of more specific serological procedures for detection of hepatitis B antigen or hepatitis A or B antibodies in serum, and hepatitis A virus in feces.

Hepatitis A. Patients with typical short incubation period "infectious hepatitis" excrete the 27 nm hepatitis A virions in their feces for a week or so prior to the development of dark urine, and this is how the causal agent was first identified by immunoelectron microscopy (Plate 23-1). Since excretion of the virions falls off greatly at the time of appearance of the jaundice, the technique is not sufficiently sensitive to have much diagnostic value. However, the same technique can be applied in reverse to identify, and even to quantitate in an approximate fashion, the IgM antibodies that develop during the icteric phase of the illness and the IgG antibodies which thereafter persist for life.

Hepatitis B. HB$_s$Ag becomes detectable in serum before the urine darkens, and it remains demonstrable for several weeks at least, and, in carriers, for years and perhaps for life. Increasingly sensitive serological techniques for the detection of HB$_s$Ag in serum have been developed and supplanted during the past few years. Successively, immunodiffusion and complement fixation were replaced by counter-immunoelectrophoresis which in turn has given way to radioimmunoassay and reverse passive hemagglutination.

Radioimmunoassay is an extremely sensitive procedure capable of detecting HB$_s$Ag in 10^{-5} dilutions (or greater) of some sera (though it should be stressed that even such a sensitive probe as this misses some infectious sera, as transmission experiments in human volunteers have demonstrated at least one serum to be infectious at a dilution one hundredfold higher than that determined by radioimmunoassay). As with immunofluorescence (see Chapter 14) there are a number of technical variants of radioimmunoassay, including methods for typing HB_sAg. The best known is the solid-phase "sandwich" system in which plastic test tubes or beads are coated with anti-HB_s and the patient's serum is incubated in the tube to allow adsorption of HB_sAg. After washing, ^{125}I-labeled anti-HB_s is added, and the amount of radioactivity binding to the solid substrate via the antigen in the anti-HB_s–HB_sAg–[125]I-labeled anti-HB_s sandwich is counted. The principles of radioimmunoassay for the detection of antigen or antibody were set out in Chapter 14.

Reverse passive hemagglutination utilizes formalin-tanned erythrocytes coated with purified antibody against HB_sAg. Such red cells, which may be stored freeze-dried, are agglutinable by HB_sAg, hence the HB_sAg titer of a carrier's serum may be rapidly determined in a simple HA assay on dilutions of the serum. Passive HA is at least a hundred times more sensitive than the counter-immunoelectrophoresis procedure widely employed by blood banks but is still slightly less sensitive than the latest modification of radioimmunoassay. HB_sAg can be typed by coating the erythrocytes with anti-*d* or anti-*y* sera.

Most of these serological procedures can be applied in reverse to detect antibodies to HBAg. Demonstration of anti-core antibody (anti-HB_c) is a good indicator of current infection. It probably plays no role in the recovery process, but appears soon after the development of clinical hepatitis and persists as long as HB_cAg continues to be synthesized. A high titer of anti-HB_c in an asymptomatic individual is generally diagnostic of the HBAg carrier state. Anti-surface antibody (anti-HB_s), on the other hand, is a neutralizing antibody which is generally not found in carriers, but correlates well with recovery from infection, appearing in all infections about 3 weeks after anti-HB_cAg and persisting for life, or at least for many years, in about half of all infections.

Epidemiology

Hepatitis A. Like poliovirus, hepatitis A virus is spread principally via the fecal–oral route. As might be expected, therefore, the disease is endemic in those parts of the world where sanitation is still primitive, namely Asia, Africa, and South America. More puzzling is the remarkable increase in incidence of the disease which has occurred since about the time of World War II in a number of Western countries where standards of hygiene were continuously improving. The analogy with poliomyelitis is probably valid. Decreasing opportunities for the acquisition of subclinical or anicteric infection in early childhood leave many older children and some adults without any acquired immunity. First infection with the virus at these ages has more serious clinical consequences. Certainly the average age incidence of clinical hepatitis in Western communities is higher than in more backward areas, with two peaks, at 5–15 years and in the young adult group. Epidemics occur every few years, but tend to be long drawn-out, often lasting for well over a year, with more cases in winter than in summer. Such epidemics may originate from and center around a focal point of community living, such as a mental hospital, military camp, children's home or comparable institution. Quite commonly, the epidemic begins explosively, indicating a common source of infection. Sometimes this can be traced to a food-handler who is a fecal carrier; food- and milk-borne epidemics may originate in this way. Water-borne epidemics, traceable to town water supplies polluted by human excreta, can be causes of major disasters; the 1956 New Delhi outbreak produced 1100 diagnosed cases in 1 week. An insufficiently publicized danger is that of oysters and clams from river estuaries close to the outlet of human sewage effluents. Family outbreaks are also very common. Typically, a child contracts the infection at school, and about a month later several of his extended family come down with the disease.

Hepatitis B. During World War II, United States troops destined to operate in zones where yellow fever was endemic were immunized with yellow fever live virus vaccine that had been "stabilized" with normal human serum. Almost 30,000 of these men contracted hepatitis, and the cause was traced to hepatitis B virus present in the serum. In Western countries today the greatest hazard is transfusion of blood, plasma or serum. Fortunately, purified globulin or albumin is safe, because the virus is removed or inactivated by the physicochemical procedures used during fractionation of the plasma. Patients requiring massive transfusions, e.g., in the course of open-heart surgery, are quite likely to become infected. In renal-dialysis units, staff as well as patients are at a considerable risk.

The second major mechanism is by transference of blood from one patient to the next via a syringe and needle that has been inadequately sterilized. In the past, doctors, dentists, nurses, and especially public health officers conducting mass immunization campaigns, have all been known to commit the sin of inoculating a succession of people with the same syringe and needle. Considerably less than 0.001 ml of infected serum can pass on the disease in this way. Little wonder then that the infection can even be transmitted by blood-sampling pipets and lancets, or tattooist's needles. Drug addicts are particularly vulnerable, and a considerable proportion become virus carriers. As a somewhat exotic rarity in modern society, the disease has been contracted in the course of the more "manly" contact sports involving mutual laceration, abrasion or other forms of blood-letting. Such minor traumatic events may be important in producing high infection rates in some primitive communities, e.g., in Melanesia (see below).

Hepatitis B virus infection is much more widespread than previously recognized, the great majority of infections being subclinical. Serum surveys for anti-HB$_s$ provide an estimate of the incidence of past infection, though they tend to underestimate the true incidence because antibody fails to develop in some infections. Such surveys reveal that in certain racial groups, e.g., Melanesians in the Southwest Pacific area, the majority of the population becomes infected. Even the HB$_s$Ag carrier rate (which is always lower than the anti-HB$_s$ rate) can be as high as 5–20% in many of the developing countries of Asia and Africa, compared with about 0.1% in North America, Europe, and Australia. Within western countries the antigenemia is much more prevalent among populations of lower socioeconomic status and in institutions such as prisons and, in particular, homes for the mentally retarded. The carrier rate exceeds 50% among some groups of U.S. "street people" addicted to hard drugs and accustomed to sharing nonsterile syringes and needles ("hippie hepatitis"). The common HBV subtype found in carriers in America and Europe is *adw* while *adr* is prevalent in Southern Europe and Asia, and *ayw* in Asia and Africa.

The data indicate that the primary mechanism of transmission of hepatitis B in most parts of the world is other than by deliberate transfusion or inoculation, but they do not shed much light on what the route is. HB$_s$Ag has been detected in saliva, feces, urine, and semen and intrafamilial spread is much more frequent between marital partners than between other members of the household. Mechanical transmission by mosquitoes is a theoretical possibility, since HB$_s$Ag has been recovered from these insects, but epidemiological evidence indicates that arthropod transmission is not important. Minor trauma and contact infection may be important. Sometimes hepatitis B virus may be transmitted vertically, either across the placenta or perinatally, from mother to baby. HB$_s$Ag is readily transmitted to babies of mothers acutely infected late in pregnancy (though less readily from chronic carriers, except in some racial groups, e.g., the Taiwanese, in whom it is common). But in most populations studied the majority of seroconversions occur postnatally—during adolescence and young adult life in western countries and earlier in childhood (especially among urban males) in countries with lower standards of hygiene and a high overall infection rate. The striking racial and even familial variation in incidence of HB antigenemia seen, for example, in certain Melanesian communities and the very high incidence in children with Down's syndrome may be associated with behavioral traits predisposing to trauma and contact infection or perhaps to inherited susceptibility.

Prevention and Control

Hepatitis A. As hepatitis A is transmitted by the fecal–oral route, control (in the absence of an effective vaccine) rests on heightened standards of public and personal hygiene. Those employed in the dispensing of food must be subject to special scrutiny. Reticulated sewerage should be the objective of every local government, and public bathing, as well as the commercial growth or private harvesting of molluscs, should be forbidden near sewerage outlets.

Contamination of water supplies by sewage is a continuing problem in advanced as well as developing countries. Sedimentation and filtration removes over 90% of virions from sewage but new physicochemical procedures (e.g., precipitation with aluminium hydroxide or with polyelectrolytes such as PE60 or adsorption onto fiberglass or cellulose nitrate membrane filters) are being developed for large-scale application to ensure the removal of viruses from treated effluents. Chlorination of water supplies cannot be relied upon to destroy relatively stable viruses such as hepatitis and the enteroviruses, particularly if large amounts of organic material are present.

Passive immunization against hepatitis A is a well established procedure which has been in use for many years. For instance, all U.S. troops in Vietnam received normal human immunoglobulin on arrival and every 6 months thereafter to ensure complete protection in the first month and protection against icteric disease as immunity waned thereafter.

Development of a vaccine against hepatitis A must await the isolation and then mass production of the agent in cell culture. If one can take poliomyelitis as a precedent, there is good reason to be optimistic about the likelihood that a live or an inactivated vaccine would be effective.

Hepatitis B. Until the major route of natural transmission of this ubiquitous agent is discovered we can do little more than take steps to minimize its more obvious parenteral spread. As most of this is iatrogenic we have the solution to a large extent in our own hands. Doctors, dentists, nurses, and we should add, tatooists and drug addicts must be educated regarding the crucial importance of sterilizing instruments that may have come into contact with blood (or feces). Autoclaving, or some other form of sterilization by heat, is the only satisfactory method; chemical disinfection is unreliable against this highly resistant virus. Disposable syringes and needles, used only once, are the ideal.

Blood banks nowadays routinely screen all blood for HB_sAg and discard material giving positive results. Since relatively insensitive tests (e.g., counter-immunoelectrophoresis) have been used until comparatively recently, such screening eliminated only grossly contaminated blood. It has not led to the dramatic reduction in post-transfusion hepatitis that had been anticipated, leading some workers to suggest that there remains another as yet unidentified agent, "hepatitis C virus."

Doctors, nurses, and technicians working in blood banks, virus diagnostic laboratories, and especially renal dialysis units are particularly exposed to virus-contaminated blood, and in these and other such environments rigid rules must be enforced to protect working personnel. Regular monthly screening of all staff and renal dialysis patients for HB_sAg and transaminase is essential, both with a view to excluding carriers and to detecting outbreaks. Spillage of even a few drops of blood must be taken seriously and the area should be regularly disinfected, e.g., with Clorox, Wescodyne, or formaldehyde. Eating, smoking, and pipetting by mouth are strictly prohibited, washing and disinfection of hands is undertaken regularly, gowns and gloves are changed at intervals, and disposable plastic equipment is used wherever feasible. In the event of an accident with HBAg-positive blood, hyperimmune human globulin is administered to all concerned.

Passive protection against hepatitis B can be transferred by human immune globulin prior to or shortly after exposure, but routine prophylaxis of hepatitis B with human gamma globulin has not proved to be nearly as successful as with hepatitis A. Clinical trials using graded doses of hepatitis B immune globulin specific for various subtypes are currently under way; preliminary results show protection against the homologous subtype and partial protection against the heterologous. Hyperimmune anti-HBV globulin should be administered to babies born of HB_sAg-positive mothers and to medical or laboratory personnel following accidental exposure, e.g., needle-prick, major blood spills in renal dialysis units, etc.

Following well-established precedents with generalized viral infections (see Chapter 12) effective control of viral hepatitis of all varieties may be anticipated when a vaccine (live or inactivated) has been developed. Usually this is contingent upon the mass production of virus in cell culture, but in hepatitis B another possibility exists at least for limited production of a vaccine. The extraordinarily high titers of HB$_s$Ag found in the serum of hepatitis B carriers presents research workers with a unique opportunity to produce an antiviral vaccine before discovering how to grow the virus in cell culture. HB$_s$Ag purified by gradient centrifugation to be free of Dane particles comprises the relevant (surface) antigens of the virus but contains no nucleic acid and is therefore noninfectious, even without inactivation. Such purified HB$_s$Ag preparations (formalin-inactivated for safety) have been shown to protect chimpanzees against HBV challenge, and will shortly be tested in man.

RUBELLA

Introduction

Rubella, or "German measles," is a trivial exanthem that would not concern us greatly were it not for the disastrous teratogenic effects it produces in the unborn child of a mother infected during pregnancy. The discovery of this association in 1941 by an Australian ophthalmologist named Gregg constitutes a striking illustration of the way in which contributions to medical knowledge have been made by observant physicians. Gregg noticed an unusual concentration of cases of congenital cataract among newborn babies in his practice, suggestive of an "epidemic of blindness." A diligent search of his records revealed that most of the mothers had contracted rubella in the first trimester of their pregnancy. Further investigations showed that these unfortunate children suffer a variety of other congenital defects, including deafness, mental retardation, and cardiac abnormalities.

In 1962 the long search for the causative virus ended with its successful cultivation in cell culture. In 1969 the first vaccine was licensed, and eradication of congenital rubella is now a realizable objective.

Properties of the Virus

The rubella virus is a typical togavirus in all respects except that it is not arthropod-borne. The virion, 60 nm in diameter, consists of a 30 nm icosahedral core enclosed with a tightly fitting envelope (Plate 23-4A). It is not related antigenically to any other viruses, and only a single serotype is known.

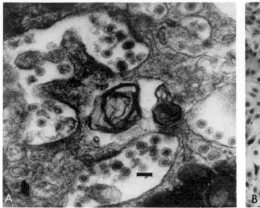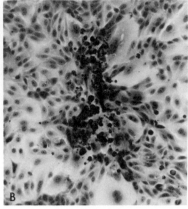

PLATE 23-4. *Rubella. (A) Electron micrograph showing budding of rubella virions into cytoplasmic vesicles. Bar = 100 nm. (Courtesy Dr. I. H. Holmes.) (B) CPE produced by rubella virus in RK-13 cells (hematoxylin and eosin stain; ×60). CPE is minimal, being initially restricted to microscopic "plaques." (Courtesy I. Jack.)*

Clinical Features

Rubella itself is usually a trivial disease; so much so that most women are uncertain about whether they have ever had it. Lymph node enlargement, followed by minimal prodromal symptoms and mild conjunctivitis, herald the development of the fine, pink, discrete macules of the rubelliform rash, which appears first on the head and spreads to the trunk and limbs. Fever is usually inconspicuous, and the rash fades after 48 hours in a fully developed case. In some however, the rash is more fleeting or even absent. Lymphadenopathy is characteristic, involving the suboccipital, postauricular, and cervical lymph nodes in particular. Arthritis is a fairly frequent feature of the disease, especially in adult females and is sometimes accompanied by myalgia and parasthesia. Postinfectious encephalitis and thrombocytopenic purpura are rare.

The *congenital rubella syndrome* (Plate 23-5) is characterized by any of the following teratogenic effects: partial or total blindness (cataracts, retinopathy, glaucoma), partial or total deafness (cochlear degeneration), congenital heart defects (especially patent ductus arteriosus sometimes accompanied by pulmonary stenosis), and mental retardation (often with microcephaly). Nonteratogenic effects include retardation of growth, osteitis, thrombocytopenic purpura, hepatosplenomegaly, hemolytic anemia, and interstitial pneumonia. Despite the diversity and severity of this pathology, the rubella syndrome is sometimes missed at birth. About 10–20% of rubella-infected babies die during the first year of life.

PLATE 23-5. *Rubella syndrome: showing severe bilateral deafness; microphthalmia, cataract, corneal opacity, strabismus, and severe bilateral visual defect. (Courtesy Dr. K. Hayes.)*

Pathogenesis

The virus enters the body by inhalation, to multiply in the upper respiratory tract and spread via the cervical lymph nodes to the blood stream. Thereafter, it seems likely that the mousepox model (Chapter 6) depicts the chain of events that occupy the 2–3 weeks comprising the incubation period. Virus can be demonstrated in the blood and nasopharynx for several days before the rash develops, and may be recovered from the nasopharynx for several days afterward. Antibody is demonstrable in the serum just after the rash appears.

When rubella infects a woman during the first trimester of pregnancy, there is a high chance that the baby will suffer teratogenic effects. The incidence of fetal damage falls off as gestation advances, but is considerably higher than previously realized. Severe damage (blindness, deafness, heart or brain defects) results from about half of all infections occurring during the first month, but rarely after the fourth month. Minor abnormalities are more frequent, and virtually all infected embryos excrete virus throughout the pregnancy as well as after birth. Abortion or stillbirth may also occur; autopsy then reveals virus in practically every organ.

Current research is focused on the basic question: what makes rubella virus teratogenic? One possible reason appears paradoxical at first sight: rubella virus is teratogenic because it is relatively noncytocidal. More cytocidal viruses may kill the fetus, leading to spontaneous abortion, but rubella virus allows many infected cells to go on dividing, at a diminished rate, and the virus is not eliminated by maternal antibody because spread occurs between contiguous cells rather than via the circulation. The hypothesis is supported by the finding that infected tissues from aborted fetuses, when explanted *in vitro*, become carrier cultures in which all cells are noncytocidally infected and continue to divide at a slower rate than uninfected controls. However, it must be added that there are many other relatively noncytocidal viruses which are nonteratogenic.

In vivo, maximum damage would be expected when viral infection occurs early in pregnancy, before the vital steps in the embryonic differentiation of eye, ear, and heart have occurred and before the embryo has acquired the capacity to synthesize interferon. Development of these organs is impaired, presumably because viral inhibition of cell division leads to a reduction in cell numbers and interferes with differentiation. In other organs inflammatory changes may result from a more usual type of cell destruction, for rubella virus is not completely noncytocidal. The persistence of virus throughout gestation and into the first months of postnatal life may continue to inflict damage, e.g., progressive development of cataracts and mental retardation after birth. We are still far from understanding why rubella virus, of all known viruses, is so prone to cause congenital defects.

Immunity

Acquired immunity to rubella lasts for many years, often for life, since there is only a single serotype, and neutralizing antibody persists for the prolonged period characteristic of systemic infections in general. Reports of second clinical attacks are often attributable to diagnostic confusion with clinically similar enteroviral exanthemata, but second infections with rubella are not uncommon. These are usually subclinical and can only be recognized by laboratory tests.

The rubella syndrome in infants provided an opportunity to look for human immunological tolerance to a congenitally acquired virus. In fact, rubella babies show no immunological tolerance. They are capable of synthesizing rubella antibodies quite normally, and they begin to do so even before birth (Fig. 23-1). The IgG found in the fetus is of maternal origin, but the high levels of IgM must have been synthesized by the fetus itself, as IgM does not cross the placenta. After birth, titers of IgM and IgG remain unusually high, as the infant continues to synthesize both classes of antibody. Nevertheless, this antibody does not eliminate the virus from its intracellular habitat, and the baby continues to shed virus for up to a year or even longer.

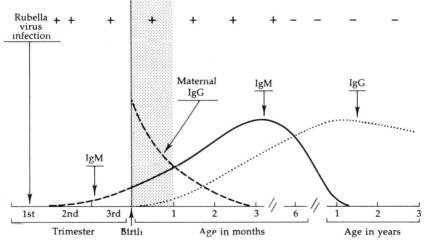

Fig. 23-1. *Schematic illustration of the pattern of viral excretion and antibody response in congenital rubella. (From S. Krugman and R. Ward, "Infectious Diseases of Children and Adults," 5th ed., p. 245. Mosby, St. Louis, Missouri, 1973.)*

Laboratory Diagnosis

A nose or throat swab is the specimen most likely to yield virus, but in the case of the congenital rubella syndrome, virus may also be readily isolated from urine, cerebrospinal fluid, leukocytes, or at autopsy from bone marrow, lens, or, indeed, virtually any organ.

The original isolation of rubella virus depended on the practical application of a well-known laboratory phenomenon, that of interference. In primary cultures of African green monkey kidney cells the virus was found to multiply well, without producing CPE; however, viral growth was recognized by demonstrating several days later that the cultures had become resistant to challenge by any one of the rapidly cytopathogenic picornaviruses. Today we have available a number of continuous cell lines in which the virus produces satisfactory CPE: BHK-21 (baby hamster kidney), Vero (green monkey kidney), RK-13 (rabbit kidney), and SIRC (rabbit cornea) are the most suitable. The CPE begins in isolated foci, spreads slowly, and rarely involves the whole cell sheet (Plate 23-4,B). CPE can be augmented by serial passage, and its cause identified by immunofluorescence or neutralization.

Evidence of past infection can be reliably obtained by serology. Titers of HI antibody remain so high for so long that this is the method of choice, although care must be taken to remove nonspecific inhibitors of rubella hemagglutination and nonspecific agglutinins from the sera before testing. A number of procedures are available for this purpose; among the most commonly employed are adsorption of the serum with kaolin, or treatment with heparin-manganous chloride or with dextran sulfate–calcium chloride. Only certain species of erythrocytes are agglutinable by rubella virus, e.g., goose, pigeon, day-old chick, or trypsin-treated human RBC, all at 4°. In addition, procedures are available to enable rubella IgM to be measured independently of IgG; an indirect immunofluorescence technique using an anti-IgM serum is one method; others require the centrifugal, chromatographic, or electrophoretic separation of IgM from serum. A high titer of rubella IgM at birth is diagnostic of congenital rubella.

There are three distinct situations that commonly call for the diagnostic facilities of the laboratory today:

(a) Nonimmunized women coming into contact with possible rubella during the first trimester of pregnancy and wishing to know urgently whether they should have an abortion,

(b) Teenage girls or newlyweds wishing to know whether they have ever had rubella, hence whether they can safely decline immunization,

(c) Newborn babies born of mothers with a history of a rash during early pregnancy or babies whose appearance gives rise to suspicion of the rubella syndrome.

The standard method for diagnosing recent rubella in a pregnant woman is to demonstrate a rising titer of HI antibodies in two specimens of serum taken 5–10 days apart. Sometimes such a delay has elapsed since the rash appeared that a high titer of antibodies may already be present in the first available serum sample. In this event a high CF titer or a high level of IgM, rather than IgG, in an HI test is strongly suggestive of recent infection (but not absolutely diagnostic, as IgM occasionally continues to be synthesized for some months).

Rubella virus can be isolated from the throat for not more than a few days after the rash appears, but it can be recovered from the amniotic fluid at any stage during the pregnancy if it is felt that the somewhat hazardous procedure of amniocentesis is justified—perhaps in the case of an extremely anxious patient recalling a rash months after the event.

The rubella syndrome is diagnosed in a newborn infant by demonstrating rubella IgM in cord blood and/or isolating virus from the throat or urine.

Epidemiology

Even though rubella is demonstrably highly transmissible within families, epidemics occur more sporadically than with measles, usually in spring; major outbreaks occur only at intervals of several years. Accordingly, most children escape infection until primary or secondary school, and about 20% of women in most Western nations reach marriageable age without immunity. The figure may be as high as 50% in some island communities.

Spread is primarily via the respiratory route. Patients are infectious for others from a few days before the rash appears until about a week after. Rubella babies are a particular danger, their saliva, urine, and other secretions being highly infectious for months after birth.

Prevention and Control

Before vaccination was available mothers encouraged their daughters to acquire immunity to rubella by exposing themselves deliberately to infection; even "rubella parties" were not unknown. Once pregnant however, women must avoid contact with rubella, especially during the first trimester. Rubella babies are a particular hazard and should be nursed in strict isolation, away from maternity hospitals, because of the extreme risk of transmitting the infection via nurses, doctors, or visitors to pregnant women attending the antenatal clinic. Immunoglobulin is of no value in the prophylaxis of the rubella syndrome.

Clearly, the solution to this problem lies in active immunization of potential mothers. Live attenuated vaccines must fulfil not only the usual requirements of immunogenicity and safety but must not themselves produce teratogenic effects (because of the uncertainty of an immunization program that aims to exclude pregnant women), and should not be transmissible to contacts.

Three live vaccines have been in general use since 1969. All have the following characteristics in common: (a) They are administered as a single subcutaneous injection. (b) Multiplication of the virus in the recipient is usually symptomless, but may cause transient arthritis (especially in adults), lymphadenopathy, low-grade fever, headache, and rarely a rash. (c) Small numbers of virus particles are shed from the throat, but do not spread to contacts. (d) When a pregnant woman is immunised in error the vaccine strain is able to reach the fetus but there is no evidence for its teratogenicity. (e) Rubella antibodies develop in about 95% of recipients, reach titers four- to eightfold lower than those following natural infection, and persist for several years. (f) Immunity to challenge with virulent rubella virus is solid initially, but after a few years declines to the point where subclinical reinfection with wild virus occurs quite commonly.

The three main contenders for widespread use are designated HPV-77, Cendehill, and RA27/3. HPV-77 was developed in the United States by passage of rubella virus 77 times in cultures of green monkey kidney cells, and then several times in duck embryo cells, to reduce the risk of carry-over of simian viruses. It is extensively used in the United States, but is not licensed for administration to adults, in whom it produces an unacceptably high incidence of arthritis.

The Cendehill vaccine, developed in Belgium by passage of rubella virus in primary cultures of kidney cells taken from a closed colony of specific-pathogen-free rabbits, is widely used in many different countries. Being more attenuated than HPV-77 it produces arthralgia in adults only rarely and lymphadenopathy occasionally.

RA27/3 is a vaccine produced in the United States by growth in the diploid strain of human embryonic fibroblasts, WI-38. It can be administered directly in the nose, where it takes quite satisfactorily, but like other vaccines is usually given subcutaneously. HI and neutralizing antibody titers are somewhat higher following RA27/3 and immunodiffusion reveals that the two principal species of precipitating antibody detectable following natural infection (anti-*theta* and anti-*iota*) are induced in considerably higher titer than by the other two vaccines. This may explain why protective immunity lasts longer with RA27/3.

There are two opposing schools of thought on rubella immunization policy. The official view in the United States is that all preschool children (male and female) should be immunized in an attempt to reduce the circulation of rubella in the community and so, indirectly, to reduce the exposure of unimmunized as well as immunized pregnant women to the risk of infection. This policy, which was introduced in 1969, has been demonstrably effective in reducing the incidence of rubella and the rubella syndrome. Theoretically at least, the virus itself could be eradicated from a country that vigorously pursued a policy of universal immunization of all preschool children. Another view, propounded in England and Australia, for example, is that immunization should be concentrated on prepubertal girls (say, on entering high school at 12–13 years), together with any nonimmune older women of child-bearing age. Since it has yet to be unequivocally established that the vaccine itself is not teratogenic, it must not be given to pregnant women and care must be taken that the vaccinated woman does not conceive in the two months following immunization. The simplest way of achieving this is to immunize immediately postpartum; otherwise the woman is advised to practice contraception for 3 months.

About 80% of adult women have serological evidence of prior natural infection with rubella (clinical or subclinical), and are therefore solidly immune for life. In some communities it is standard practice to determine the titer of HI antibodies and to immunize only the 20% or so who are seronegative. In the long run however it is less costly and less time-consuming to immunize all women of child-bearing age.

The main argument in favor of the policy of immunizing 12–13 year old girls rather than all infants is that naturally acquired immunity is very much more effective than that following immunization with any of the vaccine strains. HI antibody titers are up to tenfold lower following immunization, and more importantly, protection against challenge does not appear to be solid for more than 5–10 years. A policy of immunizing all preschool children could reduce greatly the circulation of wild virus in the community and so deprive adult women of natural subclinical boosters, leaving them nonimmune at the very time they need protection.

There remains the awkward problem of what to do with the woman who does in fact develop a rash in the first trimester of pregnancy. The first essential is to verify the diagnosis by the laboratory methods described above. If the diagnosis of rubella is established, the risk to the baby is so substantial that the attending physician should recommend a therapeutic abortion.

SLOW VIRUS DISEASES OF THE BRAIN

Introduction

The extraordinary findings with scrapie in sheep triggered an explosion of research into the possible role of "slow viruses" in the etiology of degenerative diseases of the human central nervous system (CNS). Though this subject has been discussed in detail in Chapter 8, it is appropriate to consider here the clinical and epidemiological characteristics of the slow virus diseases of the human brain that have been discovered in the last few years, the spongiform viral encephalopathies: kuru, and Creutzfeldt–Jakob disease.

Kuru

Kuru is a disease of the central nervous system characterized by cerebellar degeneration, leading to ataxia, tremors, incoordination, dementia, and certain death within a year. It has never been reported outside a single tribe of Melanesians and is confined to an area of 1000 square miles in the South Fore region of the New Guinea highlands (Plate 23-6). According to the recollections of the village elders, kuru was unknown until about 1920. By 1957, when the disease was first described, kuru was responsible for 80–90% of all deaths in South Fore women. Indeed, it had so decimated the female population that the tribe had come to contain only one third as many women as men. As most of the deaths were in young women of childbearing age, kuru had left a legacy of orphans, who presented a tragic social problem. Children of both sexes were also afflicted by the disease, but older males were largely spared (except insofar as 20% of all male deaths resulted from ritual murder on suspicion of sorcery in connection with kuru). This led to the hypothesis (now discredited) that kuru was a sex-linked genetic disease, in which the mutant gene had become widely disseminated throughout this isolated tribe by centuries of inbreeding. In order to explain how such a common lethal gene had not exterminated the tribe long ago, it was postulated that, by analogy with diseases like favism in man or scrapie in sheep, the mutant gene merely predisposed the individual to kuru, and that the actual precipitating cause was perhaps a plant or a virus first introduced into the area around 1910–1920.

PLATE 23-6. *Kuru. Irreversible cerebellar degeneration in child from the Fore region of the New Guinea highlands. (Courtesy Sir Macfarlane Burnet.)*

In 1965 Gajdusek and his colleagues working at the United States National Institutes of Health demonstrated that a fatal kuru-like disease could indeed be transmitted to chimpanzees by intracerebral inoculation with brain extracts from patients who had died from the disease. Since that time they have shown that the causative agent is filterable, that it is present in high titer in a variety of human tissues in addition to the brain, and that the infection can be passed to other chimpanzees and monkeys, in which the condition can be more readily studied. Characterization of the etiological agent has led many virologists to believe that it may be a membrane-associated "viroid" (see Chapter 8), rather than an orthodox virus.

The epidemiological pattern of kuru has changed dramatically in the years that have elapsed since Gajdusek first described the situation in 1957. Kuru has become much less common, with no new cases at all in children, and the average age of onset in women rising by about one year every year. The change coincided in time with the advent of the Australian colonial administration, which first penetrated into the area in the 1950's and promptly put an end to the practice of cannibalism. Closer interrogation has since revealed that the women of the tribe were introduced to cannibalism around 1910–1920 and had ever since pursued the custom, accompanied by elaborate ritual, of consuming deceased relatives. The men took no part in these ceremonies but the children were involved because tribal custom required that they lived exclusively with the women prior to their initiation. The peculiar sex and age distributions of kuru thus had a social rather than a genetic basis.

The incredible story of kuru is far from solved. Not the least of the remaining mysteries is where the virus (if it be a virus) first came from, and what its ecology is in its more "natural" environment (?Creutzfeldt–Jakob disease).

Creutzfeldt–Jakob Disease

This rare presenile dementia, in which the patient, usually middle-aged, becomes uncoordinated and demented and dies within 9–18 months, presents a histological picture so similar to kuru as to suggest that the two may be caused by the same or very similar agents. In both there is a "spongy degeneration" of the brain (status spongiosus), characterized by vacuolation of the axons and dendrites of neurons accompanied by proliferation and hypertrophy of astroglia, and a rather striking electron microscopic picture revealing a tangled web of membranes. The total absence of inflammation is consistent with the belief that the etiological agent does not elicit an immunological response, perhaps because it does not contain an immunogenic coat. A comparable disease with the same histopathology has been successfully transmitted to chimpanzees, monkeys, and cats by inoculation of filtrates from affected human brain. The recent accidental transmission of Creutzfeldt–Jakob disease to the recipient of a corneal transplant implies that the virus is not restricted to the brain.

Other Degenerative Diseases of the CNS

In addition to the rare conditions of kuru and Creutzfeldt–Jakob disease discussed in this chapter there are several other viruses that have now been implicated in more common degenerative diseases of the CNS. Measles virus has been unequivocally shown to be the cause of subacute sclerosing panencephalitis (SSPE) discussed in detail in Chapters 8 and 21. A papovavirus has been associated with progressive multifocal leukoencephalopathy, discussed in Chapter 22. EB virus has been implicated in the Guillain–Barré syndrome (acute polyradiculoneuropathy), Bell's palsy, and transverse myelitis, while another herpetovirus has been recovered from cases of subacute myeloopticoneuropathy (see Chapter 16). Influenza is sometimes followed by polyneuritis. It may well transpire that other viruses are also causally involved in these various diseases.

The most important of the demyelinating diseases of the CNS are multiple sclerosis, the common presenile dementias, and Parkinson's disease—all still of unknown etiology. The epidemiology of multiple sclerosis suggests an infection acquired in childhood. Over the years hopes of identifying the agent have been raised, and subsequently dashed, by reports of a series of different viruses isolated from cases of this disease, including most recently parainfluenza type 1 and measles. All associations suggested so far appear to be irrelevant, and the search goes on.

Research into the etiology of these diseases is slow because human volunteer studies are out of the question, and transmission to primates is both uncertain and expensive. Reasonable suspicion of a viral etiology of such diseases rests on the induction in primates of a clinical and histopathological picture identical to that seen in man, using a cell-free filtrate of material taken from a human case. Serial passage in primates is then necessary to establish the present of a propagating agent. Koch's postulates cannot really be fulfilled until a particular virus is consistently isolated and passaged in cell culture.

FURTHER READING

HEPATITIS

Blumberg, B. S., Sutnick, A. I., London, W. T., and Millman, I. (1972). "Australia Antigen and Hepatitis." Chem. Rubber Publ. Co., Cleveland, Ohio.
Feinstone, S. M., Kapikian, A. Z., and Purcell, R. H. (1973). Hepatitis A: Detection by immune electron microscopy of a viruslike antigen associated with acute illness. *Science* **182**, 1026.
MacCallum, F. O. (1972). Viral hepatitis. *Brit. Med. Bull.* **28**, 103.
Prier, J. E., and Friedman, H., eds. (1973). "Australia Antigen." Univ. Park Press, Baltimore, Maryland
Purcell, R. H. (1975a). Current understanding of hepatitis B virus infection and its implications for immunoprophylaxis. *Perspect. Virol.* **9**, 49.
Purcell, R. H. (1975b). Current status of diagnostic tests for viral hepatitis. *In* "Immunological Aspects of Anesthesia and Surgical Practice" (A. Mathieu and B. D. Kahan, eds.). Grune and Stratton, New York.

Symposium. (1975). Viral hepatitis. *Amer. J. Med. Sci.* (in press).
Vyas, G. N., Perkins, H. A., and Schmid, R., eds. (1972). "Hepatitis and Blood Transfusion." Grune & Stratton, New York.
World Health Organization. (1973). Viral hepatitis. *World Health Organ., Tech. Rep. Ser.* **512.**
World Health Organization. (1975). Viral Hepatitis. *World Health Organ., Tech. Rep. Ser.* **570.**
Zuckerman, A. J. (1975). "Human Viral Hepatitis: Hepatitis-Associated Antigen and Viruses." 2nd ed. North-Holland Publ., Amsterdam.

RUBELLA

Banatvala, J. E. (1970). Rubella. *Mod. Trends Med. Virol.* **2,** 116.
Horstmann, D. M., Leibhaber, H., Le Bouvier, G. L., Rosenberg, D. A., and Halstead, S. B. (1970). Rubella: Reinfection of vaccinated and naturally immune persons exposed to an epidemic. *N. Engl. J. Med.* **283,** 771.
Horta-Barbosa, L., Fuccillo, D., and Sever, J. L. (1969). Rubella virus. *Curr. Top. Microbiol. Immunol.* **47,** 69.
Huygelen, C., Peetermans, J., and Prinzie. A. (1969). An attenuated rubella virus vaccine (Cenderhill 51 strain) grown in primary rabbit kidney cells. *Progr. Med. Virol.* **11,** 107.
Krugman, S. (1969). Proceedings of the International Conference on Rubella Immunization. *Amer. J. Dis. Child.* **118,** 1.
Krugman, S., and Katz, S. L. (1974). Rubella immunization: A five-year progress report. *N. Engl. J. Med.* **290,** 1375.
Parkman, P. D., and Meyer, H. M. (1969). Prospects for a rubella virus vaccine. *Progr. Med. Virol.* **11,** 80.
Plotkin, S. A., Farquhar, J. D., Katz, M., and Buser, F. (1969). Attenuation of RA 27/3 rubella virus in WI-38 human diploid cells. *Amer. J. Dis. Child.* **118,** 178.
Rawls, W. E. (1974). Viral persistence in congenital rubella. *Progr. Med. Virol.* **18,** 273.

SLOW VIRUS DISEASES OF THE BRAIN

Brody, J. A., Henle, W., and Koprowski, H. (1967). Chronic infectious neuropathic agents (CHINA) and other slow virus infections. *Curr. Top. Microbiol. Immunol.* **40,** 1.
Fucillo, D. A., Kurent, J. E., and Sever, J. L. (1974). Slow virus diseases. *Annu. Rev. Microbiol.* **28,** 231.
Gajdusek, D. C., and Gibbs, C. J. (1973). Subacute and chronic diseases caused by atypical infections with unconventional viruses in aberrant hosts. *Perspect. Virol.* **8,** 279.
Gibbs, C. J., and Gajdusek, D. C. (1971). Transmission and characterization of the agents of spongiform virus encephalopathies: Kuru, Creutzfeldt-Jakob disease, scrapie and mink encephalopathy. *Res. Publ., Ass. Res. Nerv. Ment. Dis.* **49,** 383.
Hotchin, J. (1974). Slow virus diseases. *Progr. Med. Virol.* **18.**
Weiner, L. P., Johnson, R. T., and Herndon, R. M. (1973). Viral infections and demyelinating diseases. *N. Engl. J. Med.* **288,** 1103.
Zeman, W., and Lennette, E. J. (1974). "Slow Virus Diseases." Williams & Wilkins, Baltimore, Maryland.
Zu Rhein, G. M. (1969). Association of papova-virions with a human demyelinating disease (progressive multifocal leukoencephalopathy). *Progr. Med. Virol.* **11,** 185.

CHAPTER 24

Common Viral Syndromes

INTRODUCTION

The preceding nine chapters may be said to have dealt "horizontally" with individual families of viruses and the diseases they cause. At this juncture it may be helpful to examine the scene in the "vertical" dimension, by taking a bird's eye view of the important clinical syndromes and the viruses that cause them.

A brief clinical description of each syndrome will be given together with a table listing the causative viruses. As some sort of guide to the relative contribution of each agent, viruses have been designated as "common" or "less common" causes of the syndrome in advanced western communities. This rather arbitrary division should be regarded as nothing more than approximate, because the prevalence of individual viruses is constantly changing, and often differs quite significantly from country to country.

For simplicity, the common viral syndromes will be classified according to the major systems of the body. This policy creates minor problems when it comes to systemic diseases with generalized symptomatology referable to many different parts of the body. Nevertheless, we can think of most of the important viral diseases in terms of infections limited to the respiratory or alimentary tract, or as systemic infections with primary involvement of particular target organs such as the brain, the liver, the joints and muscles, the eyes or the skin.

Throughout this chapter it should be borne in mind that most of the syndromes to be described are not peculiar to viruses; the differential diagnosis will usually include many types of microorganisms, as well as noninfectious causes. For example, pneumonia can result from infection by viruses, bacteria, rickettsiae, chlamydiae, mycoplasmas, fungi, yeasts, protozoa, or helminths. Quite respectable descriptions of rashes in the "clinical literature" of 2000 years ago could be taken today to apply equally well to certain exanthemata of viral, bacterial, rickettsial, or fungal origin. By this we do not mean to imply that such diseases are clinically indistinguishable, but merely that there is not always a definitively "viral" character about a given viral syndrome. Where helpful differentiating features favoring a viral diagnosis are apparent, they will be mentioned in the text. For more detailed descriptions of the clinical manifestations of infection by an individual virus, the reader is referred back to the relevant chapter, and more particularly to textbooks on infectious diseases such as those listed in the Further Reading lists. The present chapter aims to do no more than draw together the threads that run through Part II of this book.

VIRAL DISEASES OF THE RESPIRATORY TRACT

Upper respiratory tract infection (U.R.T.I.) is the most common disease of man, and most of it is caused by viruses. It has been calculated to account for an average of six episodes per person per annum, up to one third of all calls on general practitioners, and countless millions of lost working hours. Attempts have been made to assess the relative contributions of the various respiratory viruses to the total spectrum of respiratory disease. The *rhinoviruses, coronaviruses, parainfluenza,* and *respiratory syncytial viruses* are the most common agents, usually producing trivial disease of the upper respiratory tract, especially when they infect adults. On the other hand, most of the respiratory infections in young children that are severe enough to require hospitalization are attributable to *respiratory syncytial, parainfluenza,* and *influenza* viruses. Seriously ill adults are more likely to be suffering from pneumonia caused by bacteria or *Mycoplasma pneumoniae* than viral infections.

Most of the respiratory viruses are highly infectious, spreading as droplets generated during coughing, sneezing, talking, etc. Respiratory viral infections have short incubation periods, of the order of 2–5 days. Hence, they regularly produce epidemics, especially in the winter and spring. A striking example is the respiratory syncytial virus, which precipitates an abrupt epidemic in infants almost every winter. More widespread epidemics, which may close schools and factories, are produced every 2–3 years by influenza A and somewhat less frequently by influenza B (see Fig. 10-1). Adenovirus types 3, 4, 7, 14, and 21 and parainfluenza type 1 occasionally cause limited outbreaks in military camps and other semiclosed communities of young people. The other important respiratory viruses, namely the rhinoviruses, coronaviruses, parainfluenza viruses, adenoviruses 1, 2, 5, and 6, and

TABLE 24-1

Viral Diseases of the Respiratory Tract

DISEASE	VIRUS	
	COMMON	LESS COMMON
Upper respiratory infection (including the common cold and pharyngitis)	Rhinovirus, >100 types Coronaviruses Parainfluenza 1–3 Respiratory syncytial Influenza A, B Herpes simplex 1	Adenovirus 1–7, 14, 21 Coxsackie A21, 24; B2–5, etc. Echovirus 11, 20, etc. Parainfluenza 4 EB virus
Croup (laryngotracheobronchitis)	Parainfluenza 1–3 Influenza A, B Respiratory syncytial	Adenoviruses Measles
Bronchiolitis	Respiratory syncytial[a] Parainfluenza 1, 3 (2)	Influenza A
Pneumonitis	Respiratory syncytial[a] Parainfluenza 1, 3, (2) Influenza A	Adenovirus 3, 4, 7 (14, 21) Measles Varicella

[a] In infants only.

coxsackieviruses A21, A24, B3, and B5, tend rather to be endemic in the community, causing only sporadic cases or occasional small outbreaks. Respiratory syncytial and parainfluenza viruses are so prevalent that the first infections with them usually occur in infancy or early childhood. In contrast, some of the enteroviruses and adenoviruses may not be encountered until adult life. Subclinical infections do not occur nearly as commonly with respiratory viruses as with viruses entering the body by other routes.

Acquired immunity to the respiratory viruses tends to be relatively short-lived. Interferon confers limited protection against all viruses for a few weeks, and specific immunity against the homologous serotype, mediated principally by IgA, persists for perhaps a few years then wanes. Hence reinfection with the same agent sometimes occurs. Even more significant, however, is the fact that a given syndrome may be caused by a large number of serologically distinguishable agents, showing little or no cross-immunity (see antigenic drift, Chapter 11).

Many of the viruses listed in Table 24-1 can, on occasion, produce disease at any level in the respiratory tract—coryza, pharyngitis, laryngitis, croup, bronchitis, bronchiolitis, or pneumonia. For example, respiratory syncytial virus or parainfluenza virus type 3 can cause a potentially lethal pneumonitis, bronchiolitis, or croup in infants, but usually little more than a sore throat or common cold-like illness in adults. Indeed, the syndromes merge into one another as the infection moves progressively down the respiratory tract (Fig. 24-1). In general, the disease becomes more serious the lower the virus goes.

Upper Respiratory Infection Including the Common Cold

We will discuss first a group of mild upper respiratory tract infections loosely labeled "U.R.T.I." The term embraces everything from the common cold to pharyngitis. The classical common cold is characterized by rhinitis, with copious watery discharge from the nose, and little or no fever. On the other hand, uncomplicated pharyngitis presents as a sore throat with cervical lymphadenitis and fever. It must be appreciated that, although these syndromes sometimes present in relatively "pure" form, they often overlap and merge. Table 24-1 and Fig. 24-1 indicate that the *rhinoviruses* are the most common, and *coronaviruses* the second most common cause of the common cold. Rhinoviruses are prevalent in children the year round, but coronaviruses are responsible mainly for winter colds in adults. "Sore throats" (pharyngitis) more frequently follows infection by *para-*

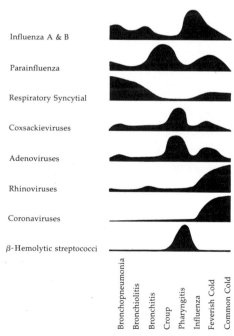

FIG. 24-1. *Diagram showing the frequency with which particular viruses produce disease at various levels of the respiratory tract.* (*Courtesy Dr. D. A. J. Tyrrell.*)

influenza viruses or *adenoviruses,* or during epidemics, *influenza* viruses. The vesicular pharyngitis and gingivostomatitis attributable to *coxsackie* and *herpes simplex* viruses come into a different category. These two conditions are marked by the occurrence of vesicles (then ulcers) on the mucous membrane of the throat and mouth respectively. *Infectious mononucleosis,* caused by the EB virus, usually presents as an exudative pharyngitis with cervical lymphodenopathy ("glandular fever"), which may need to be differentiated from diphtheria.

Secondary bacterial invasion may complicate any of these viral infections and delay recovery by a few days. Sinusitis or otitis media may supervene. However, bacteria are not often the primary pathogens in upper respiratory infections. *Streptococcus pyogenes* is the most important, being responsible for something less than 10% of the pharyngitis, but most of the acute exudative tonsillitis.

Influenza

Influenza must be considered separately (see Chapter 20 for details). The severe case is a distinct clinical entity, with fever, headache, generalized aches, and some degree of laryngotracheobronchitis. Extensive epidemics involve a high proportion of the nonimmune members of the community. The mortality rate is significant in debilitated adults, especially old people who are already weakened by chronic pulmonary or cardiac ailments. Influenza viruses are responsible for most of the statistically observed "excess deaths" among these groups during winters when they are epidemic.

Laryngotracheobronchitis (Croup)

Croup is one of the most serious diseases of young children, and is caused almost exclusively by viruses (Table 24-1). The child presents with hoarseness, cough, and partial respiratory obstruction, resulting in breathing difficulties, rib retractions, inspiratory stridor, and in severe cases, cyanosis. Tracheotomy may be required to save the child's life. The most common causative agents are the *parainfluenza* viruses, with the *respiratory syncytial virus* second in importance. During epidemics of *influenza,* this virus becomes a major cause of severe croup. A rare bacterial cause today is *Corynebacterium diphtheriae.*

Bronchitis

Influenza, parainfluenza, and *respiratory syncytial viruses* are the most important causes of acute bronchitis. The role of viruses in chronic bronchitis has been debated for years. There is no persuasive evidence that any virus is etiologically involved, though smoking, air pollution (smog), and perhaps bacteria are. Some data suggest that the disease may be exacerbated by intercurrent viral infection.

Bronchiolitis

The *respiratory syncytial virus* is the most important single cause of bronchiolitis (Plate 21-5). The syndrome is largely confined to infants under 12 months. At this age their narrow airways are easily blocked by mucus and edema. The disease can develop with remarkable speed. Breathing becomes rapid and labored, and is accompanied by a persistent cough, expiratory wheezing, cyanosis, a variable amount of atelectasis, and marked emphysema, visible by X-ray (Plate 21-6). The infant may die overnight. Maternal antibody fails to provide any protection against the disease; indeed, the peak incidence of bronchiolitis is in the first 6 months of life when maternal IgG is present. *Parainfluenza* viruses and *influenza* (in epidemic years) are less common causes of this potentially lethal disease.

Pneumonitis

Respiratory syncytial and *parainfluenza* (especially type 3) viruses may also produce pneumonitis in young children; each is responsible for just over 10% of all pneumonia in children. The "sudden infant death syndrome" in which babies are found dead in the morning after being put to bed "perfectly well" the night before ("cot deaths") are generally considered to be of physiological or biochemical causation. Lung pathology is usually minimal but respiratory syncytial, parainfluenza or influenza viruses are often isolated. During influenza epidemics this virus becomes a significant cause of pneumonia; pure "viral" pneumonitis is rare, but superinfection by pneumococci, staphylococci, or streptococci leads frequently to lobar or bronchopneumonia which may cause death, particularly in old people already debilitated by chronic pulmonary or cardiac disease.

Now that immunosuppressive drugs are in widespread use we are seeing increasing numbers of lethal viral pneumonitis due to *measles, varicella, cytomegalovirus,* or *adenoviruses,* e.g., in leukemic children. The "giant-cell pneumonitis" of measles (Plate 24-1) is more common than generally thought, but only rarely presents as the overwhelming disease that kills immunologically deficient invalids without producing a rash. Varicella also occasionally terminates fatally with inter-

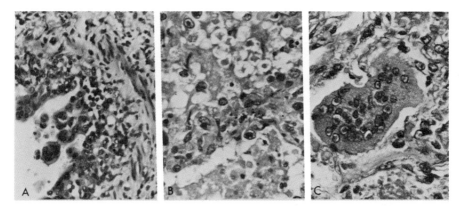

PLATE 24-1. *Viral infections of the lung. Sections from fatal cases of pneumonitis in children. (A) Respiratory syncytial virus, bronchiolitis (hematoxylin and eosin, ×135). Note multinucleated giant cells being desquamated into bronchiole, which is surrounded by leukocytes. (B) Adenovirus, pneumonia (hematoxylin and eosin, ×126). Note characteristic intranuclear inclusions. (C) Measles, giant-cell pneumonia (hematoxylin and eosin, ×216). Note multinucleated giant cell with intranuclear inclusions. (Courtesy I. Jack and A. Williams.)*

stitial pneumonitis in immunologically compromised children, including those infected perinatally from acutely infected mothers, or even in adults.

Viral pneumonitis differs from the more lethal bacterial pneumonias in that it is usually confined to diffuse interstitial lesions, rather than the more uniform consolidation of classical lobar pneumonia or the streaky consolidation of bronchopneumonia. The radiological findings are not striking; they often show little more than an increase in hilar shadows or, at most, scattered areas of consolidation (Plate 24-2). Differential diagnosis from the "primary atypical pneumonia" caused by *Mycoplasma pneumoniae, Rickettsia burneti,* the chlamydia causing psittacosis, or certain rare fungi is not always easy.

VIRAL GASTROENTERITIS

Gastroenteritis is second only to respiratory infection as a cause of human morbidity, and in many parts of the world is a major killer of undernourished children. Symptoms include diarrhea, with a variable degree of nausea, vomiting, malaise, cramping abdominal pain, and fever. The incubation period is 24–96 hours (average 2 days), the dura-

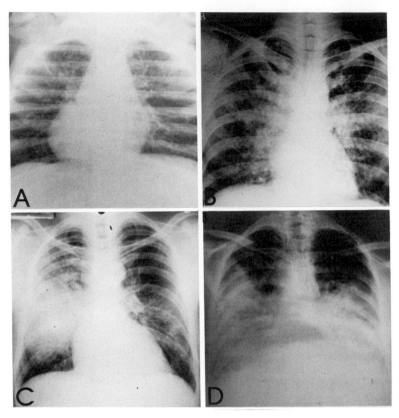

PLATE 24-2. *Radiographs of viral pneumonitis. Note streaky, patchy, or nodular consolidation. (A) Respiratory syncytial virus. (B) Varicella. (C and D) Influenza. (Courtesy Dr. J. Forbes.)*

tion of illness 12–48 hours, and complete recovery is the usual outcome. Until recently the etiology of most cases remained obscure. It has been recognized for years that the common bacterial causes, the salmonellae, shigellae, and pathogenic *E. coli* are between them responsible for only a small minority of the total. Even allowing for the fact that some others are of allergic or psychogenic origin, the great majority of cases are clearly viral. This was established some years ago by transmission of the disease to volunteers using stool filtrates.

Recently two important new viruses have been discovered by utilizing the novel technique of immunoelectron microscopy on ultracentrifuged fecal preparations. The *Norwalk agent* is a 27 nm icosahedral

TABLE 24-2
Viral Gastroenteritis

PROVED	POSSIBLE
Rotavirus	Coronaviruses
Norwalk agent	Adenoviruses
Hawaii agent	Echoviruses 11, 14, 18, 22, and others
	Coxsackie A viruses
	Reoviruses

virus morphologically resembling a parvovirus (see Chapter 22). It seems to be a major cause of the disease variously known as "epidemic diarrhea and vomiting," or "winter vomiting disease." This agent occurs predominantly in outbreaks of acute infectious nonbacterial gastroenteritis during the colder months (September to March) in both adults and children throughout the world. It shows no serological cross-reaction with the *Hawaii agent,* which may or may not be a member of the same family.

The *rotavirus* is a member of the Reoviridae family. It appears to be the major cause of "infantile enteritis" in young children throughout the world, infecting most infants in the first or second year of life with a low but significant mortality, but is relatively uncommon in adults (see Chapter 22 for details).

Additional viruses will doubtless be implicated in gastroenteritis before long. *Coronaviruses* and *adenoviruses* which currently defy *in vitro* cultivation are commonly visualized by electron microscopy in feces from patients with gastroenteritis. The role of *echoviruses* and *coxsackieviruses* is still unclear. Certain serotypes, e.g., echoviruses 11, 14, 18, and 22, have been repeatedly recovered during epidemics, but the evidence for their causal relationship to the disease must still be regarded as circumstantial in light of the frequency with which echoviruses are excreted by well people (Table 24-2).

VIRAL DISEASES OF THE CENTRAL NERVOUS SYSTEM

Most meningitis and almost all encephalitis is of viral etiology (Table 24-3). Many viruses infect the CNS as an occasional complication of an infection established elsewhere in the body which has fortuitously spread to the brain via the bloodstream. For simplicity, we shall discuss meningitis and encephalitis separately; it must be understood, however, that there is a meningitic component in many of the encephalitides, and vice versa.

Aseptic Meningitis

The term "aseptic" refers to the type of meningitis in which the cerebrospinal fluid (CSF) is nonpurulent (therefore not turbid). *Mumps, coxsackieviruses B1–5 and A9,* and *echoviruses 4, 6, 9, 11, 14, 18,*

TABLE 24-3
Viral Diseases of the Central Nervous System

SYNDROME	VIRUS	
	COMMON	LESS COMMON
Aseptic meningitis	Mumps Coxsackievirus B1–6, A9 Echovirus 4, 6, 9, 11, 14, 18, 30	Other enteroviruses Poliovirus 1–3 Herpes simplex Lymphocytic choriomeningitis Many other viruses occasionally
Paralysis	Poliovirus 1–3	Coxsackie A7, possibly others
Encephalitis	Several arboviruses (see Table 19-4) Herpes simplex Enterovirus 71	Mumps Rabies Adenovirus 7 Herpes B virus
Postinfectious encephalomyelitis	Measles	Vaccinia Varicella Mumps Influenza Rubella
Encephalopathy, incl. Reye's syndrome		Influenza Varicella-zoster
Infectious polyneuritis Guillain–Barré syndrome Transverse myelitis Bell's palsy Radiculomyelitis		Influenza EB virus?
Subacute opticoneuropathy	New herpetovirus?	
Subacute sclerosing panencephalitis	Measles	
Progressive multifocal leukoencephalopathy	Human papovavirus	
Creutzfeldt–Jakob disease	Viroid?	
Kuru	Viroid?	

and *30* are the common causative agents (Table 24-3). Other *enteroviruses* (see Chapter 18), including *polioviruses* and many other viruses, are more rarely involved. Exotic viruses include *lymphocytic choriomeningitis (LCM)*, acquired from rodents (see Chapter 22).

The patient presents with headache, vomiting, fever, and neck or back stiffness. The Kernig and Brudzinski signs are often positive. Lumbar puncture reveals a clear CSF under moderate pressure, with a slightly elevated protein (40–100 mg/100 ml), normal glucose, and a moderate number of white cells (10–2500, usually 50–500/mm³), mainly lymphocytes after the first day or two. Differential diagnosis is from nonpurulent meningitis due to such organisms as *Mycobacterium tuberculosis, Treponema pallidum,* Leptospirae, or the yeast *Cryptococcus neoformans (Torula histolytica),* or from partially antibiotic-treated infections with the more common pus-producing organisms, namely *Neisseria meningitidis, Hemophilus influenzae, Diplococcus pneumoniae, Escherichia coli,* and a host of rarer possibilities. All of the bacteria produce much more serious disease with a substantially higher incidence of death and neurological sequelae. The viral disease is relatively mild, with a negligible mortality and rare sequelae.

Viral meningitis is not necessarily accompanied by any other indication of viral infection elsewhere in the body, even in the case of mumps. However, when it is, diagnosis is greatly facilitated. For instance, meningitis due to echovirus types 4, 9, or 16 or coxsackievirus types A9 or A23 is sometimes accompanied by a maculopapular rash. Moreover, enteroviral meningitis is usually seen during the course of an established epidemic, chiefly during the summer and early autumn months, and in all age groups; mumps meningitis, on the other hand, is more often seen in the winter and spring, principally in children and adolescents. Most of the other victims of the epidemic, including contacts in the same family, will not be suffering from meningitis but perhaps from an inapparent infection, a mild "undifferentiated U.R.T.I.," or a rash, etc. Paralysis is rarely seen except in frank poliomyelitis. However, it has occurred following aseptic meningitis due to coxsackievirus A7, while other enteroviruses not infrequently produce an inconsequential and fleeting paresis.

Encephalitis

Encephalitis is the most serious of all the viral syndromes. A large number of *arboviruses* are endemic to restricted areas of the world, particularly the tropics, and cause occasional outbreaks also in more temperate regions under particular ecological circumstances (see Chapter 19). Fortunately, even then, encephalitis is a relatively rare complication of what is usually an inapparent infection. In most temperate countries the most common identified cause of sporadic encephalitis is *herpes simplex* virus (usually type 1, except the neonatal cases which are type 2). This disease, perhaps more correctly described as a meningoencephalitis, has a mortality of about 50% and a distressingly high frequency of severe permanent sequelae (see Chapter 16). The disease is only sometimes heralded by herpes labialis. It often develops insidiously with psychological abnormalities and may progress to aphasia, hemiparesis, and visual field defects, which invite confusion with a cerebral abscess or tumor. Heroic therapy with substituted nucleosides is a matter of considerable controversy (see Chapter 13). *Rabies* is a rare cause of encephalitis in countries in which it is endemic (see Chapter 22).

Viral encephalitis (Plate 24-3) is characterized by fever, headache, vomiting, and neck stiffness, followed by stupor, confusion, ataxia, and sometimes convulsions, coma, and death. Residual effects such as mental retardation, epilepsy, paralysis, deafness, or blindness are quite common in those who recover. The encephalitogenic arboviruses are listed in Table 19-4 and discussed in detail in that chapter. Some tend to produce relatively mild encephalitis; others confer a very high mortality.

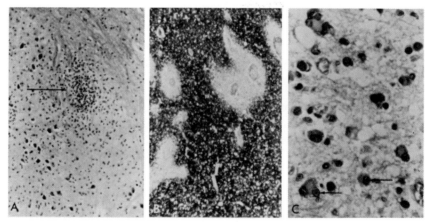

PLATE 24-3. *Viral infections of the brain. Sections from fatal human cases of encephalitis.* (A) *Arbovirus, Australian encephalitis (hematoxylin and eosin, ×55). Arrow indicates focal concentration of leukocytes.* (B) *Measles, postinfectious encephalitis (luxol fast blue stain for myelin, ×20). Note areas of demyelination around vessels.* (C) *Measles, subacute sclerosing panencephalitis (hematoxylin and eosin, ×220). Arrows indicate intranuclear inclusions. (Courtesy I. Jack, R. McD. Anderson, and A. Williams.)*

Postinfectious Encephalomyelitis

A quite different type of encephalitis is the "postinfectious" encephalomyelitis which occasionally complicates certain common childhood diseases or vaccination. This serious demyelinating disease of the brain and spinal cord occurs as a complication of *measles*, commonly (almost one in 1000), *varicella, mumps,* and *rubella* or *influenza,* rarely. Symptoms, which develop any time within a week of the appearance of the measles rash or within 3 weeks of smallpox vaccination, consist of headache, fever, stupor, and convulsions; there is a significant mortality. The CSF may be normal or may show mononuclear pleocytosis and elevated protein concentration. The main histopathological changes in the brain are perivenous inflammation and demyelination. Virus cannot regularly be recovered from the affected brain, but there are recent reports of the demonstration of measles virus by cocultivation of measles encephalitis brain biopsy with monkey kidney cells.

The mechanism of the demyelination is unknown. All the viruses mentioned share a common capacity to cause fusion of myelin membranes, and this in itself may suffice to explain the phenomenon. On the other hand there is some evidence that this is an immunopathological ("auto-immune") disease, in which the virus provokes an immunological reaction directed against myelin. The histology closely resembles that seen in experimental allergic encephalomyelitis of animals, or in the allergic encephalomyelitis of man that quite commonly followed the repeated inoculation of the 10% emulsion of infected rabbit brain which comprises the Semple vaccine against rabies.

Degenerative Diseases of the CNS

A number of CNS diseases of unknown etiology have been noted to follow infection with certain viruses. Positive etiological relationships are almost impossible to establish in the absence of (a) repeated isolation of the virus in question from the brain, and (b) transmission of the disease to animals. Nevertheless, certain associations have now been reported with sufficient regularity to justify their inclusion in a textbook, if only to encourage an extended search for more data.

The *Guillain–Barré syndrome* (*acute polyradiculoneuropathy*) is characterized by acute flaccid paralysis of muscles with associated muscle pains and paresthesia. Though some deaths occur, from respiratory failure, most cases recover with at worst a little muscle wasting. There are some reports of the disease occurring in association with epidemics of gastroenteritis and with the *EB virus*. This virus has also been claimed to be a cause of *Bell's palsy* and *transverse myelitis,* but the ubiquity of the agent makes it difficult to draw firm conclusions about its role. The isolation of a *new herpetovirus* from cases of *subacute opticoneuropathy* in Japan has been mentioned already in Chapter 16. Another herpetovirus, *varicella–zoster,* has been associated with *Reye's syndrome,* a lethal *encephalopathy* associated with fatty degeneration of the liver. Several other viruses have also been implicated, notably *influenza,* especially type B. Indeed, during epidemics of influenza A or B a number of neurological complications have been observed, including *polyneuritis.*

The reader is referred back to Chapters 8, 21, and 23 for detailed descriptions of four rare viral degenerative diseases of the brain that fit the general description of "slow infections," namely *subacute sclerosing panencephalitis (SSPE)* (caused by measles virus), *progressive multifocal leukoencephalopathy* (PML) (caused by a human papovavirus), and *kuru* and *Creutzfeldt–Jakob disease,* perhaps attributable to "viroids."

VIRAL SKIN RASHES

Many viral infections involve the skin either as localized or systemic infections.

Localized Infections. These include the common benign hyperplastic conditions, *warts* and *molluscum contagiosum,* or the recurrent vesicular outbreaks of *herpes simplex* on skin and mucous membranes.

Systemic Infections. These infections, with dissemination of virus throughout the body, may lead to the development of an exanthem (see Chapter 6), which may be either maculopapular or vesicular (Table 24-4). In the case of vesicular rashes, caused by *varicella, smallpox,* or *coxsackievirus A5, 10, and 16,* the virus can be isolated from the lesions. On the other hand the maculopapular rashes of *measles, rubella, echoviruses 4, 6, 9, and 16, coxsackieviruses A9, 16, and 23, enterovirus 71, EB virus,* or *arboviruses* (see Table 19-2 and Chapter 19) may result from hypersensitivity reactions in some cases (see Chapter 7).

The clinical differentiation of maculopapular viral rashes is much more difficult now than it used to be in the days when we knew of only rubella and measles. The terms "rubelliform" and "morbilliform" were coined to distinguish the fine, discrete, macular rash of rubella

TABLE 24-4
Viral Skin Rashes

| SYNDROME | VIRUS | |
	COMMON	LESS COMMON
Maculopapular rash	Measles	Echovirus 2, 5, 11, 18, and others
	Rubella	Coxsackie A2, 4; B1, 3, 5, and others
	Echovirus 4, 6, 9, 16	Several arboviruses (see Table 19-2)
	Coxsackie A9, 16, 23	EBV (infectious mononucleosis)
	Enterovirus 71	Reovirus 2
		Adenovirus 3, 7
Vesicular rash	Varicella–zoster	Smallpox
	Herpes simplex	Vaccinia
		Coxsackie A4, 5, 9, 10, 16
		Enterovirus 71
Hemorrhagic rash	Several arboviruses (see Table 19-3)	Smallpox
	Arenaviruses	
Localized lesions	Herpes simplex	Cowpox
	Warts	Milker's nodes
	Molluscum contagiosum	Orf

from the larger, blotchy, semiconfluent macules of measles (Plate 24-4C). Now we are aware of more than a dozen enteroviruses that may produce macular rashes essentially indistinguishable from one or the other (see Chapter 18). Enteroviral rashes, like those due to measles and rubella, are usually seen in children, but tend to occur during the summer months, often in epidemic form with an incubation period of 3–8 days separating the cases in a family. The "Boston exanthem," caused by echovirus 16, features a rubelliform rash spreading from the face downwards, accompanied by fever, sore throat, headache, and myalgia. Echovirus type 9 (coxsackie A23, the most common enterovirus of all) often produces a blotchy purplish morbilliform rash on the face, but a rubelliform rash on other parts of the body, including the palms and soles. Something of a curiosity is the condition known as hand, foot and mouth disease, caused by certain coxsackieviruses, in which a generalized maculopapular rash is accompanied by vesicles and even bullae on the palms, soles and buccal and pharyngeal mucosae ("vesicular pharyngitis").

In general, vesicular rashes do not present a diagnostic problem. Localized lesions are usually the recurrent form of *herpes simplex*, or when they correspond with a particular dermatome, *herpes zoster* (see

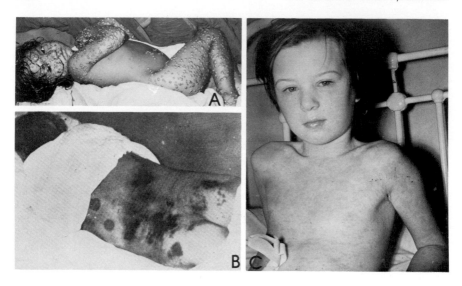

PLATE 24-4. *Three distinct types of viral rash.* (A) *Vesiculopustular rash of* smallpox. (B) *Hemorrhagic rash of hemorrhagic fever.* (C) *Maculopapular rash of* measles. (A, *courtesy Prof. A. W. Downie; B, courtesy Dr. A. Shelekov; C, courtesy Dr. J. Forbes.*)

Chapter 16). Recognition of secondary spread of *vaccinia* from a vaccination site is not difficult, but the differentiation of *chickenpox* from *smallpox* in a vaccinated traveler can test the most seasoned physician. Classically, varicella lesions are relatively superficial, concentrated on the trunk, and develop asynchronously, whereas those of smallpox involve the deeper layers of the dermis, are prominent on the head and limbs, and develop synchronously (Plate 24-4A). These differences may be obscured in vaccine-modified cases where the distribution of lesions and their mode of development may be quite atypical. Fortunately, rapid diagnostic techniques are available in the laboratory (see Chapters 14 and 17).

The hemorrhagic fevers produced by certain *arboviruses* (see Chapter 19 and Table 19-3) and *arenaviruses* (see Chapter 22) comprise a distinct, though not homogenous, group. Petechiae and ecchymoses are present on the skin (Plate 24-4B) and all the mucous membranes, hence, extensive bleeding occurs from the gastrointestinal tract and other surfaces. The patient becomes severely shocked and may die. Differential diagnosis from thrombocytopenic purpura of other etiology is not always easy, but may be aided by the presence of protean clinical manifestations of a systemic arboviral infection together with evidence of the epidemic nature of the disease and its limited geographical distribution.

GENITOURINARY INFECTIONS

Herpes simplex type 2 is now second only to the gonococcus as a cause of venereal disease in most western countries (see Chapter 16). Vulvovaginitis with or without cervicitis in females can be primary or recurrent. HSV2 infection has been closely correlated with subsequent development of cervical carcinoma but there is no proof that the association is a causal one (see Chapter 9). Other viruses, notably echovirus type 4, have also been reported to be associated with cases of cervicitis.

Virus is regularly shed in the urine during generalized infections by many paramyxoviruses, herpetoviruses, adenoviruses, togaviruses, arenaviruses, hepatitis viruses, rubella, and the human polyomavirus. As far as is known however, viruria is not an indication of impairment of renal function. Only in *cytomegalovirus* infections is there clear evidence of kidney pathology (Plate 16-6). A temporary decrease in creatinine clearance has been observed in *mumps*.

Glomerulonephritis, which is an occasional feature of *hepatitis B* infection (see Chapter 23) is attributable to the accumulation of viral antigen-antibody complexes in glomeruli (as well as in arterioles). Such immune complex disease is a frequent manifestation of chronic viral infections in animals (see Chapter 8) and may turn out to be much more common in man than we yet appreciate. A careful search for viruses, or immunoglobulin deposits, or complement depletion in human glomerulonephritis may reveal that some "idiopathic" cases are in fact due to persistent infection with known or unknown viruses.

Acute hemorrhagic cystitis characterized by hematuria, frequency, and dysuria has recently been associated with *adenovirus types 2, 11, and 21* (Table 24-5).

TABLE 24-5
Genitourinary Infections

DISEASE	VIRUS
Vulvovaginitis, cervicitis	Herpes simplex 2
Glomerulonephritis	Hepatitis B
Cytomegalic inclusion disease	Cytomegalovirus
Acute hemorrhagic cystitis	Adenovirus 2, 11, 21

VIRAL CONJUNCTIVITIS

It is not generally appreciated how frequently viruses can involve the eyes (see Table 24-6). The conjunctiva can be quite a sensitive indicator of systemic disease in the case of *measles,* certain *enteroviruses,* or *dengue* and *phlebotomus fevers.* The epidemic *adenoviruses* 3, 4, and 7 commonly produce a harmless pharyngoconjunctival fever (Chapter 15). A pandemic of *enterovirus 70* recently alerted the medical world to a newly recognized disease, acute hemorrhagic conjunctivitis, sometimes associated with radiculomyelitis (see Chapter 18).

Keratoconjunctivitis without systemic involvement is a feature of infection by *adenovirus 8* ("shipyard eye," epidemic keratoconjunctivitis), *herpes simplex, vaccinia,* or *Newcastle disease virus.* The latter two are rare accidental infections, arising by autoinoculation from a smallpox vaccination lesion, or from infected fowls, respectively. Herpetic keratitis, on the other hand, is quite common and may occasionally lead to blindness. The basic lesion is the "dendritic" ulcer, a creeping, tree-like ulcer of the cornea, which may go on to form a "geographic" ulcer, or a deep "metaherpetic" ulcer, then discoid keratitis. As with any other herpetic infections recurrent attacks often occur and the final result may be gross scarring and blindness. Fortunately, this is one viral disease which is amenable to specific chemotherapy (see Chapters 13 and 16).

Chorioretinitis is occasionally seen in *Rift Valley fever,* or more commonly as part of the *rubella* syndrome in babies who may or may not also be born with congenital cataracts, glaucoma, or a wide variety of other abnormalities.

TABLE 24-6
Viral Conjunctivitis

COMMON	LESS COMMON
Adenovirus 3, 4, 7, 8	Dengue
Herpes simplex	Sandfly fever
Herpes zoster	Echovirus 9, 16
Enterovirus 70	Vaccinia
Measles	Newcastle disease virus
Rubella	

VIRAL HEPATITIS

"Infectious" and "serum" hepatitis between them represent perhaps the major remaining challenge to medical virology. The disease has increased dramatically in incidence during the last 30 years or so. The symptomatology, pathology, and laboratory diagnostic procedures have been described in detail in Chapter 23 together with a full discussion of the major etiological agents, *hepatitis viruses A and B.*

It must be remembered that several other viruses may involve the liver as part of a more general disseminated disease process (see Table 24-7). Hepatic damage can be quite as severe as with classical infectious hepatitis. Differential diagnosis is facilitated by the concurrence of other clinical features characteristic of the disease in question. Lymphadenopathy and pharyngitis suggest *infectious mononucleosis,* which usually involves the liver and produces jaundice in 10% of cases; diagnosis rests on tests for heterophil agglutinins and EBV antibodies (see Chapter 16). *Arboviruses* must be considered in endemic areas. *Yellow fever* is the classical example, but several of the arboviruses causing hemorrhagic fever also have pronounced effects on the liver (see Chapter 19).

Neonatal hepatitis presents a special diagnostic problem. The baby is jaundiced at birth or develops the condition in the immediate postnatal period. Viral causes include *cytomegalovirus* and *rubella* acquired *in utero* or *herpes simplex* and *coxsackie B,* more probably picked up during or after parturition. In each case, hepatitis is but one manifestation of a generalized infection of great severity; careful examination will always reveal other abnormalities of diagnostic value (see chapters dealing with individual viruses concerned). Such babies often die. Affected organs all over the body show characteristic histopathology and, in the case of the two herpetoviruses, inclusion bodies are prominent. From all four, the causative virus is readily recovered from almost every organ.

TABLE 24-7

Viral Hepatitis

PRIMARY	SECONDARY	PRENATAL AND PERINATAL
Hepatitis A	EB virus (infectious mononucleosis)	Rubella
Hepatitis B	Togaviruses (e.g., yellow fever)	Cytomegalovirus
	Arenaviruses	Herpes simplex
	Echovirus 19	Coxsackie B
	?Adenoviruses	

The reader hardly needs reminding that the possible causes of jaundice are legion. They include everything from bacteria, protozoa, and helminths to a battery of noninfectious causes like hepatotoxic chemicals or physical obstruction of the biliary system. Whole books have been devoted to the difficult problem of differential diagnosis which will not be pursued further here.

MYOCARDIOPATHY

Coxsackieviruses show a special predilection for muscle, as is apparent in their effects on infant mice (see Chapter 18). In particular, acute myocardiopathies, with varying degrees of endocarditis, pericarditis, or myocarditis, are attributable to infection by the coxsackieviruses B1–5 in particular, and to a lesser extent the coxsackie A viruses (e.g., 4 and 16) and echoviruses (e.g., 9 and 22). The fatal late summer disease of infants, "myocarditis of the newborn," has been known for many years to be due to coxsackie B viruses, but more recent work has shown that infants who recover from these severe perinatal infections may develop chronic heart disease, and that coxsackievirus infections of adults can also cause more insidious heart damage. The possible role of coxsackieviruses in congenital heart disease, e.g., aortic and mitral incompetence, or calcific pancarditis and hydrops fetalis, is under investigation.

Polymyositis and dermatomyositis (polymyositis with skin rash) can follow congenital or postnatal infection with coxsackievirus A9 and probably other coxsackieviruses (Table 24-8).

The typical *togavirus* infection takes the form of a febrile generalized myositis and arthritis with or without an associated skin rash. The predilection of these viruses for muscles and joints was discussed in Chapter 19.

TABLE 24-8
Viruses Involving Other Systems

DISEASE	VIRUS
Arthritis	Togaviruses, including rubella
	Hepatitis B
Myositis	Togaviruses
	Coxsackieviruses
Carditis	Coxsackieviruses, especially B
Parotitis, pancreatitis, orchitis, etc.	Mumps

NEONATAL INFECTIONS

Intrauterine Infection. During prenatal life intrauterine infection can have either, or both, of two types of effect on the fetus: (a) severe disseminated disease, or (b) teratogenic effects. Either, if sufficiently severe, can be fatal, leading to spontaneous abortion or stillbirth.

Rubella and *cytomegalovirus* are by far the most important known causes of developmental abnormalities (Plate 24-5, ranking well ahead of toxoplasmosis and syphilis. *Herpes simplex* has only recently been shown to produce cerebral abnormalities akin to those produced by cytomegalovirus, while coxsackie B viruses may possibly be responsible for some congenital heart defects. The teratogenic capacity of influenza virus is uncertain. Differential diagnosis between the so-called "Torch" agents (toxoplasmosis, rubella, cytomegalovirus, herpes simplex) is quite difficult since all may produce an underdeveloped, jaundiced anemic baby with pneumonia and retinopathy. Demonstration of specific viral IgM in cord blood is diagnostic. The pathology, clinical manifestations, diagnosis and management of cytomegalic inclusion disease and the rubella syndrome were discussed fully in Chapters 16 and 23.

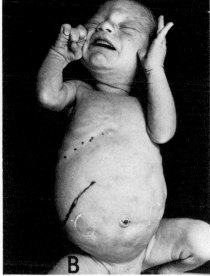

PLATE 24-5. *Viral embryopathies.* (A) *Rubella syndrome.* (B) *Cytomegalic inclusion disease.* (*Both plates courtesy Dr. K. Hayes.*)

TABLE 24-9
Neonatal Viral Diseases

TIME OF INFECTION	SYNDROME	VIRUS
Prenatal	Teratogenic effects	Rubella
		Cytomegalovirus
		Herpes simplex
		Coxsackie B
		Influenza?
Perinatal	Disseminated disease	Coxsackie B
		Herpes simplex
		Hepatitis B
		Varicella
		Smallpox
Postnatal	Pneumonitis	Respiratory syncytial virus
		Influenza
		Varicella
		Adenovirus
		Measles
	Enteritis	Rotavirus

Perinatal Infection. This can be defined as infection occurring shortly before, during or after birth. *Herpes simplex type 2* and *coxsackie B* viruses are the most important causes of overwhelming infections acquired during the perinatal period. Neonatal HSV2 (see Chapter 16) is usually picked up during passage through an infected birth canal. Coxsackievirus may also be first encountered during parturition or shortly thereafter; the resulting encephalomyocarditis neonatorum (see Chapter 18) usually develops 1–2 weeks after birth. *Hepatitis B, varicella,* and *smallpox* viruses can all cross the placenta from an acutely infected mother to cause severe generalised neonatal disease.

Postnatal Infection. Infants tend to be postnatally protected against most infections by maternal antibody. If this is lacking, or if the baby is premature, sickly, or immunologically deficient, it is very vulnerable to infection with the common viruses in its environment, even (or particularly) in hospitals. Table 24-9 lists the viruses most frequently responsible for the death of premature babies during the first few weeks of life.

FURTHER READING

Chanock, R. M., Mufson, M. A., and Johnson, K. M. (1965). Comparative biology and ecology of human virus and mycoplasma respiratory pathogens. *Progr. Med. Virol.* **7,** 208.

Christie, A. B. (1974). "Infectious Diseases, Epidemiology and Clinical Practice," 2nd ed. Churchill, London.

Ciba Foundation. (1973). "Intrauterine Infections." Elsevier, Amsterdam.

Craighead, J. E. (1975). Viral induced pancreatic lesions and diabetes mellitus in man and animals. *Progr. Med. Virol.* (in press).

Eichenwald, H. F., McCracken, G. H., and Kindberg, S. J. (1967). Virus infections of the newborn. *Progr. Med. Virol.* **9,** 35.

Hoeprich, P. D., ed. (1972). "Infectious Diseases." Harper, New York.

Jackson, G. G., and Muldoon, R. L. (1975). "Viruses Causing Common Respiratory Infections in Man." Univ. of Chicago Press, Chicago, Illinois.

Johnson, K. M., Halstead, S. B., and Cohen, S. N. (1967). Hemorrhagic fevers of South-East Asia and South America: A comparative appraisal. *Progr. Med. Virol.* **9,** 105.

Kapikian, A. Z. (1974). Acute viral gastroenteritis. *Prev. Med.* **3,** 535.

Knight, V. (1973). "Viral and Mycoplasmal Infections of the Respiratory Tract." Lea & Febiger, Philadelphia.

Krugman, S., and Ward, R. (1973). "Infectious Diseases of Children and Adults," 5th ed. Mosby, St. Louis, Missouri.

McLean, D. M. (1967). Aseptic meningitis. *In* "Recent Advances in Medical Microbiology" (A. P. Waterson, ed.), pp. 54–84. Churchill, London.

Miller, D. L. (1973). Acute respiratory virus diseases. *Postgrad. Med. J.* **49,** 747.

Perkins, F., ed. (1974). "The Importance of Viral Agents in Man and Animals." Int. Ass. Biol. Stand.

Top, F. H., and Wehrle, P. F., eds. (1972). "Communicable and Infectious Diseases," 7th ed. Mosby, St. Louis, Missouri.

Tyrrell, D. A. J., ed. (1968). "Acute Respiratory Diseases." College of Pathologists, London.

Utz, J. P. (1974). Viruria in man. *Progr. Med. Virol.* **17,** 77.

Ward, T. G. (1973). Viruses of the respiratory tract. *Progr. Med. Virol.* **15,** 126.

Wenner, H. A. (1973). Virus diseases associated with cutaneous eruptions. *Progr. Med. Virol.* **16,** 269.

World Health Organization. (1969). Respiratory viruses. *World Health Organ., Tech. Rep. Ser.* **408.**

Zuckerman, A. J. (1975). Human Viral Hepatitis," 2nd ed. North-Holland Publ., Amsterdam.

Subject Index

A

Abortion, 111, 188, 442, 448, 473
Abortive infection, 75–76
Abortive transformation, 163
Acrylamide gel electrophoresis, 11, 49, 62
Actinomycin D, 49, 66, 96
Acute respiratory disease (ARD), 291
Adamantanamine, *see* Amantadine
Adaptation, of viruses, to new hosts, 32, 70, 75
Adenine arabinoside, 250–251, 305, 313
Adeno-associated viruses, 20, 75, 295–296, 423–424
Adenosatellovirus, *see* Adeno-associated viruses
Adenoviruses, 21, 289–296
  classification and properties, 21, 289–291
  clinical features, 291–292, 454–459, 461, 469–471
  epidemiology, 195, 294–295
  genetic interactions, 76–81, 295
  laboratory diagnosis, 293–294
  multiplication, 58
  oncogenicity, 165–168
  pathogenesis, 292–293
  vaccines, 295
Adenovirus-SV40 hybrids, 79–81, 164–165, 232, 295, 296
Adjuvants, 235
Adsorption, *see* Attachment, of viruses

African horse sickness, 219, 407
Age, effect on infection, 135–136, 336, 342
Air travel, 219
Alastrim, 323
Aleutian disease of mink, 151
Alimentary infections, *see* Intestinal infections
Allergy, *see* Hypersensitivity
Alphaviruses, 358, *see also* Togaviruses
Amantadine, 251–252
Animals
  in evolution of viruses, 203–216, 379
  as reservoirs, 188–191, 219, 379, 414
  for virus isolation, 36, 135, 267–268, 326
Antibiotics, 243, 262
Antibodies, 116–129
  fluorescent, *see* Immunofluorescence
  IgA, 118–119, 339
  IgG, 117–122
  IgM, 117–118, 422, 443–445
  immune complexes, 126, 150–153, 159–160
  maternal transfer of, 122
  measurement of, 194, 268–284
  neutralization of viruses by, 120–121, 275–278
  passive immunization, 235–236, 308, 328, 396, 414, 437–438
Antigenic drift, 119, 210–214, 377–379, 383
Antigenic shift, 210–213, 377–379, 383
Antigens, 116, 257–261, 268–284, 428, 439
Antiviral agents, *see* Chemotherapy

C 7
D 8
E 9
F 0
G 1
H 2
I 3
J 4
  5